Sex, Evolution, and Behavior

SECOND EDITION

Martin Daly • Margo Wilson
McMaster University

Wadsworth Publishing Company   Belmont, California

A Division of Wadsworth, Inc.

Library of Congress Cataloging in Publication Data

Daly, Martin
  Sex, evolution, and behavior.

  Bibliography: p.
  Includes index.
  1. Sexual behavior in animals. 2. Sex. 3. Reproduction. 4. Evolution. I. Wilson, Margo. II. Title.
QL761.D34 1983   591.56   82-23193
ISBN 0-87150-767-6

ISBN 0-87150-767-6

Printed in the United States of America
    13 14 15 16 17 18 19 20

This book was set in English Times and Korinna by Caron LaVallie of PWS Publishers Composition Dept. Interior design and production were provided by Bywater Production Services. The cover was designed by Hannus Design Associates.

Preface to the second edition

Theodosius Dobzhansky once entitled a lecture, "Nothing in biology makes sense except in the light of evolution." Students of behavior have taken that maxim to heart only rather recently and the result has been the "sociobiological revolution." Organisms are "designed" by natural selection, and if we are to understand their behavior, we must ask what that behavior is designed to achieve, or in other words, how it contributes to fitness.

Five years ago, when we wrote the first edition of this book, behavioral research was dominated by studies of proximate causation and ontogeny. We then felt obliged to proselytize for the integration of such studies within an evolutionary functional perspective. As sex and reproduction are central to an understanding of behavior, we focused our book on an interdisciplinary study of this area. Today the tide has turned, and the research journals are full of tests of evolutionary hypotheses. Recent progress necessitates both an update and a reorganization.

We have expanded our introductory treatment of the essentials of evolutionary theory and genetics, including new discussions of fitness, reproductive effort, and nepotism. Hamilton's classic analysis of the consequences of haplodiploidy in the social insects is now featured, and there is a new chapter covering recent progress in the analysis of sex ratios and sex allocation. Life history strategy, the analysis of evolved schedules for the adaptive expenditure of reproductive effort over the lifetime, has also been accorded a new chapter, and we have reorganized our presentation of sexual selection and the adaptive logic of sex differences to include recent empirical and theoretical studies of female choice. Instructors using the book as an introduction to sociobiology will find the second edition more complete than the first.

In chapter 1, we review the process of evolution by natural selection and distinguish evolutionary explanations of biological phenomena from other modes of explanation. Chapters 2 and 3 outline basic sociobiological theory: the concepts of fitness and reproductive effort are introduced and then extended to become inclusive fitness and nepotistic effort. Chapter 4 concerns the adaptive significance of sexual reproduction; chapters 5 and 6, the processes and consequences of sexual selection. Chapters 7 and 8 focus on the adaptive rationale for species differences in social organization and life history, respectively. Chapter 9 treats the allocation of reproductive effort between the sexes as a budgeting problem. Chapter 10 is devoted to sexual development and differentiation: how individuals become females and males. In the last two chapters, 11 and 12, the human animal is examined in the light of sociobiological theory, with the emphasis on empirical research results rather than speculation.

Most of the illustrations in this second edition are new, and most of these are based on data published since the first edition appeared in 1978. Our bibliography contains more than 300 references published in 1979 or later, and we have had to overlook many good papers to keep the list manageable. New theory and research are featured in every chapter from 3 through 12. These are lively times for students of sex, evolution, and behavior.

We are primarily familiar with the literature on mammals and birds, but we have tried to reduce our mammalocentric bias, with what we hope are salutary effects. The implications of internal fertilization, for example, only come clear when the several vertebrate classes are contrasted. Much of the best life history research has been conducted with invertebrates. And of course the haplodiploid Hymenoptera have inspired insights that could never have been gained from study of diploid organisms. (We have included only a very few botanical examples, for fear of straining the good will of readers from the social sciences; to general biologists, we apologize for this zoological chauvinism.) Nevertheless, we retain a heavy concentration on a single mammalian species, *Homo sapiens,* an animal that has been increasingly the object of our own research efforts. There are still many scientists who would exclude *H. sapiens* from sociobiological analysis. Some argue that other animals pursue evolved fitness interests, while people pursue acquired psychological goals; some argue that language or culture or volition or self-awareness exempts us from the evolutionary principles applying to all other creatures. We are unmoved by these arguments, which seem to us to be founded on bankrupt dichotomies of biology versus culture and nature versus nurture. *All* behavior requires ontogenetic as well as phylogenetic explanation, and *all* animals pursue proximate "psychological" satisfactions whose connections to fitness are indirect and disruptable. People are obviously very special animals, but we are convinced that the same body of evolutionary theory is applicable to all creatures. We hope our readers will also be convinced. But convinced or not, please appreciate that we do not zoomorphize people (or anthropomorphize animals) frivolously.

Valuable criticisms of draft chapters of this edition were provided by Ric Charnov, Daniel Q. Estep, Carl R. Gustavson, William Irons, Jeffrey A. Kurland, and John Tooby, all of whom we thank. Thanks also to expert word-processors Bev Pitt, Wendy Tasker, and Cindy Montana; to artist Catherine Farley; and to editor Jean-Francois Vilain.

Martin Daly and Margo Wilson
Brookfield, Prince Edward Island
August, 1982

Preface
to the first edition

Reproduction is of central importance in the biological and social sciences. Other functions, eating and breathing for example, are vital, to be sure, but animals eat and breathe in order to reproduce. Each living organism is a product of evolution by natural selection, a process that favors only those characteristics that pay off in reproduction. And reproduction, leaving aside certain exceptions, is a matter of sex. Thus there is a logic by which the subject of sex holds center stage in the behavioral sciences. For example, all social organization is in principle interpretable as the outcome of the sexual strategies by which animals try to reproduce themselves.

Our approach to sex is deliberately broad, but it is also integrated. We discuss evolutionary theory, reproductive physiology, animal behavior, sex-role development, cultural anthropology, and other subjects all in one book because these apparently disparate topics are all germane to understanding behavior, and, what is more, they are mutually illuminating. The study of sex calls for an interdisciplinary approach, and we have attempted to integrate the approaches of various disciplines around the central theme of evolved reproductive strategies.

Our inclination to write an interdisciplinary book about sex can be traced to a seminar in the fall and winter of 1975–76 at the University of California, Riverside. The participants were discussing E. O. Wilson's *Sociobiology: The New Synthesis* with excitement and occasional acrimony. It seemed to us then, and still does, that the evolutionary functionalism of sociobiology offers an exciting and potentially integrative perspective for the behavioral sciences; nevertheless it arouses considerable skepticism, criticism, and even condemnation. Such reactions stem largely from a widespread lack of appreciation of the complementarity of multiple explanations of behavioral phenomena: An item of behavior cannot be fully comprehended until we grasp its adaptive significance *and* its physiology *and* its evolutionary history *and* its stimulus control *and* its ontogeny.

This book aims to synthesize various approaches to the study of sex. We begin in chapter 1 by introducing evolutionary thought and the notion of multiple explanatory levels. In chapter 2 we outline the "strategic" rationale that explains behavior in terms of the adaptive ends for which it has evolved. In chapters 3 through 6 we apply this rationale to the adaptive significance of sexual reproduction, to sex differences in behavior, to the comparative study of mating and social systems, and to the comparative study of parental behavior. We present recent sociobiological theory in these middle chapters and apply it to the comparative facts of vertebrate behavior. We also emphasize the use of a unitary theoretical framework for dealing with a broad spectrum of specific phenomena. The remaining chapters strive for synthesis. We discuss the physiology of sex and reproduction in chapter 7, concentrating upon the mammals, and argue that diversity in physiological mechanisms is comprehensible strategically. In chapter 8 we discuss developmental aspects of reproductive function and sex differences and again argue for the application of a strategic perspective. In chapter 9 we consider human behavior and attitudes from that perspective, drawing upon the discoveries of cultural anthropology, social psychology, and sexology. Biological approaches to human behavior have always met resistance, and in chapter 10 we refute certain criticisms and try to explain just what biology has to offer the social sciences.

This book should be comprehensible to an intelligent reader with no more background than the most basic high school biology, but it is not elementary in style or content. We make certain arguments that others may dispute. We present recent discoveries and recent theory from several sciences. To do so we have had to read widely in areas outside our own expertise, and no instructor using this book will be expert in all the subjects broached. We have therefore cited both original research papers and general works that we found valuable. We have tried to assure that the book will be useful to a diverse audience from undergraduates to professional scientists. It is suitable reading for courses in animal behavior, comparative psychology, and ethology and should be of interest to instructors teaching more specialized courses in sociobiology, sex and sex differences, evolution, and biological approaches to the social sciences.

Four colleagues read most of the book in draft and provided much useful criticism: Daniel Q. Estep, Jeffrey A. Kurland, Thomas L. Patterson, and George C. Williams. Others who read one or more chapters and offered valuable comments were Napoleon Chagnon, Lee C. Drickamer, W. J. Hamilton III, Edward E. Hunt, Jr., William Irons, Celia Moore, Robert K. Selander, John G. Vandenbergh, and Suzanne Weghorst.

Contents

2   Fitness   21

A gene's fitness is a matter of its reproductive success relative to alternative alleles. Fitness of whole organisms and fitness of behavior patterns can be defined similarly: reproductive success relative to other organisms and alternative behavior. Inclusive fitness refers to the organism's contribution to the replication of its genes in both descendant and collateral kin.

3   The animal as "strategist"   37

Evolution by natural selection produces organisms with reproductive and nepotistic strategies for promoting their inclusive fitness. Reproductive effort has to be allocated between pursuit of matings and parental care. The inclusive fitness interests of genetic relatives overlap, but they are not identical.

4   Why sex?   59

Why has sexual reproduction evolved? It has certain costs in comparison with asexual reproduction and must have compensatory benefits. These benefits probably lie in the genetic diversification of sexually produced offspring. And why are there usually two sexes?

5   Sexual selection: Competition for mates   77

Females typically invest more time and energy in each offspring than do males, who compete for mating opportunities. A successful male may monopolize many females and sire many young, and the competition for this fitness prize is intense.

6   Sexual selection: Mate choice   113

Males often court indiscriminately, but females must select their mates with care. Females often choose their mates on the basis of the resources that males proffer or control. Females may also select males partly for their genetic quality.

7   Comparative reproductive strategies   137

Reproductive strategies vary among related species in ways that can be attributed to ecological differences. In highly polygynous species, males tend to be large and combative. Biparental care of young is associated with monogamy, and exclusively paternal care with polyandry. Where fertilization is internal, paternity is uncertain.

8   Life history strategy   179

Organisms have evolved adaptive schedules for the expenditure of reproductive effort over the lifetime. There are tradeoffs between survival and reproduction, between growth and reproduction, and between parental nurture and fecundity.

9 Sex allocation   223

Reproductive effort must be allocated between daughter and son production. Nonchromosomal sex determination mechanisms permit parents to adaptively adjust offspring sex ratios. In hermaphrodites, the allocation of reproductive effort between female and male function is an adaptive problem analogous to that of offspring sex ratio.

10 Sexual development and differentiation   243

Mammals' body, brain and behavior exhibit masculine development if androgens are present early and feminine development otherwise. Hormones have early "organizing" effects upon brain and bodily differentiation, distinct from their later "activating" effects upon sex-typical behavior. Few activities are sex-typed consistently in all cultures, though sex roles are important everywhere.

1

Natural selection

Sex is so pervasive a factor in our lives and in the world about us that we can easily overlook the fundamental ways in which it challenges our understanding. Why sex even exists remains one of the great conundrums in evolutionary biology. The most essential property of life is reproduction, but reproduction needn't be sexual. Sex is terribly complicated—within the cell, in the physiology of individual organisms, and between individuals. Where is the utility of it all?

When we consider the two sexes as we know them, further mysteries arise. Exactly how *do* males and females differ? Which distinctions are basic and which are mere window dressing? We may be confident that the wearing of trousers or earrings by one sex or the other is an arbitrary fashion, easily reversed. What, if any, are the nonarbitrary behavioral differences? And, however we answer such questions, we cannot escape another nagging problem: *Why*? As one theorist [676] has posed it, "Why are males masculine and females feminine and, occasionally, vice versa?" In this book we intend to examine the rationale behind this duality.

"Why" has a special meaning here: Why have organisms evolved the traits they have and not others? Why, for example, do the males of so many species fight more than the females and die younger? Why are birds usually monogamous and mammals hardly ever? Why do some fish change sex, and why don't we?

Questions of this sort are most fruitfully addressed from the perspective of Darwin's theory of evolution by natural selection. Our first three chapters therefore review essential aspects of the modern version of that theory. Chapter 4 concerns the question of why organisms reproduce sexually. In Chapters 5–9, we then explore the ramifications of the male-female phenomenon, attending primarily to the evolution of social behavior. Chapter 10 concerns the developmental processes by which individuals become females or males.

Chapters 11 and 12 focus on the human animal. There is still resistance to the suggestion that evolutionary biology can help us understand ourselves, and that resistance is nowhere stronger than in discussions of the nature of the two sexes. We are convinced that a comparative, evolutionary perspective can cast a good deal of light on human behavior and on the essential nature of woman and man. One aim of this book is to make that case.

It is well over one hundred years since Darwin's theory won the minds of the world's scientists. Yet it is only in the last two decades that the theory has been powerfully applied in the behavioral sciences, and it has revitalized them.

NATURAL SELECTION

In 1858 Charles Darwin and Alfred Russel Wallace proposed to a meeting of the Linnaean Society that evolution occurs as a result of "natural selection" [138]. It was an idea that the two English naturalists had developed independently, and indeed it was an idea whose time had come. Most thoughtful biologists were already entertaining the hypothesis that animal species were not individually created but had somehow gradually evolved one from another. Such a theory seemed to provide the only unitary and

FIGURE 1-1
Hierarchical taxonomic classification

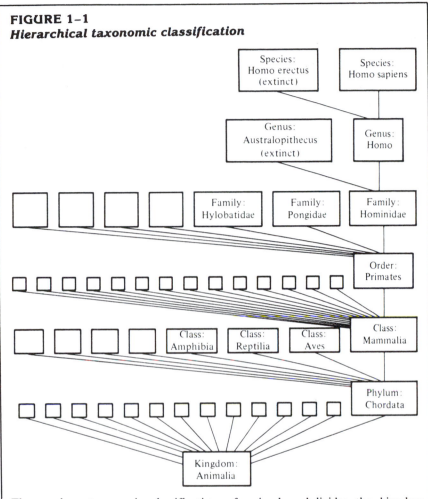

The modern taxonomic classification of animals subdivides the kingdom Animalia into more than twenty *phyla*, which are in turn divided into *classes*. The phylum Chordata includes several classes of fish and fishlike animals and the classes Amphibia, Reptilia, Aves (birds), and Mammalia. Each class is divided into *orders*, each order into *families*, each family into *genera*, each genus into *species*. The two-name label of each animal species (e.g., *Homo sapiens*) identifies first its genus and then its species.

This nested hierarchy lends itself to an evolutionary interpretation as a descent diagram. Each level represents a more distant common ancestor: The genus *Homo* emerged more recently than the family Hominidae. Further back in time is the last common ancestor of all primates; still further back is the last common ancestor of all mammals; and so on.

concise explanation for an enormous number of rapidly accumulating facts.

The first compelling body of knowledge was that of comparative morphology, the study of the form and structure of organisms. Biologists confronted with the common body plans of animals as diverse as birds and men were repeatedly led to speak of species as if they were modifications of one another, an idea that was implicitly evolutionary. Collarbones might vary among the mammals, but they were all recognizably collarbones. The skeletal structure of a bird's wing or a seal's flipper suggested that these were modified forelimbs. It was already a century since the Swedish naturalist Linnaeus (no evolutionist himself!) had used these structural commonalities to devise biology's modern classificatory system, and that system lent itself to an evolutionary interpretation (Figure 1–1).

FIGURE 1–2
Evolution of horses

Genus and geological period	Relative body size	Skeletal structure of foot
Equus Modern horses		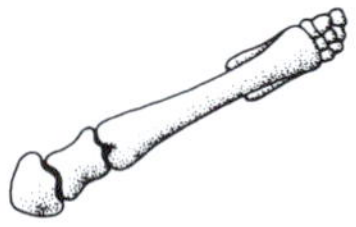
Merychippus Miocene, about 20 million years ago		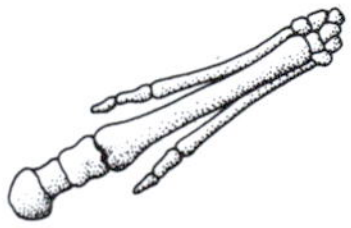
Miohippus Oligocene, about 30 million years ago		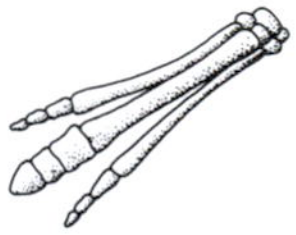
Eohippus Eocene, about 50 million years ago		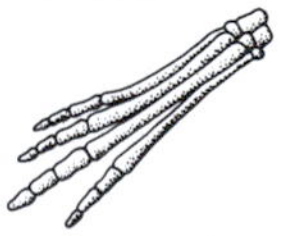

One of the most complete fossil series available is that of horses. Over the last 50 million years horses have become progressively larger, predominantly grazers rather than browsers, and highly specialized for running on hard ground. The modern horse's foot has but a single toe. (After Simpson, 1951 [572].)

A second body of knowledge, a fossil record of extinct and apparently transitional forms, had built up; and its sequential characteristics, as revealed in geological strata, were beginning to be understood (Figure 1–2). In a third field of biology, comparative embryology, mounting evidence of similar development in the embryonic forms of dissimilar animals further indicated their relatedness. Evolutionism was very much in the air.

What was lacking was a mechanism. Before Darwin no one had conceived of a comprehensible process that would make evolution happen. Perhaps, to take an example, it was true that giraffes' necks had lengthened during evolution in order to permit browsing higher in trees. Such explanations were far from satisfactory. They invoked a purposiveness that smacked of supernaturalism. *How* could such a purpose conceivably influence the animal's structure?

The answer proposed by Darwin and Wallace now seems obvious. **Natural selection** is the mechanism that causes evolution. Darwin called it natural selection by analogy with the artificial selection practiced by people who control the breeding of domestic animals. If we wish to increase the milk yield of our cattle, then for our breeding stock we should *select* those cows with the greatest milk yield (and, as Darwin noted, we should select bulls whose female relatives are good milk producers too). In such a case, there are human agents with specific goals in mind who select which animals are to breed, but nature can "select" without intent. Some animals live and others die barren. Furthermore, among those that live, some leave more surviving offspring than others.

Where animals vary, and where those variations are at least partly inherited, the less successful varieties will be eliminated, and the more successful will become relatively numerous. That was the crux of Wallace's and Darwin's reasoning. If giraffes with slightly longer necks than average are slightly more successful treetop feeders than average giraffes, they will therefore survive and grow better than other giraffes and be slightly more successful at reproduction as well. Their slightly longer-necked offspring will increase proportionately in the giraffe population and will numerically dominate future generations. In this way the observed range of neck lengths will shift, not through any purposive stretching, but through the **reproductive advantage** of those with longer necks.

Notice that natural selection is only partly a matter of differential survival abilities. If a successful animal's traits are to be passed on to descendants, that animal must of course survive to maturity, but it must furthermore *reproduce*. Imagine a population in which adults regularly raise two offspring annually. If a type arises that can successfully raise three, then that type will take over and drive the previous type to extinction, not because of any difference in survival capability, but simply because of a

---

FIGURE 1-3
Natural selection as differential reproduction

Imagine a population in which normal individuals (open circles) raise two offspring in each annual generation. Suppose a mutant form (crosshatched circles)
that is capable of raising three arises in generation 2. The mutant form will then
increase its proportional representation in the population.

Genera-tion	Proportion mutant ⊗
1	—
2	.25
3	.33
4	.43

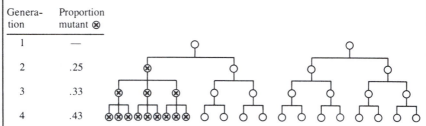

Of course, a population cannot expand indefinitely like the one illustrated
above. Increasing total numbers are not essential to the argument: Mortality
between generations may keep the population size in check, but as long as mortality strikes both forms equally, the ⊗ form will show the same *proportionate* increase. By generation 14 in the above example, the ⊗ form will constitute more
than 99 percent of the population.

---

difference in reproductive capability (Figure 1-3). All animal species *over-
produce* offspring, not all of which can survive to reproduce in their turn.
Thus, there is inevitable competition among the individuals of each species
for the means to survive and reproduce, and any inherited advantage in this
competition will be naturally selected.

The outraged response that Darwin's theory provoked is well known.
In a famous debate at Oxford, Bishop Samuel Wilberforce sarcastically inquired of the comparative anatomist T. H. Huxley whether he claimed
simian descent through his mother as well as his father. The biologist
replied that he would rather have an ape as an ancestor than a learned man
who would debate a grave issue with irrational mockery [293]. Another
famous confrontation occurred decades later between fundamentalist
William Jennings Bryan and defense lawyer Clarence Darrow in a Ten-
nessee courtroom in 1925 [143]. It is sometimes forgotten that Darrow's
brilliant defense of schoolteacher John Scopes could not prevent his convic-
tion for the crime of teaching evolution.

Nor is the battle over. As we write, the state of Arkansas is appealing a
historic ruling [474] that its religiously motivated "creation science" law
violates the United States Constitution. Creationists have presented the

issue as the opposition of two respectable biological "theories," but evolution by natural selection meets no significant opposition within biological science. The principle of natural selection has a logical necessity to it; indeed, some have argued that it is a tautology rather than an empirical law. The facts were overwhelmingly in accord with the principle, and the subsequent development of biology in major new areas like genetics and biochemistry has only reinforced Darwin's conclusion that the facts make a belief in the theory of evolution by natural selection "inescapable."

GENES AND NATURAL SELECTION

Darwin was unaware of the mechanism of inheritance when he proposed that evolution occurs as a result of natural selection. Genes were discovered by Darwin's contemporary, the Austrian monk Gregor Mendel, who sent Darwin a copy of the paper outlining his results and theory. Darwin evidently never opened it [15]. Only after both men were long dead did the scientific community rediscover Mendel's experiments on hybrid pea plants in a monastery garden and understand their importance. And it was several years more before genetics and evolution were seen to complement one another. This insight is resonantly celebrated in biology as "the modern synthesis," a wedding of genetic and evolutionary theory that has structured all subsequent thought about why we animals are as we are.

Mendel's great contribution was to perceive that **inheritance is particulate:** Parental traits are not blended in a continuous fashion but are passed along in discrete packets now called genes. In his experiments he wisely focused upon several traits in pea plants that could be categorized in a binary fashion and that then proved to be determined by single genes (Figure 1–4) [583].

Mendel elegantly explained the reappearance in later generations of apparently lost ancestral characteristics by proposing that organisms carried two full complements of hereditary factors (dubbed "genes" by the Danish botanist Wilhelm Johannsen), one acquired from the mother and one from the father, and that a "dominant" gene might mask the presence of a "recessive" gene (Figure 1–4). The alternative forms of a gene, such as Y and y in Figure 1–4, are its *alleles*. The alleles of a gene are said to occupy one genetic *locus* (pl. loci). By virtue of having its genes in pairs, whether identical (*homozygous*) or different (*heterozygous*), the organism's genotype is *diploid*. The pairs of genes become dissociated in the production of *gametes* (sperm and ova) by *meiosis* (Figure 1–5). Gametes contain only a single complement of genes and are therefore called *haploid*. Each gamete carries one or the other of the parent's two matched genes, with equal likelihood. The fusion of two gametes creates a new diploid individual with

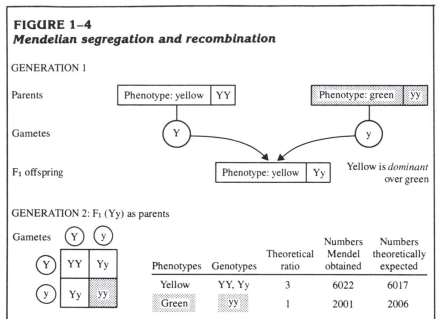

FIGURE 1-4
Mendelian segregation and recombination

GENERATION 1

Parents — Phenotype: yellow | YY   Phenotype: green | yy

Gametes — Y   y

F₁ offspring — Phenotype: yellow | Yy   Yellow is *dominant* over green

GENERATION 2: F₁ (Yy) as parents

Gametes Y y

	Y	y
Y	YY	Yy
y	Yy	yy

Phenotypes	Genotypes	Theoretical ratio	Numbers Mendel obtained	Numbers theoretically expected
Yellow	YY, Yy	3	6022	6017
Green	yy	1	2001	2006

Mendel crossed purebred yellow peas (YY) with purebred green peas (yy). All the offspring ("F₁ generation") were yellow. But when these F₁ peas were bred with one another he obtained 6022 yellow and 2001 green, almost exactly three yellow for each green.

From these results he surmised that both parental genes were present in the F₁ plants (Yy) but that only the dominant gene was expressed and that the parental genes segregated in gamete production before recombining at fertilization.

The 3:1 ratio, he concluded, reflects the fact that F₂ offspring are equally likely to inherit the genotypes YY, Yy, yY, and yy; and the first three of these all produce the yellow phenotype.

(By convention, the dominant gene is represented by an uppercase letter, hence Y for yellow; and the alternative, *recessive* allele by the same letter in lowercase, hence y for green.)

a novel genotype. This segregation and recombination of particular genetic elements was demonstrated by Mendel to proceed at different loci independently (Figure 1-6).

Alternative definitions of a gene have been proposed on the basis of biochemical considerations, but the most useful definition for our purposes remains **the basic particulate unit of inheritance that is passed intact from one generation to another.**

Every individual animal has a *genotype*—the particular combination

FIGURE 1-5
Meiosis

In a diploid sexual organism, most cell nuclei contain two full sets of chromosomes, one of maternal origin (here shaded) and one of paternal origin (here unshaded). The organism illustrated has a diploid number of $2N = 4$.

In *mitosis,* the nonsexual process of cell division that occurs continually and throughout the body, each chromosome is duplicated; one copy goes to each daughter cell, preserving the genetic identity of the diploid cells of an individual.

In *meiosis,* the duplicated chromosomes do not completely separate but remain attached at a point called the *centromere.* Homologous chromosomes from the paternal and maternal complements line up beside one another in *metaphase,* and one goes to each daughter cell. Chromosomes assort independently, so that both, one, or neither of the chromosomes received by a daughter cell might be of maternal origin.

Only after the *first meiotic division* do the paired chromosomes fully separate, and a *second meiotic division* then produces four *haploid* cells with a single set of chromosomes ($N = 2$ in the present example).These cells become four sperm cells in a male; one ovum and three disposable polar bodies in a female.

Haploid gametes from two parents later combine at conception to form a *diploid zygote,* the offspring.

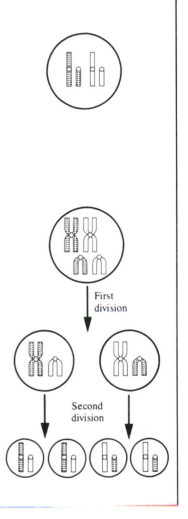

First division

Second division

of genes that have profoundly influenced every detail of its growth and development—and a *phenotype*—the manifest animal as is. The phenotype is a concept that embraces psychological and behavioral attributes as well as physical ones. Both ''genotype'' and ''phenotype'' are often used to refer to a part rather than the whole. Thus, several animals may be said to have the same genotype because they are identical with respect to a particular set of genes under consideration, despite other differences.

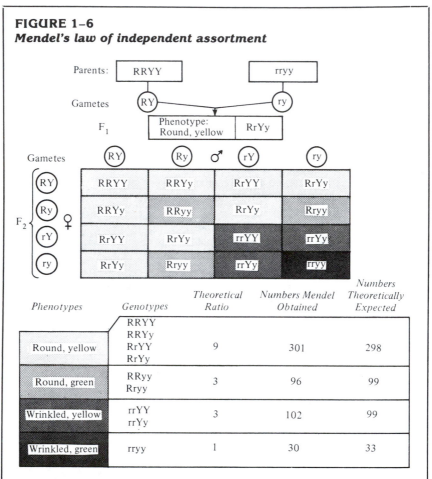

FIGURE 1–6
Mendel's law of independent assortment

Phenotypes	Genotypes	Theoretical Ratio	Numbers Mendel Obtained	Numbers Theoretically Expected
Round, yellow	RRYY RRYy RrYY RrYy	9	301	298
Round, green	RRyy Rryy	3	96	99
Wrinkled, yellow	rrYY rrYy	3	102	99
Wrinkled, green	rryy	1	30	33

Mendel crossed purebred round yellow peas (RRYY) with purebred wrinkled green peas (rryy). All F_1 peas (RrYy) were phenotypically round and yellow. In the F_2 generation all possible combinations were obtained, in the numbers shown. By this and similar experiments Mendel showed that genes determining different traits segregated and recombined *independently* of one another.

The same phenotype can arise from different genotypes, and identical genotypes can give rise to disparate phenotypes as a result of different environments and experiences during development (see, e.g., [267], p. 388). But although there is no strict one-to-one relationship between genotype and phenotype, the former always has some influence on the latter. Thus, part of the variability among individual phenotypes in any population of animals can be attributed to relevant variability in genotypes. This must be so if natural selection is to do its work. For natural selection operates on phenotypes, the actual animals that live and die and breed more or less successfully, but the resultant evolutionary change is transmitted by surviving genes. These are the repositories of the successful qualities of successful parents, the vessels of posterity. That is why **modern biology equates natural selection with the differential reproduction of genotypes.**

A fine example of an observable process of evolution in action is afforded by the phenomenon of "industrial melanism." In areas of heavy industrialization, black (melanic) forms of moths and butterflies often become predominant in species that typically have pale, mottled wing patterns. This phenomenon was long misunderstood, but research in Britain at last established a simple explanation in terms of natural selection [326, 55]. Where industrial pollution is severe, tree trunks lack lichens and may be blackened. Against such backgrounds, the melanic forms of winged insects are inconspicuous, and the robins, flycatchers, redstarts, and other songbirds that prey upon them tend to overlook the melanics. On lichen-covered tree trunks in unpolluted areas, by contrast, the pale forms are camouflaged while the melanics stand out. Melanism arises by mutation at a very low rate, and a good deal is now known of the genetics of melanism. In the best-studied cases melanism is controlled by a single dominant gene.

Industrial melanism is still appearing and spreading in various insects throughout the world. To date it has been found in more than a hundred species of British moths and butterflies. But now in certain areas an interesting reversal of the selective process has taken place. Where antismoke laws have been put into effect, melanics have become rare.

THE PHYSICAL BASIS OF HEREDITY

Mendel postulated genes, diploidy, segregation, and recombination entirely on the basis of indirect evidence, and it is testimony to his brilliance that all these entities and processes have now been directly observed. If Mendel's law of independent assortment (Figure 1-6) were strictly correct, however, evolutionary theory and population genetics would be much simpler than they are. Once Mendel's work was rediscovered in 1900, it was only four years before clear evidence of nonindependent assortment was obtained: Departures from predicted recombination ratios indicated that some Mendelian factors were "linked" (Figure 1-7). And this discovery lent support to an earlier hypothesis that the physical basis of heredity lay in the *chromosomes* in cell nuclei. Though entire chromosomes might assort and recombine independently, two genes situated on the same chromosome would not.

FIGURE 1–7
Linkage

The recessive genes *pr* (purple eyes) and *vg* (vestigial wings) are situated on the same chromosome in *Drosophila melanogaster*. T. H. Morgan crossed homozygous flies with normal red eyes (PP) and normal wings (VV) with pure recessive flies (ppvv) to produce heterozygous F₁ flies (PpVv) with normal phenotypes. He then testcrossed F₁ females with pure recessive (ppvv) males. The advantage of such a test cross is that *each resultant phenotype corresponds to one and only one genotype* (cf. Figure 1-6):

Parental: PPVV   ppvv
Gametes: (PV)   x   (pv)
F₁: PpVv females   x   ppvv males

Phenotype	Genotype	Maternal gamete	Paternal gamete	Numbers Morgan obtained	Numbers expected if assortment independent
Red eye, normal wing	PpVv	(PV)	pv	1339	710
Red eye, vestigial wing	Ppvv	Pv	pv	151	710
Purple eye, normal wing	ppVv	pV	pv	154	710
Purple eye, vestigial wing	ppvv	(pv)	pv	1195	710

But why then was linkage not perfect? In 1909, the Belgian cytologist Janssens published a description of the twisting together of pairs of chromosomes during meiosis (Figure 1-5) and hypothesized that material might be exchanged between them. This *crossing over* (Figure 1-8) was correctly interpreted by Thomas Hunt Morgan, as early as 1911, as the mechanical event responsible for his linkage results. The probability of a crossover in any particular stretch of chromosome would then depend on its length, and so Morgan and Alfred Sturtevant interpreted the degree of departure from Mendelian ratios as a measure of distance along the chromosome.

Morgan had discovered the perfect animals for genetical research—fruit flies of the genus *Drosophila*. With a generation time of only two weeks, numerous lines of flies could be bred in jars in the laboratory and

Similarly, pure strains, each with only one of the two recessive mutations, were crossed, and the F_1 females were testcrossed with pure recessive males:

Parental: PPvv x ppVV
Gametes: (Pv) (pV)
F_1: PpVv females x ppvv males

Phenotype	Genotype	Maternal gamete	Paternal gamete	Numbers Morgan obtained	Numbers expected if assortment independent
Red eye, normal wing	PpVv	PV	pv	157	584
Red eye, vestigial wing	Ppvv	(Pv)	pv	965	584
Purple eye, normal wing	ppVv	(pV)	pv	1067	584
Purple eye, vestigial wing	ppvv	pv	pv	146	584

In both experiments, the F_1 females produced gametes with more of the original parental gene combinations and fewer recombinations than would be expected if the loci assorted independently. The two loci are *linked*. The occasional appearance of recombinants is due to crossing over (Figure 1-8).

scrutinized for mutants. And mutants there were aplenty—individuals with white eyes, unusual numbers of bristles, and stubby wings, mutants called hairless, spineless, eyeless, bent, shaven, tan, cleft, forked, and bobbed. In 1913, Sturtevant, then an undergraduate student in Morgan's laboratory, produced the first *map* of six mutant genes on a single chromosome, their positions estimated from their observed degree of linkage with one another. Soon Morgan and his students had mapped dozens of genes in four linkage groups, precisely *Drosophila*'s number of chromosomes. With such results plus the confirmation of Janssens's crossing-over hypothesis by microscopy, the theory that the genes reside on the chromosomes was secure.

In recent decades, genetical research has advanced enormously—the genetic code is broken and knowledge of the biochemical intricacies of gene expression grows daily. But the molecular biology of the gene is not our

---

FIGURE 1–8
Crossing over

During meiosis, homologous chromosomes may exchange corresponding segments.

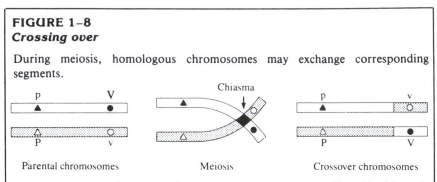

Parental chromosomes   Meiosis   Crossover chromosomes

If a crossover point (a *chiasma*) happens to occur between two linked loci (e.g., P and V in the illustration), then recombination of the parental alleles at the two loci occurs: The probability that a chiasma will occur between two loci increases as the distance between the loci increases. In the example above, the crossover has produced recombinations PV and pv of the parental genotypes pV and Pv.

---

topic. Mendel's successors had already arrived, by 1930, at an essentially correct, more molar account of the physical basis of heredity [167]. The stage was set for a modern interpretation of Darwinism. The fact of discrete genetic differences between individuals necessitated that a populational view replace typological taxonomy—a species was a gene pool, an amorphous, varying, statistical entity. Natural selection was to be understood as differential reproduction of genotypes, and evolution was to be understood as change in the gene pool, such as one allele's replacing another. And the alleles that survived were those that built the phenotypes most skilled at replicating them. This view of evolution by natural selection is still essentially unchanged, but its implications are still being explored.

MULTIPLE EXPLANATION OF BEHAVIOR

The concept of natural selection puts purpose into blind nature. It makes the question "Why?" intelligible. Biologists may wonder, for example, why a male bird sings in the spring. And they may suggest that it is "in order to" attract mates. The purposiveness of this explanation does not trouble them, nor should it. They realize that it implies neither conscious intent on the part of the bird nor the benevolent attentions of a supernatural planner. What the biologists are suggesting is simply that bird song is maintained by natural selection because of its utility in the attraction of mates and hence in promoting successful reproduction.

Other scientists confronting the question of why a male bird sings in the spring may propose a quite different answer. Experiments have shown that song is dependent upon high levels of the hormone testosterone in the bird's blood [17]. We know further that the increase in testosterone levels results from a spring growth in the size of the testes, in which the hormone is produced. And we know that the environmental factor upon which the growth depends is the spring increase in the number of daylight hours [453]. Having traced this chain of events, our scientists may feel they have a very satisfactory explanation of why a male bird sings in the spring.

Different as these two explanations are, they are not competing alternatives. Their difference hangs on a difference in meaning of the question "Why?" They illustrate the distinction between *ultimate causes* and *proximate causes*. **Proximate causation** concerns the direct mechanisms that bring something about—in this illustration how the coming of spring induces song. It is, in a sense, a structural description of the individual bird as a behaving machine, of how the animal is organized so that different environmental factors and events influence its behavior in particular ways. **Ultimate causation** concerns adaptive significance—that is, the value of spring song, its selective consequences, which must ultimately entail reproductive consequences. It is causation on a generational time scale. The claim that it is causation at all rests on the assumption that these adaptive results of behavior have been instrumental in its evolving and being maintained by natural selection.

The distinction between proximate and ultimate causation must be understood. In the pages that follow we shall often talk as if the animal computes the costs and benefits of various courses of action and then behaves in order to maximize payoffs or to minimize risks. This is nothing more than a metaphor, a convenient shorthand. These sophisticated computations need not be a part of the immediate proximate causation of the animal's "choice." Cost and benefit are implicit in the process of natural selection, for it is selection that has designed the mechanisms of proximate causation to achieve their ultimate goals as efficiently as possible. The male bird need not "decide" when to sing. Natural selection has assured that his reactions to seasonal factors will make him begin to sing in that season when song is useful to him.

Ethologists employ different terms to draw the same distinction. They call proximate causation "cause" and ultimate causation "function." Admittedly, there is a certain confusion in this terminology. Both processes are causal, although on different time scales. Moreover, there is great ambiguity in the word **function**. Ethologists generally restrict its usage to those adaptive consequences of behavior presumed to have had selective significance during the history of the species. Experimental psychologists are apt to broaden the meaning of function to include behavioral consequences on which selection has never acted. They may refer, for example, to the functions achieved by a rat's pressing a bar in an experimental apparatus. Furthermore, a physiologist's functional account of a system may be a mechanistic description that is utterly proximate; here the meaning is virtually the opposite of the ethologist's!

Proximate and ultimate levels of explanation may be further subdivided. Experimental psychologists often investigate the proximate causes of behavior by examining the associations between stimuli and responses, such as the fact that bird song is light-dependent. Other scientists, equally concerned with proximate causation, will want to investigate the physiological mechanisms underlying that association—the mechanisms of day-length detection, for example, and the links by which these mechanisms stimulate the testes. Taking a different tack, the scientist interested in ultimate causation may be concerned to show that song repels other males of the same species and thus serves to defend territory. Others will be interested in demonstrating more ultimate consequences—for example, that variations in the success with which territory is defended are indeed related to variations in reproductive success.

When all these investigations are complete, our question—Why does a male bird sing in the spring?—is not yet fully answered. If we take an egg from the nest and hatch out and rear a male in isolation, he may never be able to sing a normal adult song. In the white-crowned sparrow, for example, such confined upbringing will result in a song that is recognizably that of the parent species, and yet is aberrant: The sounds are correct, but the normal sequence of phrases is lacking [338]. The bird has been deprived of a critical experience: hearing an adult song when he was an impressionable nestling, many months before he would first open his own beak in song. This experience is crucial for many species of songbirds, though not all. Thus, a third type of explanation is invoked: the developmental. If proximate causal explanation aspires to a sort of structural description of the organism as an exquisitely complex behaving machine, then a developmental explanation is concerned with changes over time in the machine's structure. Development within an individual life span is called **ontogeny**, in contrast to the evolutionary change, called **phylogeny**, that takes place over many generations. Phylogeny may be considered a fourth category of explanation.

A phylogenetic explanation of behavior is an account of the evolutionary progression by which the behavior (or the proximate causal structure underlying it) has been formed out of some preexisting organization. We have no good phylogenetic account of bird song at present, but we can make hypotheses. For example, song might have evolved from the simpler alarm calls given by the same birds at the approach of a predator.

Explanations in terms of ultimate causation and phylogeny are intimately related, but they can be differentiated. The former is concerned with the adaptive significance of behavior—with what it is for and with the consequences that are relevant for its maintenance by natural selection. Phylogeny is concerned with the raw material out of which the behavior evolved—the preexisting behavioral organization from which natural selection has been able to sculpt the behavior in question. In practice, phylogenetic explanation of behavior has been relatively neglected, and we' too must give it less than its full due. Nevertheless, knowledge of phylogeny will often be relevant for understanding the particular adaptations available to particular animals. Internal fertilization preadapts many animals for maternal care, for example. Ancestral traits constrain an evolving species' options. Lactation, for example, makes it unlikely that any mammal will evolve uniparental male care of young, as have various birds, frogs, and fishes.

The Dutch ethologist Niko Tinbergen has suggested that there are four great problems common to all areas of biology, including behavior, and that each deserves equal attention. He refers to these as cause, function, development, and evolution [613, 614]. They correspond to our proximate causation, ultimate causation, ontogeny, and phylogeny. We decide which explanatory level is appropriate whenever we attempt to answer the question "Why?" Yet defenders of an explanation in one of these modes often imagine that their pet theory obviates explanation in another. Most of the acrimonious debates of behavioral science derive in large measure from this sort of narrow advocacy. Take the controversial subject of aggression, for example. There has been much fruitless debate among three groups: supporters of ontogenetic explanations of aggressiveness, such as the theory that attaches great importance to the behavior of parents as models for children; those who explain aggression as the proximate result of frustration; and still others who see it as an evolved aspect of behavior selected for its value in attaining useful goals. Nobel prizewinner Konrad Lorenz, in his popular book *On Aggression* [394], correctly stressed that aggressive behavior has evolved because of its adaptive value to the aggressor. He repeatedly implied, however, that aggression must therefore be impervious to variations in early experience, which neither follows logically nor accords with actual evidence. Some critics have replied with a similar non sequitur: Since aggressive behavior is learned ontogenetically, there is no validity to evolutionary explanation. Both arguments overlook the necessity for multiple explanation. Natural selection operates on phenotypic outcomes, whatever their ontogenetic histories. If aggression serves its perpetrator, it will be selected for. Natural selection is no less potent because the aggressive behavior happens to be in part the result of learning. It is quite wrong to imagine that behavior that has been learned has not also evolved. But it is equally wrong to suggest that behavior that has evolved is therefore somehow "contained in the genes," as if environmental variations during ontogeny could have no influence at all.

In a similar debate, those who would explain male and female roles as evolved features are pilloried by others who see such roles as the products of sexist child-rearing practices. It should be clear that there is no necessary incompatibility between these two views! The question of the origins of the behavioral differences between men and women is emotionally charged. Liberals and radicals often perceive biological explanations of behavior as threatening and reactionary. To suggest that sex differences have an evolutionary history and an adaptive explanation seems to many critics an attempt to justify an unjust status quo. This we deny.

Sex differences must be explained at all the levels we have discussed. Belief in nature does not preclude an appreciation of the importance of nurture. People may intentionally alter the relationship between the sexes by educational means. Sex roles as we know them may, in the future, be deliberately abolished. These are among humanity's options. Social reformers will not make their tasks easier by ignoring evolutionary biology nor by a naive insistence that our present roles are entirely products of an arbitrary sexist culture. Behavior must be explained—indeed, can only be explained—multiply.

Having stressed the need for multiple explanation, we have chosen to start by concentrating on ultimate explanation in terms of adaptive significance. Although behavior may be explicable at the proximate level of physiological systems, those systems must themselves be understood in terms of their adaptive significance. We shall see in Chapter 8, for example, that females of different mammalian species have different reproductive cycles. These cycles can be understood partly in terms of proximate causation, but a proximate causal account fails to explain why they differ as they do. Why, for example, is a mouse's ovulatory cycle so much shorter than a monkey's? Such questions demand ultimate explanations.

The same considerations apply to developmental accounts of an individual's behavior. The behavior can be explained in terms of the sequence of processes leading up to it, but developmental chronologies and processes are themselves evolved adaptations. Why is the female elephant seal reproductively mature at a much younger age than the male? The different rates of maturation can be given a proximate mechanistic explanation, but there is also an adaptive significance to the difference.

Hypotheses about ultimate causation are much more difficult to investigate experimentally than hypotheses about proximate causal mechanisms or ontogenetic processes. Evolutionary theory was therefore largely neglected by behavioral scientists until the 1960s. How the study of adaptive significance was then given fresh impetus is the subject of our next chapter, and in those that follow we shall see what light the modern evolutionary perspective sheds on the facts of sex and social behavior.

SUMMARY

Sex is a biological phenomenon, and its nature and origins can best be elucidated within a biological framework. To make proper use of that framework, one must understand the principle of natural selection: Some individuals leave relatively great numbers of offspring who inherit to some degree whatever attributes made their parents relatively successful reproducers. Satisfactory explanation of biological phenomena must be pursued at multiple levels: Any behavior typical of some animal can and should be understood in terms of its physiological control, its adaptive significance, its developmental history, and its evolutionary history.

SUGGESTED READINGS

Ghiselin, M. T. 1969. *The triumph of the Darwinian method.* Berkeley: University of California Press.

Ruse, M. 1979. *The Darwinian revolution: science red in tooth and claw.* Chicago: University of Chicago Press.

Stern, C., & Sherwood, E. R., eds. 1966. *The origin of genetics: a Mendel source book.* San Francisco: Freeman.

2
Fitness

Herbert Spencer epitomized Darwinian theory as "survival of the fittest," but this catchy phrase is doubly misleading. In the first place, survival is relevant only in the service of reproduction: Long-lived celibates are evolutionary failures whose genes die with them. The important criterion of success is reproduction. Moreover, the fittest are not necessarily those who can do their calisthenics without wheezing. Being fit in the ordinary sense of that word may have little or nothing to do with success in nature's reproductive scramble. **Fitness is now usually defined in biology to mean reproductive success.**

This definition has led some critics to suggest that the principle of natural selection in its modern form is circular and devoid of meaning: The best reproducers are the best reproducers. But it is more than just that. It is in effect a three-part statement: Variations in reproductive success exist among the members of populations; such variations are partly due to genetic differences and are therefore heritable; and these facts together explain the mechanism by which gene pools and phenotypes evolve in the direction of improved adaptation.

Fitness, in its modern sense of reproductive success, can sometimes be given precise mathematical expression and applied to particular alternative genes as well as to whole animals. For instance, among fruit flies males woo females with a stereotyped series of courtship behaviors that increase the probability that a female will permit copulation. A number of cases have been discovered in which a change in a single gene lowers the male's fitness by making his courtship behavior less effective [175]. Males burdened with one such gene vibrate their wings at a frequency other than the most seductive. Males with another are too forward in their courtship. The costs in fitness of each such gene can be worked out precisely in experiments involving breeding competition. Fitness estimates derived in this way can be used to predict the rate at which such genes will be eliminated over generations or the rate at which advantageous genes will spread.

Caution is needed here, however. A numerical value of fitness is specific to a particular population at a particular time. A "bad courtship" gene is costly only in competition with better courters. Thus, fitness is always expressed **relative to the alternatives**. (It is customary to set the fitness of either the fittest allele or the average allele equal to 1.0, and to then measure others relative to it. In Figure 1–3, p. 6, for example, the mutant type who raises three young might be assigned a fitness of 1.0, in which case the other type's fitness would be 0.67.) Furthermore, the same gene's fitness may change over generations or according to circumstances. To take another example from fruit flies, certain rare genotypes enjoy a mating advantage over commoner ones simply by virtue of their rarity: The females are more receptive to a male of novel characteristics, and this effect can halt the decline of an otherwise disadvantageous allele [175]. A gene's fitness is also a function of the environment and even of the particular set of other genes at other loci within the same organism.

We also speak of the fitness of alternative courses of action. An animal's behavior can thus be said to be **designed** (cf. p. 39) by natural selection to maximize fitness. This is generally a matter of arriving at the best compromise solution to the many demands impinging on each animal. Male songbirds, for example, aggressively defend territories from others of their species. If a bird is not aggressive enough, he may be unable to control an area adequate for his food requirements. But it is possible to be too aggressive. The costs in time and energy can exceed the benefits. If he becomes preoccupied with preventing all incursions upon his space, he may neglect to contribute adequately to the nurture of his nestlings. Clearly, there is an optimal level of aggressiveness (which may vary with the stage of the breeding cycle and other variables), a fittest level of aggression. Birds that aggress either more or less than is optimal face the penalty of reproductive failure.

Moreover, this fittest level is likely to be the usual level, thanks to natural selection, which acts to eliminate departures in either direction while preserving those genotypes that perform at the optimum. The important general conclusion is that **species-typical (or population-typical) behavior is usually fit behavior**. Constant weeding by natural selection makes the optimum the norm. This principle is elegantly demonstrated by David Lack's studies of clutch size, discussed later in this chapter.

WYNNE-EDWARDS'S CHALLENGE

With the concepts of fitness and differential reproductive success understood, we can dispel a popular and persistent misunderstanding. Many people, including some biologists, envisage natural selection as a process that somehow ensures that animals behave for "the good of the species," but this is not a necessary outcome of selection. Animals behave for their own good—or, more precisely, for the good of their own genes ("good" here meaning replication and perpetuation). Natural selection, as a competitive process within species, helps those genes that help themselves. Any gene that helps its carrier to reproduce more fruitfully than other members of the same population will gain in its proportional representation in the gene pool. Fitter genes supplant less fit genes, a fact implicit in the definition of fitness.

Consider the behavior of members of a harem-forming species. The Hamadryas baboon is a large terrestrial monkey that lives in bands in semiarid areas of East Africa. The Swiss zoologist Hans Kummer has studied these bands and discovered that they consist of many small harems of one to five females guarded by an adult male [349]. This maned overlord, far bigger than the females, herds the little harem relentlessly and punishes those that stray with a neck bite of savage appearance.

That is the way of Hamadryas society. Like it or not, a male born into such a society has little option but to behave accordingly. Let us imagine two males of this species. One jealously ensures the fidelity of his harem in the way we have described. The other is unconcerned if his females mate with other males, but instead plays the altruist, channeling his energies into pursuits of value to the group, such as sentry duty. Whose genes will be better perpetuated? And if genetic differences are indeed relevant to the behavioral differences between the two males (potentially a very important if!), which type of behavior will prevail in future generations? Selection does not favor attributes that help the species; it favors attributes that help individuals reproduce better than their neighbors. Naturally, such attributes are often good for the species too, but they needn't be.

The foregoing argument suggests that any gene that enhances its bearer's fertility, by whatever devious chain of causality, must gain in its representation in the population at the expense of alternative genes that bestow lesser fertility. The principal effect of natural selection must surely be the maximization of reproduction. Yet nature does not always present the appearance of a mad scramble to reproduce. All about us we see restraint and decorum. Why? In most species of birds, for example, each breeding female lays a predictable number of eggs and then stops. A yellow-shafted flicker, for example, usually lays a clutch of six to eight eggs, one per day, then incubates the eggs and rears the young. However, in a classic experiment one female laid seventy-one eggs in seventy-two days when her eggs were removed as they were laid [496]. Clearly, the flicker has the physiological capacity to produce many more than the half-dozen eggs to which she normally limits herself. Why does she show such restraint? Why has not natural selection decreed her replacement by less dainty breeders?

In 1962 considerations such as these led V. C. Wynne-Edwards of Aberdeen University to publish his theory, massively documented, that reproductive restraint is nature's normal state of affairs [700]. This restraint, he suggested, protects animal species from the disastrous population crashes that follow unrestrained growth and the exhaustion of resources. The evolution and persistence of such restraint, he reasoned, must depend on a process of natural selection different from the individualistic competition we have considered thus far. The process he envisaged was the extinction of whole groups of animals as a result of their profligacy, while other groups, less inclined to destroy their own resource base, survived and thrived. This process has been called group selection—or, more precisely, interdemic or interpopulational selection—in contrast to individual selection.

Wynne-Edwards argued that the workings of group selection and individual selection must be antagonistic to some degree; group selection may be expected to curb the excesses of individual competition. This is so because the calculus of individual selection assures that any gene that gives its carrier some competitive edge over other group members is likely to spread within the group. But the victory of that selfish gene will only be to its own short-lived advantage and eventual woe, if it endangers the group as a whole. So, at least, goes the argument. Variations that maximize breeding performance must repeatedly be eliminated by the extinction of the groups in which they arise.

Wynne-Edwards amassed many data that were challenging to proponents of the primacy of individual selection, but his theory has not fared well. Population biologists and geneticists were never happy with it—the processes that the theory demanded were too implausible [422]. For instance, vertebrate populations are seldom if ever divided into groups that are either as genetically isolated one from another or as numerous as the theory seemed to demand. A gene that is individually disadvantageous but of benefit to the group can become established in small gene pools only by chance (genetic drift). There would always be danger that immigrants would bring selfish genes into the gene pool and that these would replace the unselfish. Data on gene flow between local populations could not be reconciled with the demands of the theory. Nor were the frequent group extinctions that were alleged to cleanse populations of the disastrous consequences of selfish genes anywhere in evidence.

It is sometimes the way in the history of ideas that an unacceptable theory is among the most heuristic. Group selection is a case in point. Wynne-Edwards drew explicit attention to the "problem of altruism": How could a competitive process of natural selection lead to the evolution of restraint or mercy or cooperation? Biologists were goaded by Wynne-Edwards's formulation of the problem and by his inadequate solution. The result was both new empirical research and new theory.

LACK'S RESPONSE

One of the most telling replies to the group selection theory was that of Oxford ornithologist David Lack, who persuasively argued that breeding restraint is illusory [354, 355]. Take the example of the bird that lays fewer eggs than she might. Would she in fact increase her individual reproductive success by laying more? She would not, as Lack and his followers have demonstrated in a variety of studies.

The classic research on this question is Lack's own work on swifts, the most aerial of birds. Swifts feed in flight and allegedly even sleep aloft. They must, however, alight to raise their young, and this they do in nests on cliff faces, in hollow trees, or in chimneys. The European swifts studied by Lack usually lay a clutch of two eggs, occasionally three, and very rarely four. The parents then work long hours bringing food to their begging nestlings. In a wet year, when conditions for hunting insects on the wing are poor, parents with three nestlings to feed may lose them all, and fewer

FIGURE 2-1
The Lack effect

The most successful clutch size is the usual clutch size.

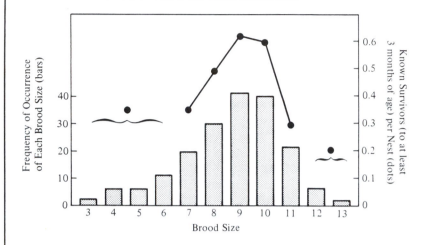

Breeding success of the great tit (Parus major) near Oxford, England, 1949–55

Unusually large broods (eleven or more nestlings) actually produce fewer surviving young than do typical broods of nine or ten. (Data from Lack, Gibb, & Owen, 1957 [358].)

young swifts are successfully fledged (raised to the age of independent flight), on average, from a brood of three than from a brood of two. Sometimes less is more (Figure 2-1).

In general, Lack argues, the most common clutch size will be the most productive clutch size, since natural selection will eliminate genotypes that tend to produce less productive clutches. The Lack effect has been widely tested, and it holds up well with certain qualifications [334, 580, 273]. For instance, it may be surprising to discover that the most common clutch size is often somewhat smaller than the most productive. This apparent paradox has at least three explanations. First, individual birds can vary their clutch size to some degree according to food availability and their own nutritional state. The result is that larger clutches are laid by the birds best able to raise them. In one study of African weaverbirds, for example, rare clutches of four eggs produced slightly more fledglings on average than typical clutches

This effect does not always hold. Sometimes, especially when conditions are good, large broods are as successful as broods of more typical size, and sometimes the largest broods are more successful.

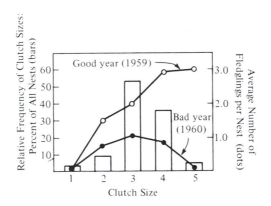

Breeding success of the boat-tailed grackle (Cassidix mexicanus) in Texas

In a "bad year" with severe flooding, the most frequent clutch size is the most successful, but in a "good year" unusually large clutches are most productive. (Data from Tutor, 1962 [625].)

of three, presumably because the only birds who laid four were those relatively able to cope with them. The hypothesis that birds who opted to lay only three eggs could not raise four was tested experimentally: When some clutches of three were artificially increased to four, fledging success was reduced [649].

The second reason the typical clutch size may be somewhat less than the most productive concerns the conservation of parental resources. The parent's goal is maximal *lifetime* reproductive success: A fractional gain in the fledgling output for one year may not be worth the price in exhaustion if the parent is so debilitated that its own prospects of surviving to nest again are jeopardized [217].

Third, some practical difficulties of measuring reproductive success should be noted. If an unusually large clutch produces an unusually large number of fledglings, that does not necessarily mean a greater share of the future gene pool. Will those fledglings breed? Large broods fledge at lower individual body weights than smaller broods, and fledgling weight is related to the probability that the young bird will survive the winter and eventually breed [355, 675]. So a big brood may sometimes produce more fledglings but fewer yearlings than a smaller one.

Although the real world always has some new complications, Lack's idea that reproduction is in fact maximal rather than restrained has been largely vindicated. The clutch size studies have been extended successfully to the explanation of litter size in mammals and have been the prototype for a variety of research that has put to the test the proposition we advanced earlier: The behavior that typifies a population is fit behavior.

In retrospect, it is not surprising that the most effective way to maximize one's personal reproductive success may not be to drop a maximum number of eggs or helpless infants into a hostile environment. Successful reproduction is a task that extends beyond the expulsion of an offspring from the parental body. The nurturing of young creates time and energy demands upon parents that may be in conflict with the goal of maximum fertility. Thus, fitness is more appropriately measured as the number of young surviving to reproduce in their turn, rather than simply as the total number produced. We shall examine the trade-off between intensive nurture and fertility maximization in considerable detail in subsequent chapters.

HAMILTON'S RESPONSE: INCLUSIVE FITNESS

An even more important reply to Wynne-Edwards's theory of group selection has been a reanalysis and extension of the concept of fitness beyond simple reproductive success. We have suggested that a fit gene is one that somehow helps itself to increase numerically relative to alternative genes. The most obvious way to do this is to facilitate its carrier's own reproduction. But is there another way? Suppose a particular "nepotistic" gene somehow disposes its bearer X, to help a sibling. X's sibling has the same one-half probability of carrying the nepotistic gene as would X's own offspring (Figure 2-2). From the gene's point of view, it is as useful for X to feed his infant brother or sister as it would be for X to feed his own son or daughter. If what is being naturally selected is the gene's contribution to its own numerical increase, assistance to collateral kin as well as to offspring can be of selective advantage. But this selective advantage—in other words, the expected average payoff in replication of the gene—will decline the more distant is the aided relative, for more distant relatives are decreasingly likely to share any particular gene.

This, in brief, is the logic behind the valuable concept of inclusive fitness, which we owe to W. D. Hamilton [249]. When we assess the fitness

FIGURE 2-2
Familial relatedness

An individual of a sexual species carries half its mother's genotype and half its father's (excluding, for simplicity, the sex chromosomes). Any maternal or paternal allele has a .5 probability of being represented by a descendant copy in the offspring. Thus, *parent-offspring relatedness* (Wright's coefficient of relatedness, designated *r*) equals .5.

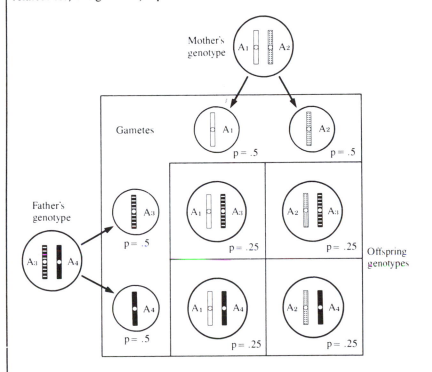

For simplicity, only one chromosome is illustrated. There are four equally likely offspring genotypes. Since each parental gene is represented by a descendant copy in half the offspring, any particular allele in an offspring has a .5 probability of being represented by a copy, *identical by descent* (*ibd*) from the same parent, in a full sibling: thus, *r* for full siblings = .5.

Half siblings share one parent only. An allele from the shared parent has a .5 probability of being represented in a half sibling by a copy *ibd*; an allele from the unshared parent has a 0 probability. The average probability that an allele will have a copy *ibd* in a half sibling is .25, and *r* for half siblings thus equals .25, *r* for cousins = .125, for grandparent/grandchild = .25, and so on.

r = the probability that any randomly selected allele in a focal individual has a copy *ibd* in the related individual.

of a gene (or of a bit of behavior), we must consider more than its contribution to the reproduction of the individual animal. We must also consider whether the gene in any way influences the reproductive prospects of kin. **Inclusive fitness** is a property of individual organisms, equal to the focal individual's own reproductive success (classical Darwinian fitness) plus the focal individual's incremental (or decremental) influences upon the reproductive success of kin multiplied by the degree of relatedness (Figure 2-2) of those kin. Similarly, the **inclusive fitness effect** of a gene is that gene's influence (relative to its alleles) upon its carrier's own reproduction, plus its influences via the carrier's phenotype upon the reproduction of kin multiplied by the degree of relatedness of those kin.

A thought experiment may help to clarify Hamilton's idea. Imagine a rare allele that inclines its carrier to act *altruistically* toward a relative. "Altruism" here refers to the actor's incurring a cost (c) in order to bestow a benefit (b) upon the recipient, both cost and benefit being measured in terms of changes in Darwinian fitness (personal reproductive success). Under what circumstances can such an allele be favorably selected and thus increase in frequency over generations? Ordinarily we would expect any allele that damages its carrier's reproductive success to be selected *against*. However, in the case of altruism, the recipient of the benefit may carry a copy of the allele by virtue of common descent from the same ancestor, so that the "altruistic allele" is really helping its own replication. The probability that the recipient indeed carries such a copy is r, the coefficient of relatedness between actor and recipient (Figure 2-2). Hamilton's condition for such an allele to be favorably selected (hence for the evolution of altruism) is that the cost to the altruist must be less than the benefit to the recipient times the probability that the recipient indeed has the gene copy: $c < rb$ (often written $r > c/b$). When this is true, altruistic acts will tend to increase the fitness of the allele. Thus, if I incur a cost of c fitness units in order to bestow b fitness units upon my full sister ($r = .5$) then each of my genes suffers the fitness cost c but may expect to gain $b/2$; if $b/2$ is greater than c, then my *inclusive fitness* is enhanced. If the recipient is instead a cousin ($r = .125$), then the expected gain is only $b/8$, and the benefits must be much greater or the costs much lower for altruism to be selectively favored. In other words, Hamilton's condition for the evolution of altruism becomes increasingly stringent as the relationship between the parties becomes more distant. The geneticist J. B. S. Haldane is said to have anticipated Hamilton's theory by remarking, "I would give my life for three brothers or nine cousins!"

Inclusive fitness is an improved concept that supplants fitness in its simpler sense of personal reproductive success. The idea that **organisms are designed by natural selection to contribute to the replication of their genes** is a more general and more powerful idea than that they are designed merely to reproduce. Before Hamilton's analysis, many instances of altruistic behavior in nature appeared to present a paradox for the theory of evolution by natural selection. How, for example, could sterile worker bees, abstaining from personal reproduction in order to raise someone else's young, evolve by natural selection? Or why should a ground squirrel risk its neck to warn a neighbor that a predator is near? As we shall see in Chapter 3, these and other examples can be elegantly accounted for by Hamilton's theory. It is now apparent that thanks to kinship, behavior other than that of personal reproduction can pay off in the replication of one's genes. The paradox has evaporated.

LEVELS OF SELECTION

Claims that organisms behave for the good of the species are now dismissed as "naive group selectionism." Arguments such as Lack's and Hamilton's have convinced most animal behaviorists that adaptation must be sought at the level of individual advantage. But the argument can also be advanced that the individual is no more appropriate a level at which to attribute adaptation than is the group [140, 142, 288].

Individuals in sexually reproducing species do not, after all, produce copies of themselves. Their genotypes are disassembled at meiosis and lost forever at death. The entity that replicates itself with sufficient frequency and fidelity to provide a stable unit for analysis on an evolutionary time scale is the gene. Adaptive attributes are therefore not "for the good of" groups nor "for the good of" individuals but "for the good of" genes. For "good," read replication and hence posterity. We can measure the fates of alternative alleles over generations, but individuals are ephemeral. Hence the gene, not the individual, is the basic "unit of selection." Richard Dawkins, the leading spokesman for this radical perspective, has suggested that "The time has come to carry [Hamilton's] 'selfish gene' revolution to its conclusion, and give up the habit of speaking of adaptation at the individual level" [141, p. 75].

One problem with this advice is that what we observe are individual organisms behaving, and it is this behavior that we wish to understand. The individual is no arbitrary or illusory level of organization upon which to focus: For one thing, the organism's fitness is the genes' only route to replicative posterity. The separate genes in an individual generally sink or swim together. This is because the gene cannot outlast the organism in which it finds itself, except by being replicated in another individual, along with a replica of a gene at each of the organism's other loci. Hence, each gene in an individual organism will almost always maximize its own fitness if the organism maximizes its inclusive fitness. From this point of view, Hamilton's inclusive fitness concept provides a way of salvaging the focus upon the individual organism instead of having to attend to the genes themselves as the level at which selection operates [424]. (There are, however, some exceptions to the general principle that an organism's constituent genes all maximize their fitnesses together. Intragenomic conflict, in which some genes gain fitness at the expense of others within the same individual, is a real possibility whose evolutionary significance is just beginning to be explored [116, 171].)

Hamilton's inclusive fitness theory is commonly referred to as kin selection theory, and many authors treat kin selection as a level that is somehow intermediate between individual and group selection. This confuses matters. Recall that natural "selection" is a metaphor that treats nature as an agent analogous to an animal breeder. "Individuals" and "groups" are real entities that an agent could "select" to breed from, but "kin" is not such an entity. Perhaps this simply indicates how far the concept of "selection" in evolutionary biology has moved from Darwin's initial metaphor. In any case, we prefer to avoid the popular phrase "kin selection."

Debate continues over the level(s) at which selection operates [see 10]. The question is probably too fuzzily formulated to be resolved, but we may sharpen it a little with this translation: At what level of organization do alternative entities "compete" for representation in future biota? Genes? Individuals? Groups, species, higher taxa, communities even? Perhaps at all these, but then at which levels is selection *effective in fashioning adaptation* [674]?

Despite a resurgence of new "group selection" models in population genetics, and despite the theoretical interest of intragenomic conflict, we think that **there is no real alternative to maintaining a focus on the individual organism as a "strategist" in pursuit of its inclusive fitness interests.** It is at that level that theory has been most heuristic.

HERITABILITY

Selection acts directly on phenotypes, not genotypes. It is not the genes that live or die, breed or help their relatives, but the realized animal. The result is that both phenotypes and gene pools evolve. An evolutionary series of fossil animals, changing in structure gradually over the course of millennia, manifests an evolution of phenotypes (Figure 1-2). It is assumed that this must reflect concomitant changes in the gene pool. For it is only by such genotypic changes that evolutionary modification of phenotypes can be retained and accumulated. The gene pool must change for evolution to occur, and it must change because of selection. Natural selection is powerless to produce evolution if phenotypic differences in survival and fertility are unrelated to genotypic differences, for in such a circumstance no systematic changes in the gene pool would accrue.

This is the crux of the concept of heritability: Phenotypic differences must have a basis in genotypic differences if selection is to operate. A classic example comes from efforts to breed laying hens for improved egg production. Breeders have been able to increase egg size by selectively breeding from those hens who laid the largest eggs. However, similar attempts to increase the number of eggs produced have been less successful. In the language of population genetics, egg size was highly heritable, whereas the number of eggs was not. Among the original group of laying hens there existed some genetic variability that was relevant to the size of the eggs they produced, and the presence of this variability meant that selection could be effective. By contrast, there was no genetic variability relevant to egg number and selection for this trait was ineffective (Figure 2–3).

"Heritability" is defined mathematically as *the proportion of the observed phenotypic variance that can be attributed to correlated variance in genotypes*. Therefore, heritability depends upon the actual variability in a particular gene pool—hence, in one particular population of animals. Heritability is not an immutable property of the trait in question. It is misleading to say that egg size is more heritable than egg number, for example, unless it is made clear that a particular population of real birds is under discussion. In a different group of hens the number of eggs might be highly heritable. Suppose, moreover, that one were to set about estimating the heritability of egg size *after* the breeders had selected for large eggs for many generations. Any remaining variability would then have little or no genetic basis, and our experimenter would conclude, in contrast to the earlier investigators, that egg size was *not* heritable. That conclusion would be correct. Every estimate of heritability is specific to a particular population, and generalizing further always risks error.

There is more. The fact that a trait is highly heritable does not imply that it could not be changed by altering the environment. All it means is that a large proportion of the observed variability is correlated with genetic differences under the particular environmental conditions that prevailed. That egg size was more heritable than egg number gives us no clue as to which of the two measures of productivity would change more if we were to alter the diet or the degree of crowding of the hens or if we exposed them to a rooster. A heritability estimate is not only specific to a real population with a particular history of natural selection but is also specific to the environment in which that population exists.

We are not arguing that heritability is a useless concept. Agricultural-

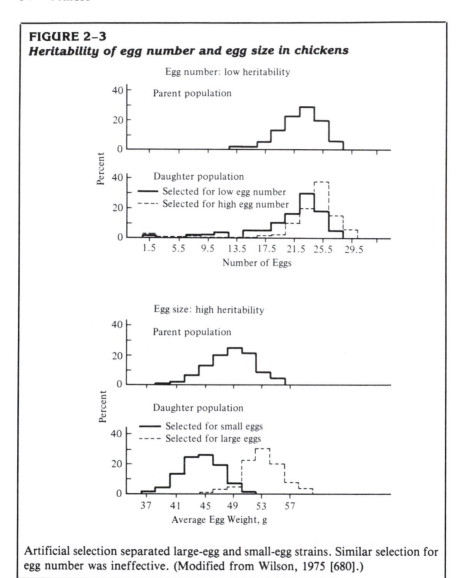

FIGURE 2–3
Heritability of egg number and egg size in chickens

Artificial selection separated large-egg and small-egg strains. Similar selection for egg number was ineffective. (Modified from Wilson, 1975 [680].)

ists and other biologists can use it effectively to describe particular gene pools and their potentialities. Broader generalizations may sometimes be possible, too. For example, if comparative studies of many bird species showed that egg number generally had low heritability, we might gain some insight into the selective pressures operating in nature: We might suppose that genetic variability in egg number has been kept down by intense selection, or that a capacity for adaptive modification of egg number in response to environmental variables has been favored, or both (cf. pp. 26–27). However, it is important to understand the limitations of the concept of heritability, for it has been much abused. In discussions of human intelligence, for instance, it has been supposed that high heritability carries the implication that education and other environmental modifications can have little effect. This inference is unfounded.

NATURAL SELECTION AND COMPETITION

The natural selection that drives evolution is a process of **competition** between genotypes within a population or species. Certain genotypes (whether alleles at single loci or larger linkage groups) reproduce more successfully than their alternatives. The result is that the species' gene pool, the set of genes that in a sense *is* the species, is altered. Competition also takes place between different species, and when one species prevails over another, this is indeed a sort of natural selection. But in itself this *inter*specific selection will not cause any species to evolve.

Starlings are hole-nesting Eurasian birds that were introduced into North America less than a century ago. In 1890 Eugene Scheifflin of the Acclimatization Society, an organization with the lofty aim of familiarizing Americans with each of the birds mentioned by Shakespeare, released eighty of these pugnacious birds in Central Park, New York City [437]. In 1891 forty more were imported, and the starlings did the rest themselves, multiplying prodigiously so that by 1971 their North American population was estimated to be 100 million [239]. One factor in their success is the starlings' extreme aggressiveness in appropriating nest holes from other birds, such as flickers and bluebirds. Several hole-nesting species, unable to compete successfully with the starlings, have suffered population declines and local extinctions.

As far as we know, the selective process by which starlings have supplanted their nest-site competitors has not directly modified any species. There is, however, a way in which such competition can make species evolve. A competitor is one sort of **selection pressure**. Starlings might interfere with the reproduction of only some flicker phenotypes (and hence, presumably, genotypes), thereby indirectly favoring others—those that respond more aggressively to the starlings, perhaps, or those that choose nest sites relatively unattractive to starlings. The starlings then become a selective agent, acting upon the flicker population to favor certain genotypes over others. In this they are analogous to other selective agents that are in no sense competitors of the flickers—environmental agents such as trees of varying hardness or characteristics of the flickers' prey.

Some further comments may be in order here about the way we are using the word *competition*. No direct confrontation is necessarily implied. Every animal species exploits limited resources, and some creatures will get more than an average share. If two animals pick about for scarce food items in the same area, they are in competition, even if each is unaware of the other. If a female mammal is pregnant, some male has impregnated her, and this precludes another male's doing so: There is competition between males for the opportunity to inseminate females, even if the males never fight or even meet.

We depart still further from everyday usage when we suggest that a flicker that can repel starlings is successfully competing with flickers that cannot, but such usage is common in biology. Many authors prefer to restrict the word to those cases where there is a scarce resource, but population biologists may use the word *competition* in any case where individuals are differentially successful. It is in this sense that we consider evolution by natural selection to be a process of intraspecific competition.

SUMMARY

A gene's fitness is a matter of its reproductive success relative to alternative alleles. Fitness of whole organisms and fitness of behavioral acts can be defined similarly: reproductive success relative to other organisms and alternative acts.

Wynne-Edwards challenged the view that natural selection tends to maximize reproduction by pointing to apparent reproductive restraint in nature. However, studies by Lack and others indicate that parents who raise few young may be raising as many as they can. Furthermore, Hamilton expanded the concept of fitness from personal reproduction to inclusive fitness, by showing that the quantity that selection tends to maximize is not simply reproduction but the replication of the organism's genes in both descendant and collateral kin.

SUGGESTED READINGS

Dawkins, R. 1976. *The selfish gene.* London: Oxford University Press.
Dawkins, R. 1982. An agony in five fits. Chapter 10 of *The extended phenotype.* Oxford: W. H. Freeman.
Williams, G. C. 1966. *Adaptation and natural selection.* Princeton: Princeton University Press.

3
The animal as "strategist"

We organisms can be viewed as machines designed by natural selection to replicate our genes. An animal's various activities can therefore be considered the elements of a **strategy** whose goal is the maximization of inclusive fitness. Clearly, we are using the word "strategy" metaphorically. We are not implying that organisms make decisions or have conscious goals. The metaphor is useful because it allows us to speak of the functional significance of behavior without long, clumsy phrases. It is much simpler to say that a male Hamadryas baboon herds his females "in order to assure their fidelity" than it is to say that the male's herding behavior increases his inclusive fitness by decreasing the probability that one of his females will be inseminated by a rival male, with the result that the herding male enjoys a selective advantage over males that do not herd.

It is often useful to elaborate the strategy metaphor in "economic" terms. Every animal has a limited amount of time or energy to budget into various activities. One may choose to invest more time in seeking food, for example, by giving up some of the time invested in advertising for mates. Alterations in such budgets lead to altered payoffs. The bottom line is inclusive fitness, since natural selection tends to adjust time and energy budgets to maximize this single quantity.

STRATEGIES AND GOALS

Although inclusive fitness is in principle the one superordinate goal, it is rarely possible in practice to measure directly the reproductive consequences of the behavior in which we are interested. Instead, we measure feeding rates and longevities and body weights and copulation frequencies. But we should not be discouraged by having to attend to measures that are imperfectly related to fitness. How, after all, do animals themselves pursue their fitness interests in a variable and unpredictable world? The answer must be that they too focus upon more proximate goals that tend on average to be correlated with fitness. Keep your energy reserves up, your ectoparasites down, your competitors away, your escape routes open. Animals clearly have strategies for the attainment of these more immediate goals. An example comes from the study of feeding strategies.

Chickadees and tits (*Parus* species) are engaging little birds common in wooded areas of Europe and North America. Unlike many small birds of temperate zones, they remain the year round in the areas of their summer breeding. In winter, tits move in small flocks through the woods in search of the insect larvae that are their staple food. They are tame birds, active and unconcerned when observed by people, so they have been the objects of much scientific study.

Searching for larvae in twigs and pine cones takes skill. The larvae are often rather rare, and they are usually well hidden and irregularly dispersed. Tits cannot waste their time and effort in a barren area, but when they find a rich one, they want to get full value from it. They shouldn't keep going over areas they've already searched. In a word, they want to search efficiently.

In order to assess the little birds' foraging efficiency, we have to compare their behavior with some theoretically derived conception of what the behavior is designed to achieve. A common first hypothesis in foraging studies is that the behavioral control mechanisms are designed to maximize the rate of net energy gain. That is clearly a rather abstract goal. An animal who is to achieve it needs some decision rules or *tactics*. The basic decision rule that chickadees seem to follow in foraging on pine cones is a "giving-up-time" rule: If you haven't found a larva for a certain time, move on to the next cone. This "certain time" is flexible, varying with the birds' experientially based assessment of overall habitat quality. Alternative decision rules they might use, but do not, include searching each cone for a fixed total time, regardless of the number of larvae found, or always departing after a fixed number of larvae have been discovered. Because the larvae are irregularly distributed, the strategy used is indeed the most efficient, as the British ethologist John Krebs and his colleagues have shown by simple calculations and experiments [343]. The giving-up-time rule maximizes the number of food items located in a given time period, in comparison to the less efficient alternative decision rules.

Ethologists have not generally drawn a sharp distinction between "tactic" and "strategy." The former is a little closer to a direct description of behavior and its causation; the latter is a more complex kind of functional hypothesis. The hypothesis that a chickadee obeys a giving-up-time rule is relatively tactical and descriptive, and it is relatively easy to agree how to check it against the data. The more strategic hypothesis that this rule serves an energy-maximization function is a little trickier to verify. What one can do is to show, as above, that other rules are less effective routes to the hypothesized goal. Here, and indeed in assessing functional hypotheses in general, our confidence in the hypothesis is based on an argument from **design** [674]: This particular decision rule seems ideally designed to maximize this particular quantity; hence that is its presumed function. The logic is exactly the same as that used by a paleontologist in deciding that a pterodactyl was a flying reptile because its forelimb structure was clearly designed to support a wing.

Functional hypotheses can be made more or less plausible by arguments from design, but they cannot be proven. Some biologists therefore object to the functional approach to behavior, considering it to be an untestable kind of storytelling. We disagree. Like all good scientific hypotheses, a strategic model is formulated in such a way that it is vulnerable to *dis*proof. Sometimes, for example, foraging birds clearly do *not* behave so as to maximize energy intake but instead "waste time" with energetically unprofitable prey [220]. In such cases, we conclude that the decision rules controlling behavior are *not* part of an energy-maximization strategy, and we may then try to formulate and test other hypotheses about the forager's strategic goals. Perhaps, for example, it is not energy that the bird is trying to maximize but calcium. Or perhaps foraging rules aim at a more complex balance of nutrients. Notice, however, that when the animal fails to obey the decision rules that we have predicted, we do not conclude that the animal is ill-designed but rather that we have guessed wrong about what it is designed to do. Although it is at least logically possible that we have had the bad luck or bad judgment to seize on some aspect of behavior with no adaptive function at all, we would be foolish to start from that hypothesis. Whereas specific functional hypotheses can be tested and disconfirmed, the hypothesis of functionlessness has no specific, testable implications. Thus, it seems far more heuristic to view organisms as evolved strategists.

REPRODUCTIVE STRATEGISTS

Foraging strategies, antipredator strategies, and the like are clearly amenable to study. But, as we noted earlier, these must be elements in a larger fitness-promoting strategy. Considering one class of subordinate goals in isolation from others is a little artificial, if for no other reason than because animals must budget their time and attention. Several theorists have therefore endeavored to develop a more encompassing strategy theory.

One intuitively appealing concept is **reproductive effort**, which may be partitioned into **mating effort** and **parental effort** [25, 401]. A male, for example, may benefit both by courting females and by protecting infants, but these activities compete for his time. There is presumably some optimal distribution of each male's efforts. With a finite amount of time, energy, and resources that can be devoted to reproduction, each animal should strive to distribute its reproductive effort so as to maximize inclusive fitness. As we saw in Chapter 2, natural selection automatically tends to produce maximally fit behavioral strategies by eliminating maladaptive alternatives.

Reproductive effort is something that animals *expend* or *invest*. If we are to apply this economic analogy at all rigorously, we shall need to have a scale for these outlays. In particular cases, we may speak of animals investing their time or their energy reserves, but neither of these is a general enough "currency." The common denominator of all such investments is **the expenditure of residual reproductive potential**.

Consider the killdeer, a common bird of American pastures, which lures predators away from its chicks with an elaborate pretense of broken-winged vulnerability (Figure 3-1). Such predator distraction displays may cost little in time or energy, but the risk to the parent may be important. That risk is an investment. Occasionally such a ruse must cost a parent bird its life. In principle it should be possible to determine the display's cost in expected reproductive success, that is to say, the average diminution of future breeding incurred by each use of the display. In principle one could

FIGURE 3-1
Predator distraction display of the killdeer, Charadrius vociferus

(Photograph by M. P. Kahl.)

also determine its fitness payoffs. Where the risk is too high or the gain uncertain, the display is not worth the cost: More young are likely to be left by the parent that saves herself to breed again than by one who dies in defense of this year's brood. By equating reproductive effort with the relinquishing or risking of future reproductive prospects, we are able to include predator distraction displays in a unitary theoretical scheme along with fighting sexual rivals and feeding the babies.

Within this framework, an animal's expected future contribution to its fitness is its **reproductive value** [191]. This quantity varies with the animal's age, sex, condition, and other circumstances. Old animals, for example, tend to have low reproductive value. When an animal makes a reproductive effort, it generally suffers a reduction in reproductive value in exchange for present fitness returns. The currency of both investment and return is fitness units.

The preceding paragraphs set out a conceptual framework to which we shall often refer. **A reproductive strategy is a program for the allocation of reproductive effort**. When to reproduce? With what expenditure of effort? And how shall the effort be distributed? As we shall see, both diversity among species and differences between the sexes can be characterized in terms of reproductive effort budgets.

PARENTAL EFFORT AND DISCRIMINATIVE SOLICITUDE

Many aquatic animals—oysters and certain fish, for example—make no parental contribution after spawning. All their reproductive effort is *pre-zygotic*. Most familiar animals, however, invest in offspring beyond the zygote stage. Or at least one sex does. It is by no means unusual for females to make a substantial *post-zygotic* investment and for males to make none at all. Causes and consequences of this sexual asymmetry will be the subject matter of Chapter 5. What we should like the reader to remark here is that two individuals, the parents, gain fitness as a result of only one's parental effort. The other, the male in this case, is getting a free ride. There is an interesting general point here: One animal's parental effort is a sort of resource that one or more others may exploit [619]. Fobbing off the work load on somebody else can be a profitable strategy, and a number of animals have evolved adaptations to parasitize the nurture of quite unrelated creatures.

Brood parasites are birds that lay their eggs in another bird's nest and then depend upon the duped foster parents to do all the work of incubating and rearing their offspring [489]. Some eighty species of birds are obligate brood parasites, which is to say that they are totally committed to this strategy and completely lack the behavioral repertoire of parenting—everything from nest building to feeding the nestlings. Most brood parasite species are cuckoos of one sort or another (though American cuckoos are not parasitic); one is a duck; altogether, the strategy is believed to have evolved independently seven times in seven different avian taxa.

Now it behooves the "host," as the victim is ironically labeled, to recognize the brood parasite's egg and it behooves the parasite to overcome the host's discrimination mechanisms. Thus, victim and parasite have often evolved considerably together. European cuckoos, for example, generally lay eggs that visually mimic the eggs of their hosts in color, in speckling or streaking, even in size. Since a single cuckoo species parasitizes several songbird species with eggs of different sizes and colors, the female cuckoos are specialists: Each picks only upon the host species that her own egg mimics.

After hatching, the deception continues. Young birds generally gape at their parents in order to be fed, and in many species the gape displays a visual pattern in the beak and throat of the helpless infant, which serves as a specific visual stimulus for parental feeding. These patterns are mimicked in remarkable detail by parasitic widow birds. Brood parasites employ still further ruses besides these mimetic ones. Some host species will abandon a nest if they find an extra egg in it; parasites counter by removing one of the host's eggs for each one that they lay. Even so a parasite's egg may provoke the host to abandon the nest or, in some cases, to build a new floor over the eggs and lay a fresh clutch. This may be repeated several times as the parasite persists in leaving its unwelcome addition.

Once the egg is established warmly under the brood patch of an unsuspecting foster parent, a crucial aspect of the parasitic strategy comes into play. The parasite's egg generally requires less incubation time than its hosts' eggs and hatches first. Some parasites then exhibit a remarkable and

---

FIGURE 3-2
An adult hedge sparrow, Prunella modularia, *feeding a much larger parasitic nestling cuckoo,* Cuculus canorus

(Photograph by J. Markham.)

ruthless adaptation. They dispose of their nestmates and thereby gain the exclusive attentions of their foster parents. Newly hatched cuckoos, though bald, blind, and in most ways helpless, go through a specialized and complex series of contortions that serve to laboriously throw other eggs and young out of the nest. In another group of parasitic birds, the African honey guides, hatchlings have a lethal hooked bill that they use to murder their nestmates. Other species do not directly attack their foster siblings but simply grow faster, gape more persistently, and starve them out. The parasite is often a relatively huge bird. Once the foster parents have become attached to their voracious ward, it no longer seems to matter that he look like one of their own, and they may continue to feed him till he reaches grotesque proportions (Figure 3-2).

Brood parasitism is a strategy that frees a mother from one burdensome aspect of reproductive effort—parental care—and thus permits increased effort elsewhere—egg production, in fact. A parasitic cuckoo commonly lays twenty or more eggs in as many different nests within a season, whereas her nonparasitic relatives produce clutches of only about two to four [488]. This strategy thus illustrates the trade-offs in reproductive effort budgeting. It also illustrates a quite different point: From the host's point of view, the parasite is a selection pressure, and what it evidently selects for is increasingly discriminative stimulus control of parental solicitude. Parents should evolve to avoid squandering nurture on unrelated young, and this should apply to young of one's own species every bit as much as to baby cuckoos.

Discriminative parental solicitude is conspicuous in nature [130]. It seems to be best developed in species in which the breeding situation is such that there is some real risk of misdirecting parental care to nonrelatives. Thus guillemots, marine birds whose nests are typically situated only a few centimeters apart, recognize their eggs and newly hatched chicks, and will reject experimental fosterings. The related razorbill, on the other hand, nests more dispersedly and is oblivious when an experimenter swaps eggs or hatchlings [53]. A nanny goat forms an individualized attachment to her own kid immediately after birth and will reject other kids who approach her. A territorial burrow-dwelling rodent, on the other hand, whose nurslings are incapable of roaming, is so lacking in parental discrimination as to accept any conspecific pups and even other species. At an age where pups become mobile, and misdirected parental investment becomes a real risk, maternal discrimination emerges.

So organisms may be said to have strategies for the discriminative channeling of post-zygotic investment to their own offspring. And offspring, as W. D. Hamilton [249] reminded us, are just a subset of the larger class of genetic relatives in whose welfare organisms have a benevolent interest. Hamilton thus broadened our conception of the organism from a **reproductive strategist** to a **nepotistic strategist**.

NEPOTISTIC STRATEGISTS

In Renaissance Rome, priestly celibacy seems not to have been a major bar to paternity. The bestowal of patronage upon the "nephews" of Vatican officials was so rife as to warrant its own name: *nepotismo*. The modern English meaning is not so very different, and biologists have borrowed the term to epitomize the inclusive fitness concept. "Nepotism" thus refers to favoritism toward kin, including offspring. Natural selection should favor the expenditure of reproductive effort only where it enhances the reproductive value of relatives. As Richard Alexander [9, p. 46] has said, "According to inclusive-fitness theory, then, we should have evolved to be exceedingly effective nepotists, and we should have *evolved* to be nothing else at all."

Parental nepotism is conspicuous and abundant, but what of benevolence to collateral kin? The likeliest candidates are the closest relatives, namely siblings. In many species of birds, for example, young adults stay home for a year or two, helping to feed nestlings who are not

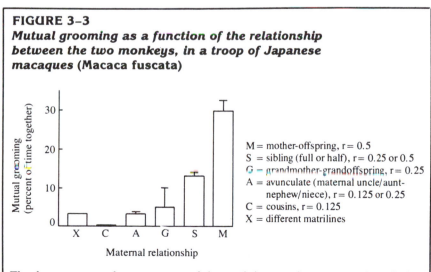

FIGURE 3-3
Mutual grooming as a function of the relationship between the two monkeys, in a troop of Japanese macaques (Macaca fuscata)

M = mother-offspring, r = 0.5
S = sibling (full or half), r = 0.25 or 0.5
G = grandmother-grandoffspring, r = 0.25
A = avunculate (maternal uncle/aunt-
      nephew/niece), r = 0.125 or 0.25
C = cousins, r = 0.125
X = different matrilines

The data represent the percentage of the total time a pair spent together, during which either groomed the other. Relationships are reckoned through maternal links only, since paternity is unknown. (After Kurland, 1977, Figure 6 [351].)

FIGURE 3-4
A female Belding's ground squirrel alarm calling.

(Photograph by George D. Lepp.)

their own offspring but are most often their full siblings and hence just as good inclusive fitness vehicles ($r = .5$) as offspring would be [71, 178]. In group-living monkeys, sibling solidarity is well documented, both in peaceful activities such as mutual grooming, and in coming to the aid of brothers and sisters under attack [351, 416, 434]. Monkeys, in fact, show finer discriminations too, favoring close relatives over cousins and the like, but favoring these more distant kin over nonrelatives (Figure 3-3).

One of the most intensive studies of nepotistic discrimination has been conducted in a high mountain meadow in the Sierra Nevadas of California, by Paul Sherman of Cornell University [558, 559, 560]. Belding's ground squirrels hibernate for most of the year and hurry through their foraging and reproductive activities during the brief alpine summer. For several years, Sherman and a small army of field assistants have observed a population of distinctly marked (courtesy of Lady Clairol hair dyes) individuals. The observers have found a matrilineal sociospatial organization: Daughters settle and breed on territories near their mothers, whereas males disperse annually and do not maintain contact with relatives.

   Ground squirrels have both terrestrial and aerial predators, and can often be seen standing up on their hind legs looking alert. If a predator is spotted, the squirrel keeps it in sight and may emit a loud "alarm call" (Figure 3–4) alerting other squirrels foraging in the grass. Alarm calling is a clear example of the sort of "altruistic" behavior that inspired the development of inclusive fitness theory. Observations indicate that the behavior at least occasionally draws the predator's attention, provoking an attack on the caller. So why not keep your eye on the predator and your mouth shut? The plot thickens: Some squirrels do just that, and we are immediately led to inquire what determines *whether* a squirrel calls. It takes a lot of data to answer such a question. In almost 4,000 observation hours, 119 spontaneous interactions were observed between terrestrial predators and marked squirrels with known matrilineal kinship links.

   Is alarm calling effectively nepotistic? If so, then females, who live among relatives, should be likelier to call when a predator appears than should males, who are isolated from kin. Sherman [558] was able to con-

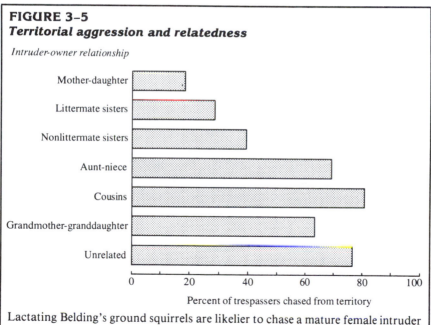

FIGURE 3–5
Territorial aggression and relatedness

Intruder-owner relationship

Percent of trespassers chased from territory

Lactating Belding's ground squirrels are likelier to chase a mature female intruder from their territory when that intruder is a distant relative or nonrelative than when the intruder is a close relative. (After Sherman, 1980, Figure 3b [559]. Reprinted by permission of Westview Press. Copyright © 1980 by the American Association for the Advancement of Science.)

firm this prediction. Perhaps more interesting than the simple sex difference was the fact that females called or not according to their circumstance. Reproductive females with living mothers, sisters, daughters, or grand-daughters nearby were likelier to call than reproductive females without such living relatives!

Besides the relatively rare, dramatic emergency of a predator's appearance, female squirrels also exhibit nepotistic discrimination in their more mundane interactions with one another [559]. Each reproductive female has a territory around the burrow where she lives alone, but she defends that territory less aggressively against relatives than against unrelated females (Figure 3-5). In fact, related animals sometimes cooperatively defend their adjacent territories and even run over to help defend one another's pups. Unrelated neighbors do nothing of the sort, even if they have lived next door to one another for two years or more.

The ground squirrel study shows how nepotistic networks can be elucidated by painstaking fieldwork. Nepotistic solicitude, beyond that of a parent to a dependent offspring, is not easy to document. We should like to have information on the genealogical relationships among individuals, and there is no quick and easy way to acquire it. Fatal fights, predation incidents, and other selectively significant crises are rarely observed and are often over before we know what is happening. But in spite of all these problems, recent field studies are demonstrating that collateral kin are indeed significant vehicles for nepotistic investment in many animal species [e.g., 439, 694, 434, 661].

STERILE WORKERS AND NEPOTISM

We people have a quite understandable tendency to view ourselves as the acme of social evolution, but the social insects must then be granted a summit of their own [679]. The effective cooperation of thousands of honeybees is an awesome spectacle: collecting nectar and telling one another where they found it, building brood cells, controlling hive temperature, feeding pupae, cleaning and defending the hive. And all this labor is conducted by sterile *workers*, while a single *queen* lays all the eggs. The queen is the mother of all the workers and of the female and male reproductives as well. The female reproductives are potential queens in hives of their own; the males serve only as mates during the queens' single youthful reproductive flight. After the mating flight, each queen settles in her hive with a lifelong supply of sperm and no further need of a mate.

The challenge that the lifelong sterility of worker bees presented to his evolutionary theory was not lost on Darwin, nor was the relevance of kinship to its solution. He argued that the sterile individual labors on behalf of her own "stock," and that natural selection between families could resolve the apparent paradox of worker sterility evolving by survival of the fittest. With no satisfactory theory of heredity, he could take the argument no further.

More than a hundred years later, in formulating the concept of inclusive fitness, W. D. Hamilton was still primarily concerned with the curiosity of a social organization in which some individuals are routinely sterile (a "eusocial" society), and he had a completely novel insight. There is something peculiar about the genetic system of insects of the order

---

FIGURE 3-6
Haplodiploid reproduction

In several animals, including insects of the order Hymenoptera, *females are diploid* in all their body cells, like ourselves, but *males are haploid*.

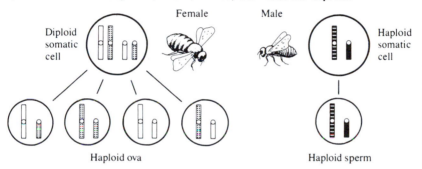

The female produces haploid eggs by meiosis, and her ova contain different subsets of her genotype, just like a female in an ordinary diploid species (cf. Figure 1-5). The male, by contrast, produces sperm by mitosis. Each of his sperm carries his complete genotype, and all his sperm are *genetically identical*.

When an ovum is fertilized, the diploid zygote always develops as a female. An unfertilized ovum develops as a haploid male.

The haplodiploid system produces asymmetries of relatedness (*r*):

	Relative:					
Focal individual	Mother	Daughter	Sister	Father	Son	Brother
Female	.5	.5	.75	.5	.5	.25
Male	1.0	1.0	.5	—	—	.5

r = (probability that an allele in the focal individual will be represented by a copy, identical by descent, in the relative).

Hymenoptera, to which the eusocial bees, ants, and wasps belong: The males are haploid. The consequence of **haplodiploid reproduction** (Figure 3-6) is that females are more closely related to their sisters than to their own daughters! This is because their father, being haploid, produces sperm by mitosis, so that each spermatozoan carries an identical, complete complement of paternal chromosomes. The diploid queen, on the other hand, produces ova by meiosis, just like females in a familiar diploid species such as ourselves. The result is that sisters are, like ourselves, half alike on average in their maternal inheritance, but they are identical in their paternal inheritance, and hence 75 percent alike over their entire diploid genome. So while a female's sister carries three quarters of her genes, her daughter bears just half. If animals have evolved to maximize *inclusive* fitness, then a female might prefer to invest her effort in raising sisters rather than daughters from the egg stage to reproductive maturity. Hamilton [249] suggested that the peculiar elevated relatedness of Hymenopteran sisters constitutes a sort of preadaptation for the evolution of eusociality. The sterile worker may not be such an altruist after all, but a sensible nepotist.

There is a complication, however: What the worker gains in raising sisters she may lose in raising brothers. From the point of view of a female faced with the choice of either reproducing or raising siblings, a son, like a daughter, is a product of a haploid ovum and hence a bearer of one half of the maternal genotype. But a brother, who is on average half like his sister in maternal inheritance, has no paternal genes in common, since he has no father at all! He therefore carries just a quarter of the sororal genotype. Hence, from the point of view of a female, one sister and one brother would represent on average one complete copy of her genotype (.75 plus .25), the same as one daughter and one son (.5 plus .5). If that were all there were to it, we should be hard pressed to perceive any link between haplodiploidy and eusociality. However, the asymmetrical relatedness of siblings allows the sterile worker to find an inclusive fitness edge: All she has to do is bias her nurture toward sisters and away from brothers. This small point has some fascinating ramifications, as we shall see in Chapter 9.

Not all haplodiploid insects are eusocial, and not all eusocial insects are haplodiploid, but the two phenomena do seem to be associated. Eusociality has evidently evolved about 11 different times in the haplodiploid Hymenoptera and only once in all the much more numerous taxa of diploid insects [679]. That one case, incidentally, is no obscure oddity but a successful social experiment that has radiated to become a large order: the termites (Isoptera). Unlike the Hymenoptera, termites exhibit ordinary diploidy, and the sterile workers and soldiers are of both sexes. The basis for eusociality in this group must be altogether different from that in the Hymenoptera. Perhaps termite eusociality is related to the necessity for social transmission of the symbiotic creatures that inhabit termite stomachs and break down cellulose there. But, the termites' alternate route to eusociality notwithstanding, it would appear that haplodiploidy and the powerful sisterhood that it can create must be given some of the credit for the astonishing evolution of social forms in bees, ants, and wasps.

KIN RECOGNITION

Organisms may well be evolved nepotists with benign inclinations toward their relatives, but how do they know who those relatives are? Of course individuals need not recognize relatives for inclusive fitness effects to be important; in a species with little dispersal, for example, neighbors will generally be related, and "altruistic" alarm calling might therefore be selectively favored in the absence of kin recognition. Wherever animals behave discriminatively in direct social interactions, however, some basis for the discrimination is necessary. Kin recognition is no trivial problem. One mistake can be devastatingly costly—a parasitic cuckoo fledged and a season or even a whole reproductive career wasted.

We have already mentioned the discriminative solicitude of a newly maternal goat toward her kid. That the discrimination is normally established in a brief period immediately after birth has been demonstrated experimentally [335]. Brief postnatal experience with her own kid will lead a goat to reject all alien kids, but if the alien kid is substituted for her own immediately after birth, then the mother transfers her affections to the adoptee. Her genetic offspring can later be returned to her, and it will be indignantly rebuffed like any other uninvited milk thief.

Clearly, it is not just simple familiarity that predicts beneficence. What is evidently important is contact at a particular time or situation where it is reliably indicative of kinship. Recall the ground squirrels who remained aloof from long-familiar next-door neighbors. Sherman was in fact able to show that females cooperated more with relatives they had recently met (younger half sisters) than with unrelated neighbors they had known longer [559].

This is not to deny that there is sometimes tolerance of individually known but unrelated individuals. In several songbird species, for example, ingenious experiments, in which tape-recorded songs are played back to territorial males, have demonstrated that these males recognize by voice their unrelated territorial neighbors, with whom they have a sort of noninterference pact; it is only a stranger's song that provokes investigation and aggression [69]. It would be simplistic to conclude, however, that mere familiarity produces fellow feeling [708]. Organisms are subtler than that. A certain tolerance of familiar neighbors and a quite distinct benevolence to familiar relatives can coexist in a single breast. Animals evidently rely upon the circumstances of their exposure to others in order to draw that distinction. The pup that emerges from one's womb is assumed to be one's own. A less reliable but common inference is that the egg in one's nest is likewise one's own; the fallibility of that inference, however, has evidently provided a sufficiently intense selection pressure for many species to have evolved individual recognition of eggs by their mottling. Collateral kinship is inferred by circumstance too. Nestmates may be presumed to be siblings. And in Sherman's ground squirrel study, some limited but intriguing observations [559] suggest that a mature female will behave nepotistically toward her younger sisters only if they were first seen in company with the mother, and have thereby come to be recognized as siblings.

Besides such circumstantial evidence, an animal might be able to assess relatedness on the basis of the **phenotypic resemblance** of the in-

---

FIGURE 3-7
Electrophoresis

Electrophoresis of a tissue or blood sample is a method for separating electrostatically distinct forms of enzymes, corresponding to alternative alleles of the genes coding for those enzymes.

The sample is poured into a well in an electrolytic "gel" (a solid or semi-solid material through which the enzyme molecules can pass). A strong current is applied across the gel, and the various components of the sample migrate toward the electrode at different rates according to their different electric charges.

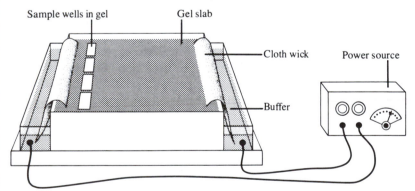

After a certain interval, the electric current is removed, and the gel is *stained* to reveal the particular chemical substances of interest. Illustrated below is the result of such staining for hemoglobin in three people: (1) a normal

dividual in question either to oneself or to a known relative [360]. Recent evidence indicates that Sherman's squirrels also employ this sort of phenotype matching. When a female squirrel comes into estrus, she typically mates with several males in rapid succession, and her subsequent litter is usually multiply sired [255]. Consequently, littermates may be either full siblings (r = .5) or half siblings (r = .25), and there is presumably no circumstantial cue that can help a squirrel distinguish her full sisters from half-sister littermates. By the electrophoretic analysis (Figure 3–7) of blood enzymes, however, Sherman *could* make that distinction [277], and the remarkable fact is that half-sister littermates were more aggressive to one another as adults than were full-sister littermates! It seems that the squirrels must assess phenotypic resemblance to themselves, probably by odor, in order to discriminate between the two classes of littermates.

If animals can use phenotypic cues in this way, then the question arises whether organisms might ever "recognize" relatives they have never met. It

---

homozygote, (2) a heterozygotic carrier of the sickle-cell trait, and (3) a homozygote afflicted with sickle-cell anemia:

Phenotype	Genotype	Hemoglobin electrophoretic pattern Origin	+4	+2	0	Hemoglobin types present
1. Normal	Hb^AHb^A				●	A
2. Sickle-cell trait	Hb^SHb^A			●	●	S and A
3. Sickle-cell anemia	Hb^SHb^S			●		S

Note that electrophoresis of several enzymes can provide information about genotypes and hence about relatedness. The method is used in paternity testing. If a child exhibits genotype (2) for example, and the mother genotype (1), then the Hb^s allele must have been inherited from the father, and any man with genotype (1) is *excluded*. Genetic analysis cannot conclusively establish paternity; it can establish only nonpaternity. When several rare alleles are involved, however, the likely father can sometimes be identified with a probability approaching certainty. Or if all possible fathers are known (as, for example, when a female squirrel has been watched continuously throughout her estrous period), then all candidates but one may be excluded, establishing the paternity of the last candidate. (After *An introduction to genetic analysis,* 2nd ed., by David J. Suzuki, Anthony J. F. Griffiths, & Richard C. Lewontin, W. H. Freeman and Company. Copyright © 1981, Figures 4–1 & 4–2 [590]. Figure 4–2 was adapted by W. H. Freeman from Rucknagel & Larus, New York: Hober 1969, Figure 2.)

seems clear that this sort of naive recognition occurs at least occasionally. The best-studied example concerns a small, dark North American bee, *Lasioglossum zephyrum*. This species is "primitively eusocial" in that the differentiation of worker and queen is neither extreme nor irrevocable; many females indeed pass their lives as nonreproductive laborers, but the degree of ovarian development is variable. *L. zephyrum* nests in a burrow that is founded in the spring by a single queen. Several of her daughters mate and return to the natal nest, and unrelated females may try to gain entry too. The single burrow entrance is guarded by one or more workers who block the tunnel with their bodies and apparently check the identity of would-be entrants by exploring them with chemically sensitive antennae. In summer, the foundress dies, but her daughters carry on. By late summer, the burrow is complex and is occupied by several laying females and workers, who presumably vary in their relatedness to one another [31]. It would seem that an inclusive fitness premium might accrue to the bees who could exercise some kind of kin recognition ability. That they indeed have

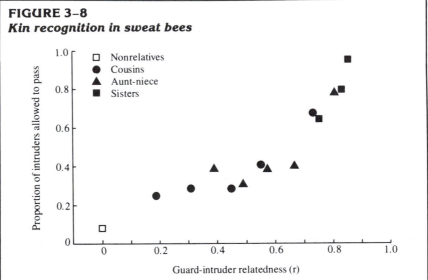

FIGURE 3–8
Kin recognition in sweat bees

Does a guard bee permit a stranger to enter the nest? Fourteen groups of bees were tested: nonrelatives ($r = 0$), cousins of five different degrees of relatedness (inbred to varying degrees), aunt-niece of five degrees, and sisters of three degrees (outbred, $r = .75$; inbred, $r = .80$; highly inbred, $r = .86$). The more closely related the strange intruder was, the likelier the guard bee was to permit her to enter. The figure is based on 1,586 guard-intruder interactions. (After Greenberg, 1979, Figure 1 [228]. Copyright 1979 by the American Association for the Advancement of Science.)

such an ability has been demonstrated in ingenious experiments by Les Greenberg at the University of Kansas [228].

L. zephyrum turns out to be a tractable species in captivity. By controlled matings starting from initially unrelated stocks, Greenberg was able to breed bees with several specified degrees of relatedness to one another. Experimental colonies were then established with uniform soil and diet so that these environmental features could not provide discriminative cues for colony recognition. The experimental test of kin recognition was then simplicity itself: Would a guard bee let a stranger pass? A number of guard-intruder pairs were tested at each of fourteen degrees of relatedness, and the results were exceptionally clear: Relatedness is a powerful predictor of the probability that a guard bee will permit a stranger to enter the burrow (Figure 3-8).

One might suppose from Greenberg's result that these bees are well equipped to make subtle discriminations on the basis of degree of relatedness to themselves. In fact, the situation is less clear than this. Experiments in which bees were fostered at a larval stage have shown that the guard directly assesses the intruder's relatedness not to herself but to those who raised her [74]. Fostered to nonrelatives, she will later admit strangers in proportion to their degree of relatedness to her foster parents, while spurning her genetic relatives. Thus, a bee normally learns what her relatives smell like but evidently does not use her *own* odors as a basis for comparison.

PARENT-OFFSPRING CONFLICT

Relatives are to be cherished. That, in a sense, is the message of inclusive fitness theory. A and B have a commonality of interest insofar as that which contributes to A's inclusive fitness also enhances B's. The more closely related, the more common cause. Indeed, if A and B are genetically identical, their inclusive fitness interests are utterly harmonious, for there is no way either can ever gain at the other's expense.

Conversely, the degree of genetic nonidentity is a measure of the potential for conflict. A and B have a conflict of interest insofar as that which contributes to A's inclusive fitness subtracts from B's. Such conflicts of interest are most obvious where individuals compete to make the same use of a scarce resource—whether mates, food, or a particularly fine nest site. But conflict can show its face in predominantly cooperative ventures too—one parent may try to leave all the work to the other, for example, a profound conflict that we shall explore further in subsequent chapters. And, as Robert Trivers pointed out in a persuasive paper in 1974 [620], we can apply this concept of conflicting fitness interests to the parent-offspring relationship too, since parent and offspring are not genetically identical. In particular, parents wish to partition their parental investment among offspring in a way that is optimal for their own inclusive fitness, while it might reasonably be expected that each offspring would like to extract more than its share of parental care.

Consider a juvenile mammal, still under parental care but with a growing capability for an independent existence. At a certain point it becomes a bad investment for its mother. Perhaps it is still consuming her milk, although it is also processing solid foods. Or perhaps its continued attachment to her affords it some slight protection from predators but also places her at some risk. Whatever the exact balance sheet of costs and benefits, the day comes when the mother's own fitness would best be served by sending it on its way. True, she can continue to assist it and it is her offspring, but she has more to gain by getting on with the business of the next child. When that point comes, the youngster may be expected to resist. Its genetic fitness continues to increase by virtue of its mother's care, and it can hardly be expected to relinquish valuable maternal investment without protest.

Now, when we speak of offspring fitness, we of course refer to their inclusive fitness. Since their mother is close kin, they are not indifferent to her future reproduction. Her further reproduction has payoffs in the replication of their genes. There is thus some point at which the offspring is gaining relatively little in reproductive value (expected fitness) from the continuation of maternal care and would gain inclusive fitness if the mother were to invest elsewhere. At that point, the offspring may be expected to cease protesting their mother's withdrawal of care.

What this theory suggests is that there is a stage of conflict that begins at the point where the mother's investment costs her more inclusive fitness than it gains her and ends at the point where it costs the young more than it gains them. During this stage the mother should endeavor to sever the parent-offspring bond, but only at its end should the offspring be completely agreeable to that severance. It may already have struck the reader that there is indeed a stage of conflict very like this, and it has a name: weaning.

Weaning can be understood in both a narrow and a broad sense, both of which are applicable to this stage of conflict. In the broad sense it refers to a total end to dependence. In the narrow sense it refers to the termination of the feeding of milk to a young mammal. Even in this narrow sense weaning is a gradual process: Nurslings do not switch abruptly from an exclusive diet of milk to an exclusive diet of solids.

When we observe the weaning process in a mammalian species, it is readily apparent that more is involved than simple maternal rejection that begins suddenly at some point when it no longer pays the mother to nurse the offspring. The mother may reject some suckling attempts and permit others; she may continue to initiate nursing bouts even after she has begun to refuse the nipple to her young. Moreover, there is a gradual, long-term transition in the spatial relationships between mother and young—in who approaches whom and who withdraws. This is not surprising. Although we outlined Trivers's theory as involving a discrete temporal stage of conflict of interests, such a simplification would apply exactly only if the mother were to invest entirely in one offspring at a time. In any real situation, shifting one's parental investment portfolio is a more gradual affair. Many mammals, for example, such as small rodents that breed repeatedly in a single season, are pregnant while lactating (see pp. 219–220). They gradually withdraw milk from the nursing young while committing a slowly increasing proportion of their resources to the growing embryos. Even when mating occurs only after a previous offspring is fully weaned, as is true of some annual-breeding ungulates and carnivores, the mother may be recovering her condition and body weight in anticipation of the next breeding. This represents an investment in her next offspring that is purchased at the cost of withdrawal of milk from the present young.

More generally, then, there is a conflict throughout the period of parental care over the extent to which the mother will invest in particular offspring. The investment of maternal resources that is optimal for each recipient's inclusive fitness will always exceed that which is optimal for the mother's own. In its general form, Trivers's theory suggests that a child will always wish to get from its parent more than the parent is willing to give.

Trivers treats parent-offspring conflict as a symmetrical power struggle, but the parent-offspring relationship is a special one. Some of the ways in which it is special have been pointed out by University of Michigan zoologist Richard Alexander [8]. The crux of his analysis is that offspring are the vessels of the parents' reproductive strategies; they are created, nurtured, and tolerated by parents for the advancement of the parents' own fitness. This special relationship cannot be treated simply as another example of conflict between two independent agents with some specified degree of relatedness, because the offspring are at the mercy of their parents (admittedly, to a variable and declining degree) and can therefore be forced to serve the parental interests. The point is the power asymmetry. Parents may even sacrifice some of their own young to achieve a redistribution of their parental investment. Alexander suggests that parents are commonly able to manipulate offspring to parental advantage and that neither party would gain if the offspring were to resist, so that we might expect offspring to "evolve tendencies to accede to parental discipline." The young may have few devices by which they can contest parental manipulation, although there are undoubtedly circumstances in which offspring manipulate parents too.

If Alexander is correct, then we should see little evidence in nature of offspring defying their parents, at least during the period of parental care.

(Sherman's full- versus half-sister result suggests that parents cannot impose lifelong cooperativeness among their progeny, and indeed it would be hard to understand how they could.) But it is by no means simple to test whether "serious" conflict occurs, that is to say, conflict that damages the fitness of parent, offspring, or both. Where parent-offspring conflict theory produces some quantitative, testable predictions is again among the Hymenoptera; and on present evidence it appears that the mother does not always get her way, even at home. But that story will have to wait until Chapter 9. We have first to attend to the question of *why* animals produce imperfectly related offspring with whom they will sometimes disagree! Why not be a thoroughgoing nepotist and raise clonal copies of oneself?

SUMMARY

Organisms can be viewed as *strategists* designed by natural selection to promote their inclusive fitness. A reproductive strategy can usefully be described in "economic" terms: the organism budgets its efforts between mating effort and parental effort, investing its capacities and resources in pursuit of fitness returns. An effective nepotistic strategy involves benefiting relatives who will use the benefits to provide inclusive fitness returns to the strategist. The inclusive fitness interests of genetically distinct individuals are not identical, and thus there is conflict of interest even between close relatives, including parents and offspring.

SUGGESTED READINGS

Alexander, R. D. 1974. The evolution of social behavior. *Annual Review of Ecology and Systematics,* 5: 325–383.

Trivers, R. L. 1974. Parent-offspring conflict. *American Zoologist,* 14: 249–264.

Williams, G. C. 1966. *Adaptation and natural selection.* Princeton: Princeton University Press.

4
Why sex?

If you take a walk in semiarid habitat in the southwestern United States, you're likely to see a whiptail dart away from you. Whiptails are alert, slender little lizards of the genus *Cnemidophorus*. In the western whiptail, *C. tigris,* courtship involves some striking behavior, with the male first displaying to the female and then biting her legs and tail before mounting. But in the desert-grassland whiptail, *C. uniparens,* no one has ever seen a male courtship display and no one ever will. There are no males! Unfertilized eggs develop into females genetically identical to their mothers.

"Is sex necessary?" inquired James Thurber and E. B. White ruefully in a witty examination of that grand antagonism, women versus men [611].

59

Apparently not! But, as good Darwinians, we must suspect that such a widespread phenomenon as sex has some raison d'être. We are therefore led to ask, with theoretical biologist John Maynard Smith, "What use is sex?" [423].

This might seem a rather inane question, but the whiptail lizards show that it is not. Biparental sexual reproduction is a process by which two individuals each provide part of the genetic material for the creation of a new individual genetically identical to neither. This is by no means the simplest or most obvious way to reproduce. Parthenogenesis, the development of an unfertilized ovum, is one of several ways by which organisms reproduce *asexually,* with the mother providing all the genetic material for the offspring [see 40]. (For the sake of simplicity, we shall consider only that variety of parthenogenesis that produces a daughter genetically identical to the mother. For those versed in the arcane terminology of comparative sexuality, we refer to apomictic thelytoky.) The genus *Cnemidophorus* includes a number of sexual species and a number of parthenogenetic strains.

Sexual reproduction confronts evolutionary theory with one of its greatest problems: Why does sex persist? Why, for example, don't parthenogenetic mutant forms arise in sexual populations and take them over more often than they do? The reason for posing this question is that sexual females would appear to be at a selective disadvantage in competition with parthenogens. There are several components to this disadvantage [676].

THE COSTS OF SEX

Whiptail lizards lay some eight or ten eggs in a breeding season. In a sexual species, on average, half are daughters and half are sons. They won't all grow up, but a typical mother can expect, as in any sexual species, to have one of her daughters—bearing half the mother's own genotype and half the genotype of some lucky male—survive to breed. And on average, she can expect the same contribution to her fitness through her sons as through her daughters. The average individual in a stable population can expect to be represented by one descendant copy of her genotype in each future generation (see Figure 4-1). Now suppose a parthenogenetic mutant appears. She too will lay eight or ten eggs, but all will be females and all will carry her entire genotype. With the same rate of survival as sexual offspring, the parthenogenetic brood should produce an average of *two* breeding females with *two* complete copies of the maternal genotype. The parthenogenetic mutant thus replicates her genes at twice the rate of her sexual competitors, all else being equal (see Figure 4-1).

The evolutionary theorist George C. Williams dubbed the above disadvantage of sex the **cost of meiosis.** He reasoned that the sexual female

FIGURE 4-1
The twofold advantage of parthenogenesis

In a stable population the average female (circle) and the average male (square) can expect to produce one breeding daughter and one breeding son to replace themselves.

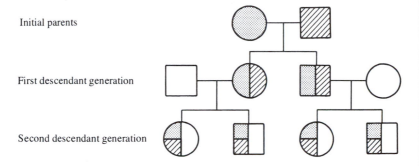

Initial parents

First descendant generation

Second descendant generation

Each parent leaves, on average, two offspring who each carry one half of the parental genotype, four grandchildren who each carry one quarter of the parental genotype, eight great-grandchildren who each carry one eighth, and so forth. In each descendant generation, the average individual leaves one descendant copy of her genotype, hence *an average of one descendant copy of each of her genes.*

Now, suppose a mutation occurs, permitting a female to reproduce parthenogenetically, and such a female is still able to rear two breeding offspring:

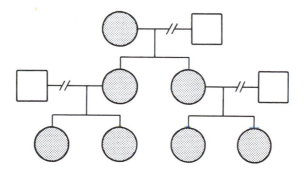

If the parthenogenetic female takes a mate, he transmits no genes to her offspring. The parthenogen leaves two daughters, four granddaughters, and so forth, each of whom carries the entire maternal genotype. *Each of her genes doubles in frequency in each descendant generation, including the mutant gene for parthenogenesis* (at least at first, while parthenogens are still rare). The mutant gene for parthenogenesis thus holds a twofold advantage over its sexual allele.

loses out by sacrificing half her genotype in meiosis (Figure 1-5, p. 9), in comparison to the parthenogen who produces a diploid ovum. Why should a female raise an offspring that carries only half her genes when, *for the same effort,* she could raise one that carries *all* her genes? (Other theorists prefer to think of this cost as a **cost of producing males,** arguing that the sexual female's disadvantage lies not in meiosis but in the fact that, by virtue of producing sons, she can only produce half as many daughters as an asexual female, so that the latter type increases proportionally in the population. But however the argument is put, it seems clear that the parthenogenetic form holds a twofold fitness advantage over its sexual competitors. Other things being equal, it should rapidly supplant them.)

As if that's not enough, sexual reproduction incurs further costs. Hard to quantify but unquestionably formidable is the **cost of recombination** (the geneticists' "recombinational load"). Sexual reproduction entails the breakup of successful genotypes and the recombination of their elements. Particularly valuable combinations of several alleles at several loci cannot be retained in such a process except by linkage. Moreover, there is a risk of new and dangerous homozygosities in recombination. A famous case is the human allele that confers protection against malaria in the heterozygous state but causes fatal sickle-cell anemia when homozygous. When two heterozygous parents reproduce sexually, one quarter of their children are likely to succumb to the anemia and another quarter will be vulnerable to malaria; only half will enjoy the advantages of heterozygosity. If the mother could just reproduce asexually, all her children would be spared the diseases. In this extreme case it is apparent that recombination exacts a terrible cost indeed.

Finally, there is the **cost of mating** [127]. Sexually reproducing organisms invest much time and energy in securing mates and in orchestrating the activities of two individuals so that they may cooperatively produce young. Much of the behavior and ornamentation that we observe serves these functions. Special chemicals are produced and broadcast. Dangerous travels are undertaken. Male attention is furthermore a mixed blessing at best. Courting males have been observed to attract predators and parasites by their conspicuous behavior, to so badger females that the latter cannot feed, to wound the objects of their affection in battles over them, and even to kill them *in copulo*! A parthenogen might be relieved to dispense with all this.

We cannot sum these costs and give them a precise value. Neither theory nor data are adequate for that. But the costs are clearly substantial. How, then, can sex persist when it incurs such heavy costs in competition with asexual reproduction? Sexual reproduction must carry some great benefit(s) sufficient to offset all these costs.

SEX AND EVOLUTIONARY POTENTIAL

Sexual reproduction recombines genetic elements. There is first the recombination produced by meiotic reduction to haploidy (Figure 1-5) and subsequent *syngamy* (fusion of gametes). Moreover, the phenomenon of **crossing-over** (Figure 1-7, p. 12) seems exquisitely designed to reshuffle the genes still more. Most biologists believe that the benefit of sexual reproduction must reside in recombination.

What is so very useful about recombination and the resultant diversification of offspring? After all, the maternal genotype must have done rather well to be mature and breeding at all. Why break up a demonstrably fit combination and incur all the costs of sex in order to gamble on novel genotypes? These questions are by no means resolved, but it will be of interest to look at some of the answers that have been suggested.

For many years, a consensus of biologists followed R. A. Fisher [191] and H. J. Muller [447] in accepting that sex's advantage is at the population level: Sexual populations can evolve faster than asexual ones. Recombination allows favorable mutants to appear together in the same genotype sooner than they would if they had to occur sequentially in a single asexual lineage, even though the mutation rates and total population sizes be similar in the sexuals and the asexuals. If environments change, asexual populations are less able to adapt and are likelier to go extinct. And insofar as the variability necessary for evolutionary change depends not on mutations but on novel combinations of alleles at several loci, sexual populations will again respond more readily to selection.

Incidentally, if a parthenogenetic mutant line cannot readily evolve further, then the idea that a sexual population pays a cost of mating relative to a parthenogen loses force. The parthenogen will generally continue to pay at least part of this cost too, because the asexuals cannot readily evolve secondary adaptations that would eliminate all such costs. Several species of parthenogenetic fish, for example, still need sperm from a male of a related species in order to initiate differentiation of their unfertilized ova [550]. Parthenogenetic dandelions still invest much energy in superfluous pollen and in yellow flowers, which no longer have any function, since the plant need not attract pollinating insects. These are evidently cases where the asexual forms lack the evolutionary flexibility to make further improvements upon their initial parthenogenetic habits.

Paradoxically, Muller [448] pointed out that sexual populations are not only able to evolve faster but are also better able to *resist* evolution. Most mutations damage fitness, and when such a mutation occurs in an asexual lineage all descendant individuals are stuck with it: Muller evoked the vivid image of a ratchet wheel, clicking inexorably forward with each deleterious mutation accumulated, with no way ever to eliminate the errors. Recombination, on the other hand, allows sexual strains to "edit" the genotype: Whereas an asexual female must transmit all her deleterious mutations to all her daughters, half the maternal genes are absent in a sexual offspring, and maternal mutations can be lost—perhaps even weeded out by natural selection among gametes or zygotes. Indeed, it is this "editing" function that is widely believed to have been the original reason for the evolution of both diploidy and sex.

> *Digression:* Alert readers may have been struck by the fact that we have discussed "Why sex?" entirely in terms of how sex is *maintained* and have shied away from the question of its *origins*. To quote Maynard Smith: ". . . There is really little alternative. Recombination probably originated some three thousand million years ago, and eukaryotic sex one thousand million years ago. Each origin may have been a unique series of events. We can speculate about such events, but cannot test our speculations. In contrast, selection must be acting today to maintain sex and recombination. We have to concentrate on maintenance rather than origins because only thus can we have any hope of testing our ideas." [424, pp. 6–7]

There is at least some evidence that "Muller's ratchet" actually operates in organisms, not just in theory. This evidence is provided by studies of various invertebrate animals that normally alternate sexual and asexual generations. When a spoilsport experimenter intervenes to ban sex, repeated asexual reproduction may be accompanied by an accelerating decline in fitness and eventual extinction of the parthenogenetic lines, apparently due to the accumulation of deleterious mutants [698, p. 147].

We cannot here review the evidence that parthenogenetic varieties are doomed to early extinction, but this really seems to be the case [see 424, pp. 52–54]. This might be due either to Muller's ratchet accumulating deleterious mutations or to a sluggish rate of evolutionary advance [563]. In either case, sex evidently serves a function, on a multigenerational time scale, for the population as a whole. But this cannot be the complete story.

SEX AND INDIVIDUAL ADVANTAGE

Sexual reproduction may keep a species healthy and flexible on an evolutionary time scale, but this group-level advantage does not dispose of the problem. Recall that we translated "Why sex?" into the question of how it is that sexual populations are able to resist invasion by parthenogens. It would seem that some more immediate fitness advantage of sex would be necessary to offset the parthenogen's immediate doubling of daughter production and the other costs that sex incurs.

George C. Williams has been the leading advocate of the view that sex must hold a fitness advantage for the reproducing individual. Many kinds of plants and animals—strawberries, aphids, gall wasps, and some grasses, for example—utilize both sexual reproduction and parthenogenesis, alternating between the two. If either form of reproduction held a net advantage in these cases, it would surely supplant the other by natural selection. The persistence of occasional sex in these species suggests that the great costs of sex must be met by great benefits in individual fitness.

What can we learn from these cases of nonobligatory sex? Williams discussed the issue in his provocative monograph, *Sex and Evolution* [676]. He began with Maynard Smith's question "What use is sex?" [423] and suggested that the answer can be induced from comparative knowledge:

> If it can be shown, for a variety of organisms that can reproduce both asexually and sexually, that they usually reproduce asexually, but use sexual reproduction in special situation s, the answer arises from inspection. Sex is an adaptaton to special situation s. This is a valid conclusion regardless of possible ignorance as to why it should be adaptive in relation to s. [676, p. 3]

Is there such a special situation *s*, a sort of common denominator to the circumstances under which optionally sexual species utilize sex? There is indeed such a situation. Sexual reproduction is used in the face of an uncertain future!

Many plants and animals invade a rich and unexploited habitat and there multiply without resort to sex, filling up a puddle or a field with genetically identical progeny—until the crunch. Resources begin to run out, and further reproduction is possible only if new offspring leave. At this point, when the young have to deal with new, different habitats, the parents turn to sex.

Parasites afford some good examples. Many parasitic species reproduce asexually during a rapid population explosion within an individual host animal. Then sex is used to produce the offspring that will disperse and invade new hosts. When parasites have complicated life cycles involving more than one host, they commonly reproduce asexually in all but one "final" host, which is generally the most mobile and hence most likely to disperse the parasite widely. Trematode worms such as liver flukes, for example, usually have a snail as an intermediate host and a vertebrate as a final host. They reproduce asexually in the snails before infesting, say, sheep, and reproduce sexually in anticipation of being dispersed in the sheep's feces.

Williams's generalization appears to hold very broadly: Animals that reproduce both sexually and asexually use sex in anticipation of dispersal. His inference that sex is therefore adapted to the task of dispersal into relatively unpredictable environments is probably sound. When we attempt to understand *how* sex is adaptive for dispersing organisms, our efforts are necessarily speculative.

Let us suppose that the genotypes that are most fit in a new environment are not those that were most fit in the parental environment. Perhaps the new locale is a little different in its soil chemistry, in its average temperature, in the presence or absence of other species. As long as it is somehow different, the genotype best adapted to live and grow and reproduce there is likely to be different from the genotype best adapted to the parental locale. By sexually producing a number of different genotypes, the parents launch a motley crew of pioneers, a few of which will land in places to which they will prove particularly suited. This variety gives them an edge over a like number of dependably similar asexual offspring in this one task of scrambling for new footholds in strange lands. Even if a majority of the sexuals are less fit than the genetically uniform asexuals, a choice few of them will by chance be preadapted to the slightly different circumstances encountered (Figure 4-2).

Suppose further that many dispersing pioneers arrive in some unexploited habitat such as a host animal or a new puddle. Each sets about reproducing itself asexually. There are then formed a number of *clones,* groups of genetically identical individuals produced by asexual division. Some clones will be fitter than others in the new environment. They will spread faster and ultimately squeeze out the less fit. When the new habitat is fully exploited, one or a few fit clones will survive and the others will become extinct. This is precisely what happens in a field of wild strawberries or other cloning plants.

Now, what of a mother sending forth such pioneer offspring? What is her best strategy to achieve genetic posterity in the winning clone(s)? She should reproduce sexually and thus have her genes represented in a variety of clones. In betting terms, this is a kind of "hedge." If she reproduced asexually, thus having identical offspring, her genes would be represented in only one clone, with a low probability of being the single winning clone.

If intense selection, high fecundity, and high mortality of offspring characterize a species, then the high cutoff point of Figure 4-2 and the consequent advantage of sexual reproduction may lead to obligate sexuality. This may be true in some cases. However, the theory works best where millions of colonists are sent out and a tiny handful with genotypes exceptionally well adapted to the particular habitat establish themselves and breed in their turn. Only then do the sexuals hold an advantage (Figure 4-2). Where a substantial proportion of the colonists survive, the average advantage swings to the asexuals. What, then, of people and other relatively infecund animals? We bear few young, and a substantial proportion survive to maturity. Is sexual reproduction therefore a poor strategy? Williams was

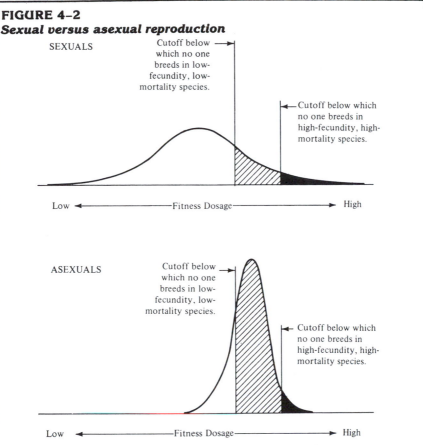

FIGURE 4–2
Sexual versus asexual reproduction

SEXUALS — Cutoff below which no one breeds in low-fecundity, low-mortality species.

Cutoff below which no one breeds in high-fecundity, high-mortality species.

Low ←———Fitness Dosage———→ High

ASEXUALS — Cutoff below which no one breeds in low-fecundity, low-mortality species.

Cutoff below which no one breeds in high-fecundity, high-mortality species.

Low ←———Fitness Dosage———→ High

A frequency distribution of "fitness dosages" of two populations of "pioneer" offspring arriving to colonize a new habitat. Fitness dosage is a sum, for each genotype, of the genetic predictors of its eventual fitness (reproductive success); it is the value of that genotype in the competition to reproduce. *On average*, asexual offspring have a higher habitat-specific fitness dosage, since each is genetically identical to a mature reproducing parent. Sexuals are far more *variable:* The majority have lower fitness dosage than the asexuals because of recombinational load, but a few lucky chance combinations have higher habitat-specific fitness dosage than *any* asexual.

In high fecundity, high infant mortality species, only a tiny proportion at the upper end of the fitness dosage scale (black) will ever reproduce. Here sexuals hold an advantage.

In low fecundity, low infant mortality species, a much larger proportion of the upper end of the fitness scale (crosshatch plus black) reproduces. Here asexuals hold an advantage. (Modified from Williams, 1975 [676]. Copyright © 1975 by Princeton University Press. Reprinted by permission of Princeton University Press.)

forced to the reluctant conclusion that sex in most vertebrates might be an inflexible vestige of the evolution of obligate sexuality by some highly fecund ancestor. But a new theory suggests that this need not be so.

PATHOGENS AND THE RED QUEEN

One of the strange experiences of Alice through the looking glass was her encounter with the Red Queen, who took her hand and led her on a wild run. When they stopped they were right where they had started, and the Red Queen explained why: "Now *here*, you see, it takes all the running *you* can do, to keep in the same place." According to Leigh Van Valen [639], things work a bit that way on this side of the looking glass too. Van Valen views the evolution of entire communities as a sort of contest among the various species, a game in which anyone's gain is somebody else's loss. Since organisms eat each other and resist being eaten, compete and interfere with one another, every evolutionary advance by one species will amount to a deterioration of the environment for one or more others. An increment in the swiftness of the gazelle is a blow to the lion. So is an improvement in the hyena's gazelle-hunting skills. And so too is a new strain of lion distemper. The point is that each species is involved in simultaneous antagonistic co-evolutionary races with its predators, its prey, its parasites or hosts, and with every species that in any way competes for any component of its ecological niche. Like Alice, it has to keep "running" just to maintain position.

Van Valen's "Red Queen hypothesis" suggests how an environment might vary enough to always select for novel recombinations in preference to perfect replicas of tried and true parental genotypes: The important thing is that the relevant aspects of the environment are themselves biotic, and are co-evolving with the species of focal concern. This is not just a matter of a predator or parasite species, for example, evolutionarily "tracking" a prey species that is trying to evolve away from it. A more subtle point is that genetic diversity per se may be valuable in confounding one's co-evolutionary antagonists, particularly if they are *pathogens* such as bacteria and viruses.

Pathogens are parasites that have short lives and large populations in comparison to the hosts they attack. Accordingly, pathogens can evolve relatively rapidly. This poses special problems for their hosts. A germ may run through thousands of generations in a single host individual, evolving in the process to exploit that individual's unique biochemistry. A parthenogenetic mother would be giving her daughter rather a poor start in the world, then, when she passes along her own genotype plus a complement of pathogens uniquely adapted to it! A sexually produced host individual with a novel genotype will not be so readily infected as will genotypes to which pathogens are already adapted.

Apart from the problem of intrafamilial infection, we expect pathogens to be adapted to the most prevalent gene complexes of their hosts, and this means that there is in general a selective advantage to rare alleles and recombinations. Particular pathogen strains often exhibit virulences specific to particular host genotypes. In other words, individual alleles in the host organism confer resistance to particular strains of pathogens while leaving the host vulnerable to slightly different strains. There may be many alleles at a single locus, each conferring resistance to a particular pathogen strain. This is the case, for example, with rust resistance in flax. Any locally prevalent strain of rust will select for resistant flax, which in turn sets the stage for other rust strains to invade and thrive until the local flax becomes resistant to them. Resistance to all strains cannot be maintained simultaneously, so that the co-evolutionary race between flax and rust has no stable equilibrium. In such a contest, sexual reproduction by the flax may be thought of as a way of "recruiting" locally successful genes [390]: "The genes recruited by mating represent a sample of the genes contained in adult individuals who have thrived in the vicinity of the sexual partner with whom they mate. Hence the recruitment of genes by mating allows a parent to track the current success of sporadically valuable genes in the surrounding population and pass these protective genes on to the progeny in the approximate proportions in which they have succeeded among neighboring individuals" [390, p. 97]. W. D. Hamilton has constructed population genetical models demonstrating that pathogen pressure can indeed give a selective advantage to sexual reproduction with high levels of recombination [253, 254; see also 510]. The models have no trouble dealing with those low-fecundity organisms whose sexuality remains paradoxical by Williams's theory.

If theoretical adequacy were all that the pathogen hypothesis had going for it, we would have been reluctant to discuss it. Fortunately, the theory has some interesting empirical implications, and initial tests of them appear promising [617]. The selective pressure exerted by pathogens should be most extreme under conditions where they thrive. Hot, wet climates, for example, are favorable for the growth and transmission of many bacteria and other pathogens. Contagion is also facilitated where there is high density or high frequency of contact among members of the host species. Islands maintain lower pathogen loads than biotically more complex mainland habitats. Considerations such as these lead to a series of predictions. If pathogens are the principal selective forces favoring recombination, then we can expect the prevalence of sexual reproduction to vary with temperature, moisture, host population density, and other geographic and sociodemographic variables. Asexual varieties should most often be found where pathogen loads and contagion are minimized.

The distribution of parthenogenetic forms seems mostly to comply with these expectations. This is particularly evident in the geographical distribution of sexual and asexual forms that are so closely similar as to have been classified as members of a single "species." (Modern definitions of "species" commonly refer to the potentiality for interbreeding among members. Obviously, this makes the concept of species a bit problematic for parthenogenetic strains! In practice, however, the working taxonomist in a museum uses morphological criteria in his classifications, and may include a sexual population and any parthenogenetic strains derived from it in a single species.) There are ticks and lice, bugs and flies, moths, beetles, grasshoppers, millipedes and more, in all of which males disappear as one moves from the tropics toward the poles. There seem to be very few examples of an opposite trend [214]. Altitudinal changes mimic latitudinal. The implication seems to be that parthenogenetic forms do better in colder climes. They also seem to do better under arid conditions. And it is clearly the case that polar and desert conditions are not optimal for pathogens.

It is worth remarking in this context that recombination is not simply all or none—sex versus parthenogenesis. It is furthermore of variable degree within sexual species. If all genetic loci assorted independently in accordance with Mendel's second law, then recombination would be total. However, they do not. Linkage mitigates the reshuffling of genes. Chromosomes, however, *do* assort independently, and thus, the larger the number of chromosomes, the more extensive the recombination. Crossing over (Figure 1-8, p. 14) also enhances recombination, and the characteristic number of *chiasmata* (crossover points) in a genome is therefore also an index of recombination rate. So is the degree of outbreeding. If the pathogen theory is correct, then all three of the indices of recombination rate should vary systematically with latitude, altitude, aridity, and so forth. Preliminary evidence tends to uphold the theory [378, 617].

What use is sex? The question is not yet resolved. The pathogen theory is promising, but we cannot predict its fate—at the least, readers may find some inspiration in this thought: Perhaps you engage in sex to confound your germs!

FEMALE AND MALE

Why sexes? The question may seem even more preposterous than "Why sex?" but nothing we have said thus far has suggested an answer to it.

The existence of sex does not necessarily imply the existence of sexes. Sex is the process of shared parenthood, where two individuals of the same species combine their genetic material to produce a novel genotype. Sexes are two different types of individuals within a species, males and females, and successful reproduction demands that the parental pair include one of each. It is not at all obvious why these two categories exist. Why not just have one type, with A capable of mating with B, B with C, and C with A? Granted the (occasional) utility of sexual reproduction, why not sex without sexes?

Such a thing does in fact exist in bacteria. Sexes are lacking, but sex is not. Once in awhile two bacteria come together to exchange genetic material and generate recombinations. The act is perfectly symmetrical. The partners defy categorization as male and female [163].

So whence the asymmetry that prevails all about us? Whence female and male? The definitional distinction may be surprising: It is a difference

---

FIGURE 4–3
Gamete size dimorphism: hamster ovum and sperm.

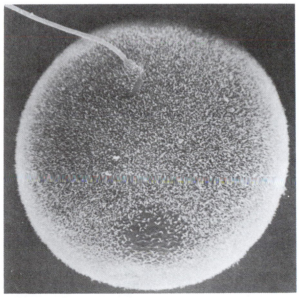

(Photograph by David M. Phillips.)

in size. The female gamete is bigger. This is not, of course, predictive of the relative sizes of the whole animals, which vary. In many creatures the male is bigger than the female, but in others, like hamsters and hawks, the female is larger. But the sperm is always smaller than the ovum, and that defines the sexes. From this trite distinction, much follows. In this and subsequent chapters we shall trace some of the paths of causality by which femininity and masculinity derive from a mere size distinction (Figure 4-3).

A sperm is an entity with a mission—search, find, fertilize. In intense competition with other sperms of similar ambitions, it has become stripped down and streamlined. As a participant in a race, it has jettisoned all nonessential baggage. The sperm must, of course, transport chromosomes, but its cellular material is otherwise minimal and is placed in the service of mobility.

The prize for which the sperms compete is a relatively enormous gamete, the ovum, which does not move to welcome the victor but sluggishly awaits him. The ovum's immobility is due to the bulky cytoplasmic mass that accompanies her share of the chromosomes. Although each parent contributes almost equally to the nuclear chromosomes of the new creature they create, not all contributions are equitable. The female provides the raw materials for the early differentiation and growth of their progeny. Here, at the very fundament of sexuality, is love's labor divided, and it is the female who contributes the most.

EVOLUTION OF THE SEXES

The origin of this primeval division of labor is lost in antiquity. Any effort to reconstruct the early evolution of the sexes is necessarily speculative. With that caution expressed, let us examine one plausible scenario, suggested by Parker, Baker & Smith [482]. Imagine a primeval organism engaged in sex without sexes. As with the bacteria mentioned earlier, any individual might mate with any other. But imagine that the organism is, unlike a bacterium, multicellular, able to devote some of the energy it ingests to the production of gametes. Each individual produces haploid gametes that contain one gene from each pair. Gametes can unite with any other gamete they encounter. Since there are no males and females, there are no ova and sperm, simply gametes. When two gametes collide, their chromosomes are united in a diploid zygote, the beginning of a new individual.

Then, as now, there would be natural selection. Some genotypes would be fitter than others—that is, more reproductively successful. What selection pressures might we expect in such circumstances? One such pressure would surely be for gametic size: There must be an adequate cytoplasmic mass for the new zygote's first divisional stages to be able to proceed. Another would be for maximal numerical productivity of gametes. But these fundamental pressures stand in opposition to one another. This is because the organism has a limited amount of material to invest in gamete production. By maximizing numbers, one minimizes size, and vice versa.

Now, suppose that there is some distribution of gamete sizes in the population—some are small, some are large, but most are of intermediate size—and that gamete size is somewhat heritable. How will selection operate? Clearly, the gametes of greater size are fitter than those of lesser size. But the critical size mass is that of the two parental gametes combined (the zygote); even a tiny gamete may be successful if it unites with one large enough to provide adequate cytoplasmic resources. Parker, Baker & Smith have shown that such a situation can give rise to *disruptive selection:* Over generations the distribution of gamete sizes becomes bimodal. Two basic types are then recognizable—a large proto-ovum and a small proto-sperm (Figure 4–4).

Once this bimodality is established, several consequences seem to follow. Whenever two of the small proto-sperms unite, there will be a fatal shortage of cytoplasm. Devices to discriminate these small gametes and to reject them as unsuitable partners should be selected for. The large proto-ova may also prefer to reject the little proto-sperms and to pair instead with other proto-ova, but the selection pressure against pairing with the little proto-sperms will be greatest upon the proto-sperms themselves. Such male-male matings will always be unproductive for want of cytoplasm. Moreover, there will be intense selection for the proto-sperms to overcome the discriminatory devices by which the proto-ova would spurn them. Sperm should then become increasingly specialized for the pursuit of ova. Ova, regularly fertilized by highly mobile sperm, should need less mobility, thus making possible the evolution of fertilization inside the female.

This reconstruction of the origins of gamete dimorphism has received a good deal of support [38, 424], but there are some problems. While it seems likely that ova and sperm diverged from an originally unimodal ("isogamous") size distribution, it is by no means certain that ancestral isogamous organisms lacked "opposite sexes" [670]. There are, for example, modern species of isogamous algae with two "mating types," arbitrarily designated + and − (since there is no real basis for labeling one female and the other male). A zygote can result only from the union of a + gamete and a − gamete. The existence of mating types in isogamous algae suggests that our question—Why two sexes?—is not solved by the Parker, Baker & Smith model. The leading hypothesis about why there are mating types in algae is that they are a way of avoiding excessive inbreeding (see p. (see p.

FIGURE 4-4
Disruptive selection of gamete size resulting in ovum and sperm

Imagine a population of organisms that reproduce sexually but without sexes: *A* can mate with *B*; *B* with *C*; *C* with *A*. There is a *distribution* of parental types producing gametes of different sizes.

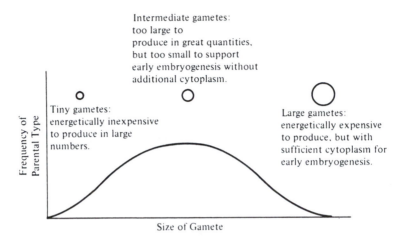

Intermediate gametes: too large to produce in great quantities, but too small to support early embryogenesis without additional cytoplasm.

Tiny gametes: energetically inexpensive to produce in large numbers.

Large gametes: energetically expensive to produce, but with sufficient cytoplasm for early embryogenesis.

Frequency of Parental Type

Size of Gamete

By mathematical models Parker, Baker, & Smith [482] have shown that this situation can lead to *disruptive selection:* Parents who produce tiny gametes or large gametes both have greater fitness than parents producing intermediate gametes. So in a subsequent generation, the distribution looks like this:

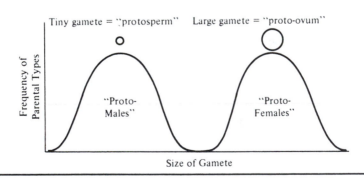

Tiny gamete = "protosperm"   Large gamete = "proto-ovum"

Frequency of Parental Types

"Proto-Males"

"Proto-Females"

Size of Gamete

305) as a result of self-fertilization. If two distinct mating types came first and the evolution of dimorphic gametes followed (perhaps according to Parker, Baker & Smith's scenario), selection should then have favored

linkage of the two traits [94] so that the appropriate union would be an ovum and a sperm.

We are speculating about an ancient series of events that may or may not have taken place in this way. Other theories are possible [see especially 116]. But whatever the historical events, the male-female phenomenon is now with us, and the female is by definition the sex that produces the larger gamete. Even at the instant of conception the roles of the sexes are differentiated, their interests imperfectly harmonious. At this first stage the female is contributing far more to the nurture of the young. The quick little sperm can be said to parasitize the parental resources that she provides. As we shall see in the next chapter, this basic difference has vast implications for female and male roles before and after conception.

SUMMARY

Reproducing sexually exacts heavy costs in competition with asexual reproduction. Sex must bestow some substantial fitness benefits, and there are several theories about what those might be. The classic view that sex contributes to evolutionary potential is probably sound, but it cannot explain the resistance of sexual populations to being taken over by parthenogenetic forms. Sexual diversification of offspring must contribute to the fitness of individual parents as well as to the fitness of whole populations.

Even given the existence of sex, the existence of the sexes still demands explanation. The female is by definition the sex with the larger gamete. Males appear to have evolved as producers of superabundant tiny gametes (sperm) that parasitize the cytoplasm of big female gametes (ova).

SUGGESTED READINGS

Maynard Smith, J. 1978. *The evolution of sex*. Cambridge: Cambridge University Press.
Williams, G. C. 1975. *Sex and evolution*. Princeton: Princeton University Press.

5
Sexual selection: Competition for mates

At conception, the mother's material contribution surpasses the father's, for it is the large ovum that provides the nutrients for early development. In the "economic" metaphor of reproductive effort budgets, the female's initial **parental investment** in each offspring exceeds her mate's. Parental investment has been defined by Robert Trivers as "any investment by the parent in an individual offspring that increases the offspring's chance of surviving (and hence reproductive success) at the cost of the parent's ability to invest in other offspring" [619, p. 139]. This definition stresses the trade-offs implicit in any behavioral choice. Parents who invest in one offspring sacrifice the opportunity to invest in others.

77

NURTURANT FEMALES

In animals such as ourselves, where parental care is an immense undertaking, the energetic cost of the initial gamete is a negligible proportion of total parental investment. The ovum-sperm disparity seems trivial. Yet nurturance is predominantly a female enterprise in many animals including almost all mammals, and the dimorphism of those little gametes is ultimately to blame. This is because the gamete size distinction has set the stage for some other evolutionary developments that have greatly amplified the initial difference in the parental investment of female and male.

Large ova bias organisms toward female nurture wherever fertilization takes place inside one parent's body because the relative immobility of the ovum assures that, whenever internal fertilization occurs, it takes place within the female. Internal fertilization protects the young from perils like desiccation and predation, thereby greatly increasing each individual ovum's prospects for fertilization and survival. But it also has a cost. By increasing her nurture of each ovum, the female can no longer invest in so many. In one season a ten-year-old cod may release as many as 5 million eggs for external fertilization. No fish with internal fertilization can match the cod's fecundity. But neither will an internal fertilizer suffer the same massive destruction from early deaths and fertilization failures as unparented cod eggs. There is a trade-off between fecundity and nurture.

Once the female is harboring zygotes within her, any further investment in nurturant or protective devices that may then evolve tends most often to be paid by the female, although both parents reap the fitness benefits. Birds and reptiles, for example, construct a protective eggshell at considerable material expense. In several types of animals, mechanisms have evolved to feed the infant while it grows within the maternal body; the mammalian placenta is perhaps the most sophisticated such device. In fact, we mammals have attained a pinnacle of concentrated maternal investment. First there is pregnancy: Prolonged internal gestation and placental nurture are enormous commitments of a female's resources that pay off in greatly reduced offspring mortality during the early stages of development. And then, at birth, the mammalian female doesn't even take a recuperative break, for we mammals have invented another major maternal investment —milk. (Just why the males don't pitch in and help with lactation is a subject we shall consider in Chapter 7.)

Internal fertilization, gestation, placentation, lactation: Each of these evolutionary developments has had the effect of concentrating female investment in a decreasing number of offspring. In the evolutionary sequence leading to modern mammals, each advance in the effective nurture of young has amplified the sex difference in parental contributions. Among vertebrate animals with external fertilization (most fish and amphibia), things are very different: If there is any postmating parental care at all, it is at least as likely to be performed by males as by females [324, 657]. In birds, internal fertilization is universal, but so is an early externalization of the developing embryo, with the result that both parents can care for it; biparental investment is the rule (see pp. 149–151). But in mammals, and indeed in most animals with internal fertilization, parental nurture is overwhelmingly female-dominated. So what are the males up to?

COMPETITIVE MALES

A single mating can generally provide a female with all the sperm necessary to fertilize all her available eggs. Once she has been fertilized, additional matings avail her nothing until she requires fertilization anew. If she is a mammal who becomes pregnant and lactates, that may not be for months or even years. If she is an insect who can store sperm, she may never need to mate again. But for the male, an additional mating may have a reproductive payoff at any time (at least within the breeding season, where such exists). These considerations provide an explanation for a widespread sex difference in the temporal patterning of sexual motivation. In many species, copulation effects a rapid decline in female sexual interest and receptivity [e.g., 86, 420, 579, 616]. Males, by contrast, are often rearoused by a novel female (the "Coolidge effect"; see [150]), as might be expected since a novel female represents a novel reproductive opportunity.

In many species, female reproductive capacity is limited by the energy and time demands of parental nurture, whereas no comparable ceiling exists for the males. The male's reproductive output is limited by his access to fertile females, whereas access to all the males in the world would not elevate the female's capacity. In our own species, for example, it is hardly possible for a woman to bear more than about 20 children in her lifetime, although the *Guinness Book of Records* credits a nineteenth-century Muscovite with 69 live births, bearing twins, triplets, or quads in each of 27 pregnancies! Astounding though that figure may be, the male record is much greater: 888 children were sired by the Sharifian emperor of Morocco, Moulay Ismail the Bloodthirsty [432]. Where one male mates with several females, we call the mating system **polygynous**; where one female mates with several males, we call it **polyandrous**. (See below for a more technical definition that treats polygyny-polyandry as a dimension.)

Most adult females in most animal populations are likely to be breeding at or close to the limit of their physiological capacity to produce offspring and raise them to successful maturity. For example, Sadleir's

FIGURE 5-1
Sexual selection in Drosophila

Male mating success is more variable than that of females, and a male's reproductive success depends on the number of mates, while a female's does not. (Data from Bateman, 1948 [30].)

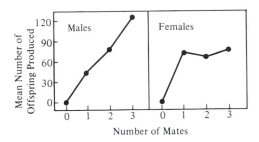

The more females a male mates with, the more offspring he sires. A female, by contrast, does not gain by accepting multiple mates.

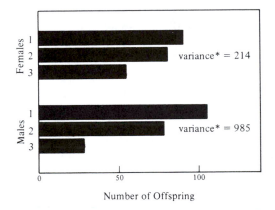

Number of Offspring

Male variance* is greater than female in the offspring produced by each individual when three male and three female fruit flies were housed together. In every one of sixty-four such experiments the males' reproductive output varied more than the females'. The data above represent one of these experiments. (Data from Bateman, 1948, Table 4 [30].)

Number of Mates

Male variance* is greater than female. When three to five flies of each sex are housed together, most females mate with only one or two of the available males.

Among males, by contrast, a substantial proportion mate with more than two females, while many others fail to mate at all.

This figure represents data from sixty-four experiments involving 215 flies of each sex.

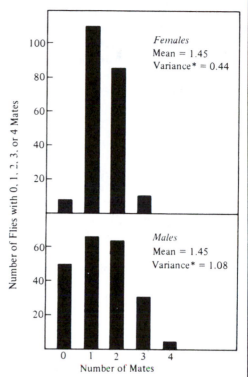

Females
Mean = 1.45
Variance* = 0.44

Males
Mean = 1.45
Variance* = 1.08

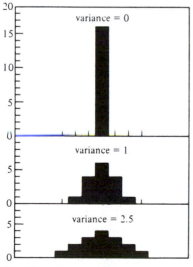

variance = 0

variance = 1

variance = 2.5

Variance is a measure of the variability in a set of scores. It is defined as the average squared deviation from the mean: $(X - \bar{X})^2/N$.

The three distributions at left all have the same mean, but their variances differ.

review of the ecology of mammalian reproduction cites studies in deer, mice, and rabbits in which more than 90 percent of adult females were pregnant under good food conditions [540]. Males, by contrast, are limited not so much by their physiology as by their competitors. Any male, quite unlike a female, has the prospect of increasing his reproductive output if he can somehow secure another fertilization. This sex difference has important implications for female and male reproductive strategies, as is exemplified in the behavior of animals as diverse as fruit flies and people.

Fruit flies

The fruit flies (*Drosophila*), which have played such an important part in the history of genetics, have also made an important contribution to our understanding of the basic sex differences in reproductive strategies, thanks to an insightful study of their mating behavior in captivity reported in 1948 by the British geneticist A. J. Bateman [30].

Bateman's experiments involved placing three to five fruit flies of each sex in each of several containers, where they bred. Offspring could be attributed to particular parents because of different "genetic markers" in the different adults, features analogous to such human characteristics as hair and eye color. By observing the sexual behavior of the flies and by subsequently counting the offspring of each parent, Bateman provided the first experimental demonstration that females gain nothing by mating with more than one male (Figure 5-1). All a female's eggs could be fertilized by sperm from the first mate. For the males, however, extra copulations continued to pay off with worthwhile gains in the number of young sired.

Figure 5-1 also illustrates Bateman's second finding: There is greater variance in the reproductive success of males than of females. Whereas virtually all females produced similar numbers of offspring, the maximum being physiologically limited by egg-production capacity, some males did far better than average and others far worse. In a penetrating discussion Bateman outlined the broader implications of his data. He noted that the "greater dependence of males for their fertility on frequency of insemination" is "an almost universal attribute of sexual reproduction" [30, p. 364]. This fact, he suggested, ensures greater intramasculine selection and therefore results in an often conspicuous competition among males as well as "an undiscriminatory eagerness in the males and a discriminatory passivity in the females" [30, p. 365]. Thus is explained the commonplace observation that it is generally males who court females rather than the reverse, a subject to which we shall return in this and later chapters. Moreover, since repeated matings afford males an advantage that does not hold for females, there is a "greater male inclination to polygamy," which we may expect to endure even in species that have evolved monogamy.

Sex differences in the within-sex variance of reproductive success can be used to characterize mating systems. Where male variance is greater, as in fruit flies and indeed most animals, the system is **effectively polygynous**. Where female variance is greater, it is **effectively polyandrous**. Thus, populations may be placed along a dimension ranging from extreme polygyny to extreme polyandry (see Chapter 7, pp. 151–152).

Since Bateman's classic laboratory study, the basic phenomenon of relatively great variance in male reproductive success has been demonstrated in a variety of animals in their natural habitats. The correct attribution of paternity remains something of a problem in field studies (but see pp. 52–53), so that we must usually rely on an indirect measure of fitness, namely mating success, which is presumably highly correlated with eventual reproductive output.

Elephant seals

Perhaps the most striking mammalian example of great variance in male reproductive success is that of the northern elephant seal [367, 368, 369], which breeds on tiny islands off the California coast. In early December the massive males haul out on the beaches, where they remain without feeding until March. Their initial concern is the establishment of dominance relationships among themselves. This is achieved mostly by threats, but there are spectacular and occasionally bloody fights, which persist for at most a quarter hour and stop short of permanent injury. Pregnant females begin to arrive in mid-December. Each delivers her pup on the beach and then nurses it for about twenty-eight days before departing. Copulation takes place during the last few days of this nursing period [371, 118]. There are females available for mating from about the end of January through February, when the competition among males is intense (Figure 5-2). Over the weeks of struggle there is generally some reordering of the male dominance hierarchy, which is constantly being tested by threats. This competition exacts an enormous energetic cost from the fasting bulls. It is not unusual for a bull to die immediately after his most successful year on the beach, presumably because of the degenerative effects of his incessant struggle to maintain his high rank. That cost is tolerated because the maintenance of high rank has huge payoffs in reproductive success.

Elephant seal reproductive competition on Año Nuevo Island has been studied over several years by Burney LeBoeuf of the University of California. Hundreds of seals have been tagged for permanent recognition.

FIGURE 5–2
Northern elephant seals (*Mirounga angustirostris*).

Males bellowing threats at one another, surrounded by females and pups. (Photograph by B. J. LeBoeuf.)

In one breeding season 115 males were present, but the 5 that ranked highest in the hierarchy performed 123 of 144 observed copulations. Hence, some 4 percent of the males present evidently sired 85 percent of the next season's pups [370]. From long-term study a picture of the lifetime reproductive success of the northern elephant seal is taking shape. The male variance is enormous. Few breed at all, for only about 10 percent of the males born survive to join the breeding competition at five or six years of age, and most of these never copulate. Beyond age six a male's chances of surviving to compete again next season are about 50 percent per year, so the probability of living to be nine or ten years old is on the order of one in a hundred. If he lives that long, a male can enjoy substantial breeding success, possibly siring 50 or more pups in a season. The most successful male in LeBoeuf's study dominated breeding at one site for four consecutive years, while evidently inseminating over 200 females [368].

Yellow baboons

Among our close relatives, the monkeys and apes, intense reproductive competition among the males is again evident. Several studies have shown a relationship between a male's rank in a dominance hierarchy and his mating success. The most careful analysis of this relationship to date was conducted by an American primatologist, Glenn Hausfater [265]. His subjects were a troop of yellow baboons (Figure 5–3) in Kenya's Masai-Amboseli Game

FIGURE 5–3
Yellow baboons (*Papio cynocephalus*) in Mikumi National Park, Tanzania. Top: two adult males in one of the frequent aggressive interactions that determine dominance. Bottom: copulation; note the considerable sexual dimorphism in body size. (Photographs by D. R. Rasmussen.)

Reserve, an area of semiarid, short-grass savanna with scattered acacia trees. The baboons' (*Papio*) basic social unit is a troop consisting of several adults of each sex and the females' offspring. (The Hamadryas baboon, *P. hamadryas*, whose basic social unit is the one-male harem, is an exception.) Males change troops at maturity and occasionally thereafter, while females usually live out their lives in their natal troops [476]. Particular males and females do not pair off within the troop, or at least not for very long, so a male may aspire to inseminate several females, and a female may mate with several males during a single estrus (the period of maximal sexual receptivity).

When Hausfater began to study his troop, it contained eight adult males, thirteen adult females, and fifteen juveniles and infants. With occasional comings and goings, births and deaths, the troop size varied during the next thirteen months between twenty-nine and thirty-six individuals, each of whom Hausfater was able to identify by sight. During mating periods there were up to fourteen sexually mature males in the troop, and at any point in time they could be ranked in a dominance hierarchy on the basis of the observed outcomes of aggressive interactions among them. As in the elephant seals, dominance rankings were frequently challenged, and individuals occasionally moved up or down the hierarchy. Overall, the rankings were remarkably stable: A male named BJ was number 1 for a ten-month stretch, for example, although he briefly fell as low as number 3; Peter occupied ranks 2, 3, and 4, each for several months; Sinister was in

---

FIGURE 5–4
Male mating success in a troop of yellow baboons

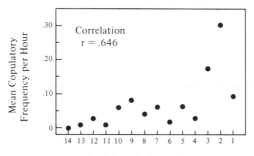

A male's copulatory frequency is related to his rank in the male dominance hierarchy. (Days D–7 to D–1 combined; data from Hausfater, 1975, Table XXIV [265].)

rank 6 for a total of eleven months, but he briefly rose to 5 and fell to 7 and even 8; and so on. And as in the elephant seals, the higher-ranking males enjoyed a higher frequency of mating (Figure 5–4).

Some mating opportunities are more important than others. The only matings that are likely to be fertile take place near the time when the female ovulates. A yellow baboon has a menstrual cycle of approximately thirty-two days. During about the first half of that cycle there is a gradually increasing swelling of the "sexual skin" surrounding the vagina. Then, quite suddenly, the swelling subsides, an event Hausfater labeled D-day (D for "deturgescence"!). Physiological studies in the laboratory indicate that the female usually ovulates on day D–3, three days before D-day, so it is then that a mating is most likely to be productive. The probability of fertilization is lower the further in time from day D–3. If we focus on the potentially fruitful matings, the advantage of high male rank is still more conspicuous. Low-ranking males occasionally copulate, but their chances are particularly low at just those times when a mating is most likely to be fertile (Figure 5–5).

FIGURE 5–5
Male mating success in relation to rank and to stage of
the female ovulatory cycle in a troop of yellow baboons

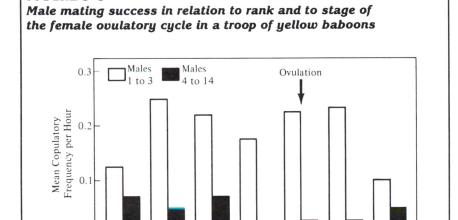

The top-ranking males mate more than lower-ranking males on all days, but especially near the time of ovulation. The low-ranking males' copulatory frequency is lowest on days D–4, D–3, and D–2, just the time when a mating is most likely to be fertile. (Data from Hausfater, 1975, Table XXIV [265].)

The extent to which high-ranking male baboons monopolize mating has sometimes been overstated: Several authors have suggested that no male outside the "ruling elite" ever gets to mate at all. Several primatologists have criticized these exaggerated accounts [535, 265, 45], but the critiques have sometimes been overstated too, giving the impression that the alleged mating advantage of high rank is fantasy. Careful examination of the data, however, permits the conclusion that there is considerable variance in male mating success in multimale primate troops and that this differential mating is strongly related to the male dominance hierarchy. The relationship between rank and mating success is perhaps weaker in primates than in other mammalian orders [151].

The selective consequences of the reproductive competition among males will not be fully elucidated until determinations of paternity are carried out in natural troops. Several recent studies have used the method of electrophoresis (p. 52) to determine fatherhood and male reproductive success in *captive* groups of monkeys [168, 574, 685]. So far, the results seem to corroborate the copulation frequency data: High rank is correlated with number of offspring sired, but paternity is not a total monopoly of the top males. No doubt, similar studies of primate troops in the wild will soon be appearing in the research journals.

People

In our own species it is again the case that males vary in their reproductive success more than do females. The Xavante are a Brazilian people whose villages have been intensively studied by anthropologists. In one sample of 184 married Xavante males 74 were married to more than one wife. Co-wives were likely to be close relatives of each other. A few chiefs had four or five wives, and these men had two to three times as many children as other men of the same age. Furthermore, just 1 of 195 women remained childless by the age of twenty years, whereas 6 percent of the men were still childless at forty. So while the average number of surviving offspring in completed families was identical for men and women—namely, 3.6 children—the male variance was more than three times as great as the female variance [543] (Figure 5-6). Human marriage is of course an imperfect reflection of the effective mating system, a fact to which we shall return in Chapter 11.

The polygynous marriage practices of these South American people are by no means unique: In fact, a majority of human societies permit a man more than one wife if he can afford them (see Chapter 11). And even in a professedly monogamous society like our own, the variance in child production is still likely to be somewhat greater for men than for women. This

FIGURE 5-6
Variance in reproductive success among Xavante Indians

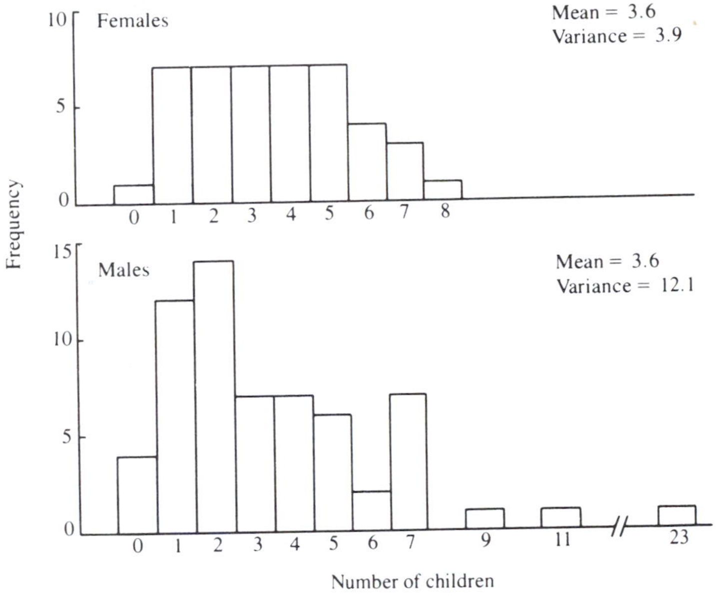

Total children produced by living adults at least forty years old plus recently deceased, in two villages in the Mato Grosso, Brazil.

Because of polygynous marriage a few men have more children than any one woman could bear. (Data from Salzano, Neel, & Maybury-Lewis, 1967, Table 8 [543].)

is due mainly to successive marriages: A man is almost certainly more likely to sire children by two wives than is a woman to have children by two husbands. (It is surprisingly difficult to find data bearing directly on this point, for census bureaus and government statistical offices routinely collect information on the fertility history of women but not of men. It is clear that the number of *marriages* has greater variance for men, who are both likelier to remarry [e.g., 93] and likelier to remain unwed. But whether the number of *offspring* has greater variance for men remains a matter for conjecture.)

So the answer to our question "What are the males up to?" is this: They are competing with one another for the opportunity to inseminate females. By apportioning a relatively large part of their reproductive effort to such competition, males of most species devote rather little to parental care. The nurture that females bestow becomes a resource for which males compete. The male who wins the right to inseminate a female also wins for his progeny a share of the female's parental investment.

FEMALE-DEFENSE POLYGYNY

In a polygynous breeding system, not all males can be harem-holders. Those who are must constantly guard against attacks and raids by the bachelors. With other males ever at the ready, even the most dominant of polygynists is liable to find his attention divided and his monopoly imperfect. Consider the bison, for example, that once dotted the Great Plains of North America. Dale Lott of the University of California observed a herd of sixty-three mature bison—twenty-six bulls and thirty-seven cows—during the three-week period of the rut. The huge bulls, at 900 kg almost twice the size of the cows, are highly aggressive, and Lott was able to rank them, just as in Hausfater's yellow baboon study, into a dominance hierarchy based on the outcomes of fights. As in the baboons, rank was predictive of mating success, but only to a degree. No bull serviced more than four of the thirty-seven cows, and only seven males (27 percent) failed to breed at all [396].

Male bison "tend" females around the time of estrus, and try to keep their chosen cows away from the other bulls. After some persuasion by the tending bull, the cow will usually copulate only once, although she may accept a second male or the same male again up to about a half an hour after the first copulation. After mating, the bull continues to accompany her for approximately that half hour, evidently guarding the cow for just about the duration of her continued receptivity [397]. He can then abandon her without fear of rival insemination, and so he does. Nevertheless, two of the thirty-seven cows in Lott's study were mated by a second male who displaced a lower-ranking bull within the half-hour postcopulatory period.

It would appear that divided attention is the main reason why the dominant bulls in Lott's herd were unable to monopolize breeding more completely. Tending an estrous cow costs time and energy, and she cannot be abandoned even after copulation without risk of losing the fertilization already won (see p. 107). So while dominant bulls are guarding cows and contesting with one another over them, lower-ranking males can be making their pitch, too. This sort of competition distributes mating among the males, especially because the rut is brief and the cows are ready to be bred simultaneously; 46 percent of Lott's cows came into estrus within a four-day period. Every competitive breeding situation has a certain **potential for polygyny** [179]. Spatial scatter of females obviously reduces the potential for polygyny: A male can be in only one place at a time. A group of females, on the other hand, is relatively defendable. But even when females are aggregated, synchrony of female receptivity and prolonged pairing are factors that will make it difficult for a high-ranking male to monopolize more than a few females, and hence will give the lower-ranking males a chance too, thus reducing the variance in male fitness.

FEMALE-FEMALE COMPETITION

Of course, females also compete among themselves. Recall that we characterized natural selection as a **fundamentally competitive process**: differential reproductive success among the members of a population (pp. (35–36). Female competition tends to be a somewhat different matter from the contests between males. Female fitness is less often limited by access to mates than is male fitness, and so females compete *for mates* relatively infrequently (although a high-quality male may sometimes be worth fighting for, especially in species where males invest parentally). Since a female can seldom profit from monopolizing the reproductive effort of several males, it is not surprising that **male-defense polyandry** is much rarer than female-defense polygyny. (Nevertheless, these rare sex-role reversals do occur, as we shall see in Chapter 7.) No, it is not access to mates that generally limits female fitness, but access to nutrients that can be converted into offspring.

Competition among females for limited resources may be more or less direct. Female gerbils defend feeding territories from which they aggressively exclude other females; a larger female occasionally manages to usurp such a territory from the resident [129]. Lionesses, wolves, and other female carnivores defend feeding territories too, and so do a host of other mammalian females. In group-living species there may be a rather delicate balance of cooperation and competition. In a wild dog pack, for example, hunting and territorial defense are cooperative, but the top female may prevent lower-ranking females (not her close kin, as it happens) from breeding or may even kill their pups [198]. In the ani, a communally nesting bird, several females lay their eggs in a single nest and young are evidently fed indiscriminately, but females about to lay may roll their rivals' eggs out of the nest first [640]. In various species of wasps several potential queens establish a nest together and then battle for the position of sole or principal egg-layer [662]; the losers are either killed or become nonreproductive workers (which still provides them some inclusive fitness returns if, as is often the case, they are sisters of the queen).

In ground squirrels, a mature female who loses her own litter to a predator commonly moves to a new area and, for the rest of a summer that is too short for her to squeeze in a full reproductive cycle, she becomes a killer [560]. The immigrant female makes sneak raids upon the burrows of her new neighbors and bites the helpless pups to death. She does not eat them. The behavior apparently functions to eliminate competitors and make a little space for the killer who, if successful, will breed there in the following summer. When female kin settle near one another, as they usually do (see pp. 46–48), they cooperate in defending their burrows against these marauding intruders, who are not related to them.

Competition can be indirect too, as when rival females simply scramble for scarce resources without challenging one another. But even in the absence of dramatic aggressive encounters, subtle social influences can be a medium of competition. Young female mice, for example, delay their puberty when they encounter a high density of residual odor cues left in the environment by mature females [417]. If the habitat is sparsely inhabited, young females mature rapidly, but if the density of adult females is indeed high, then a reproductively mature young female will be attacked by established residents. If she is not prepared to compete successfully, she is better to remain in a prepubertal state, in which her own odors do not elicit aggression from adult females, and to await a less contested breeding situation.

Female competition is a serious matter, then, but there is a simple rationale for our emphasis on male competition: Males generally compete more intensely. Intensity of competition can be measured as variance in reproductive success (cf. Figure 5–1), and the higher male variance means that the males are playing a higher-stakes, higher-risk game.

A MALE'S CAREER: SHORT AND SWEET

Male competition tends usually to be more confrontational and intense than female competition because male fitness, as Bateman first clearly perceived, depends on maximizing mating frequency. Males cannot afford to take the contest lightly. The prize is substantial, and the greater the prize, the greater will be the risk that the hopeful male should venture in order to secure it. For a big enough prize it will even be worth his while to risk death.

It may seem bizarre to suggest that natural selection often promotes early death by favoring types whose energetic and dangerous life-style can lead to an early grave, but a little consideration will show that it must be so. Imagine a bull elephant seal that has no stomach for the dominance battles of the breeding beach. Very well. He can opt out: remain at sea, never endure the debilitating months of fast and battle, outlive his brothers. But mere survival is no criterion of success. Eventually he will die, and his genes will die with him. The bull seals of the future will be the sons of males that found the ordeal of the beach to be worth the price.

A simple model proposed by Gadgil and Bossert [206] and elaborated by Trivers [619] makes clear the reason why male mortality is generally higher than female. The argument can be understood without recourse to their diagrams and equations. The critical notion is that of **reproductive effort**, the expenditure of future reproductive potential. For example, in order to breed this year our seal enters a debilitating contest that almost certainly increases his chances of dying before the next breeding season. A male's output is potentially much higher than a female's, but this potential can be reached only if he bests other males in competition. His success is therefore highly dependent on the reproductive effort expended by other males, his competitors, whereas a female's success depends more upon her own effort and less upon the effort expended by other females. This is another way of saying that males compete more than females, and the result is selection for increased male mating effort. The limit on female reproductive potential imposes a limit on the effort that it is worth her while to expend. Why should a seal kill herself for one season's single pup? For the male, on the other hand, the potential payoff of a good season is higher, and the risk worth taking is therefore higher too. Moreover, he has no real options: The competitive situation assures that he must expend a great deal of reproductive effort to get any payoff at all. So he fusses and fights, suffers stress and starvation, and perhaps even dies trying to breed.

It is hardly surprising, then, that males generally succumb at an earlier age than females. It commonly follows, especially in mammals, that adult females outnumber males despite an initial equality of numbers of each sex. (Just why it is generally a good parental strategy to produce equal numbers of male and female offspring will be considered in Chapter 9.)

In many cases the very qualities that permit males to compete for mating opportunities exacerbate their risks. Song and bright colors make males conspicuous to predators as well as to females. A gaudy tail may be of value in sexual displays and yet be a hindrance in such activities as feeding or flight.

Such potentially maladaptive features led Darwin to his theory of sexual selection [137], published twelve years after the epoch-making *Origin of Species*. Sexual selection takes place whenever individuals of one sex (usually males) compete for mating opportunities or when individuals of one sex (usually females) choose mating partners. This Darwin differentiated from natural selection, which was a matter of the environment selecting the most viable individuals. He stressed the distinction mainly because the two kinds of selection could, in principle, act in opposition to one another, so that sexual selection could explain such burdensome features as oversized tails or antlers. The implications of female choice will be explored at length in the next chapter, but we wish first to examine further the costliness of male reproductive effort.

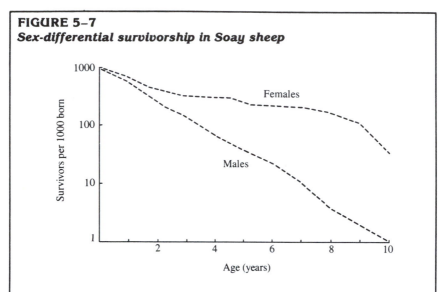

FIGURE 5-7
Sex-differential survivorship in Soay sheep

Survivorship curves for soay ewes and rams on the Scottish island of St. Kilda. There is intense sexual selection, and only one in a thousand males lives ten years. (After Grubb, 1974, Figure 10-10 [235].)

The phenomenon of excessive male mortality is especially well documented among the mammals, where it occurs very generally. The evolutionary innovations of the placenta and of milk occasioned a great initial disparity in the parental investment of the two sexes, and the females' intensive nurture is a resource for which males are willing to pay a substantial competitive price. One result of this competitive effort is great male mortality. An exemplary set of data is presented in Figure 5-7. Similarly, among deer, yearling males suffer about twice the mortality of females [530]. Stags are therefore much rarer than does, and they continue to die off faster through adulthood. These deaths are more or less directly due to the brutal competition for females. Males expend a tremendous amount of energy during the rut and enter the winter in a weakened state, weaker in fact than the does on whom has fallen the burden of pregnancy and nursing! Seasonal violence among males competing for reproductive status is a widespread mammalian phenomenon. Male rhesus monkeys [388] for example, fight in the breeding season, may be seriously wounded, and may die of those wounds (Figure 5-8).

In many other mammals, reproductive competition promotes male mortality by less dramatic avenues. Saharan gerbils, for example, are rather solitary animals. Wherever a little plant life can be found in the harsh Saharan terrain, individual females inhabit and defend small domains adequate to provide food for themselves and their pups [129]. The distribution of females is thus related to food resources, but the critical resource for males is the females themselves. The imperative of male gerbil existence is getting around to call on the females. Males strive to visit as many females as possible and as often as is necessary to catch each female on her infrequent days of sexual receptivity. Toward this end males are constantly making their rounds. They are apt to inhabit burrow sites that are too poor in food ever to appeal to a female but that are centrally located for visits to several females living in scattered food-rich areas.

These gerbils afford another example of the perils of the competitive male life-style. The female has a very good chance of living a year or more; she sticks to a small, well-known area with lots of cover, and she maintains several minor escape burrows. A male, on the other hand, repeatedly takes his life in his paws by running hundreds of meters across open country in

FIGURE 5-8
***Seasonal variation in woundings and deaths of adult
male rhesus monkeys (Macaca mulatta) on Cayo
Santiago, Puerto Rico***

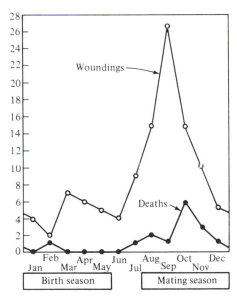

(Data from Wilson & Boelkins, 1970 [678].)

order to visit a female who is certain to drive him away, but who may just possibly be willing to mate with him before doing so. It is a dangerous quest. The entire adult male population of an area is likely to turn over every few months, while females rest secure in their territories. Despite equal numbers of each sex at weaning, adult female gerbils are sometimes two to three times as numerous as males [see also 411].

WHY DO MEN DIE YOUNG?

A sex difference in mortality is well known in people too (Figure 5-9). In *Males and Females* [292], a provocative account of human sex differences and their physiological bases, British psychologist Corinne Hutt reviewed a great variety of prenatal and postnatal causes of death to which males are more susceptible than females and concluded that "The adage of the male being the stronger sex seems to be limited very much to physical strength!" [292, p. 23]. Hutt noted that "the greater vulnerability of the male has been hitherto one of the less explicable features of development" and then went

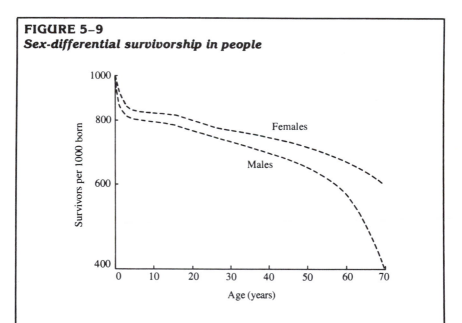

FIGURE 5-9
Sex-differential survivorship in people

Cohort survivorship from birth to age 70 among white Americans born between 1899 and 1903. (After U.S. Department of Health, Education, and Welfare, 1972, Table 1 [633].)

on to outline an explanation in terms of chromosomal differences between the sexes. Women have two X chromosomes, while men have just one X and a much smaller Y. Thus, many genetic loci are duplicated in a woman's cells but not in a man's. One result is that certain defects, such as color blindness, are more likely to be expressed in men: A man will have the defect if a certain relatively rare recessive gene occurs on his single X chromosome, whereas a woman will exhibit the defect only if the rare gene occurs on *both* of her X chromosomes, which is much less likely to occur.

This explanation is undoubtedly correct for color blindness, and it is probably sound with respect to some sex differences in vulnerability to disease. It is, however, an incomplete explanation of the sex difference in vulnerability. At best it explains only a relatively proximate mechanism of that difference, leaving out of account the ultimate causes. It is particularly apparent that the chromosome theory lacks something when we consider animals with other genetic systems. The mechanism of sex determination in birds is the opposite of that in mammals—the males are homogametic (ZZ) and the females heterogametic (ZW; see Figure 9-1, p. 225). Nevertheless, where there is severe male competition for mating opportunities, as for example in the polygynous great-tailed grackle[553], the mortality of males exceeds that of females (see also Figure 7-4, p. 154). The ultimate causes of sex differences in mortality are the different reproductive strategies of females and males [619]. The males' greater competitiveness commits them to greater risks and resultant mortality. Selection for intense male mating effort is the ultimate cause of the males' greater degree of dissolution, disease, and death (see pp. 99–100). To call human males more "vulnerable" than females obscures the fact that excess male mortality is largely a consequence of risky behavior, particularly in young men. This is a subject to which we shall return in Chapter 11 (pp. 299–301).

MALE PHYSIOLOGY AND STRATEGY

There is no end to the ploys by which males vie for fertilizations. There are resounding battles, brilliant displays of color, and fervent serenades. These and many other elements in male reproductive strategies share a common causal factor—the testicular hormones called *androgens*. (The role of ovarian hormones in female reproductive strategies will be discussed in Chapter 8.)

The location of the testes, external to the main body cavity in most mammals including ourselves, makes them especially vulnerable to accident and malice. It is therefore hardly surprising that the functions of the testes have been at least partially understood since antiquity. Castration was practiced as a punitive measure in China three thousand years ago, and it has since been performed on slaves, the conquered, sexual offenders, and young sopranos, as well as in religious rites [409, 500]. Through the ages the most commonly remarked effects of castration have included the atrophy of male reproductive organs; a reduction in the capacity for sustained penile erection, for orgasm, and for seminal emission; and a decline in aggressiveness [597].

While the effects of castration were well known, the secretory functions of the testes remained but dimly suspected. In 1889 the pioneer endodrinologist Charles-Édouard Brown-Séquard injected himself, at the age of seventy-two, with small quantities of ground-up testicles of guinea pigs. He soon reported to the Parisian Société de Biologie that he was greatly reinvigorated by the injections and was able to work in the laboratory as he had not done in years. Soon French doctors were claiming that such injections cured almost every conceivable malady, and by 1890 more than twelve thousand physicians were administering the concoction to their patients. The fad died a deserved death within a few years, as the early optimism faded [471].

Not until 1935 was testosterone, a gonadal androgen, first chemically identified and synthesized; this feat was achieved independently in two laboratories. The 1939 Nobel Prize was awarded the two principal investigators, the Austrian-born Swiss Leopold Ružička and the German Adolf Butenandt. (The latter was denied his prize by the Nazi government.) That first pure synthetic testosterone was disappointingly ineffective when injected experimentally into animals. In 1936, however, an effective chemical medium for slowing its absorption and metabolism was discovered, and the first precise experimental studies of hormone function became possible. Methods for the accurate measurement of naturally occurring hormone levels have been developed only within the last twenty years, opening the doors to an explosion of endocrinological discovery. (See [34] for a history of research on hormones and behavior.)

Androgens, and indeed vertebrate hormones in general, are secreted by endocrine glands into the circulatory system. The term *endocrine* refers to glands that lack specialized ducts for the secretion of their products but are instead richly vascularized so that their products are continually being carried away in the blood. Although a few hormonally induced events, such as the milk-ejection reflex, take place rapidly, hormonal effects generally wax and wane at a leisurely pace. (By one general definition, "hormones are chemical messengers that help to coordinate the activities of one group of cells with those of another" [596].) Androgens are responsible for sex differences in coloration, muscle growth, hair and plumage patterns, metabolic rates, and a host of behavioral inclinations. (By "responsible," we mean that androgens induce and maintain these sex differences in mature animals. The role of androgens and other hormones in the initial development of sex differences—sexual differentiation—we shall leave aside until Chapter 10.)

Hormones are constantly being released in tiny quantities that fluctuate in response to social stimuli, environmental cues such as daylength and temperature, and various events inside the animal, including the secretion of *other* hormones. Hormones are then constantly broken down by enzyme action in the liver and kidneys. The concentration of hormones in the bloodstream, then, is determined by a sort of algebraic summation of their production in the endocrine glands and their removal from circulation.

Hormones have "target tissues," highly specific parts of the body where they provoke highly specific responses. Blood-borne hormones are carried to all parts of the body, but nontarget tissues are relatively unaffected. Hormones do not initiate any novel cellular processes; they act as catalysts, regulating the rates at which certain chemical processes occur in the target tissue. By so influencing cellular events hormones modulate the rate of secretion from secretory cells; adjust neural thresholds so that particular nerve cells fire more or less; and initiate and depress the growth of all manner of somatic tissues, including sensory receptors, vaginal walls, fur and feathers, bones and breasts.

As we remarked above, castration has long been known to emasculate animals. Males lose both interest in, and capacity for, sexual behavior, though the decline may occur surprisingly gradually in view of the fact that there is virtually no testosterone remaining in the blood a few hours after castration. The sultan cuckolded by his castrated harem master is the subject of many a ribald tale, but more sober evidence suggests that human males also lose libido over the months following the operation. And not only are castrated males less interested in sex; they are also less interesting. An estrous female rat, for example, will choose to approach an intact male in preference to a castrate on the basis of odor cues alone. In a great many species castrated males become relatively unaggressive toward other males, and they also become less effective stimuli for the elicitation of aggression from intact males. These morphological and behavioral deficits can all be mitigated or completely redressed by androgen replacement therapy.

Androgens induce males to violent and risky behavior, and hence increase mortality risk. It furthermore appears to be the case that androgens contribute to male mortality even more directly by hastening degeneration and senescence! The question of the longevity of castrated men has been studied by J. B. Hamilton [246, 247]. Imperial China and the Ottoman Empire maintained palace eunuchs into the present century, and Hamilton has studied some of these men. His primary source of eunuchs, however, was American homes for the mentally retarded, in which castration was sporadically administered for several decades with the aim of rendering the men more docile and manageable.

Hamilton and Mestler [247] found that the median age attained by castrated retardates was 69.3 years, while an otherwise comparable group of intact retardates attained a median age of only 55.7. This surprisingly large difference is not solid proof of the mortal effects of circulating male hormones: The castrated group may have been special in other respects, perhaps related to the decision to castrate them. However, the comparison group of intact men was carefully chosen to be as similar as possible to the eunuchs in all ways other than the surgery, and Hamilton makes a strong case for the conclusion that testosterone has degenerative consequences. Experimental studies of animals have strengthened that conclusion: The hormone's importance in the acceleration of male metabolism has been established, and castration experiments with cats have shown that the operation can indeed prolong life [248]. It is a price few men would willingly pay!

In many seasonally breeding species the testes regress and regenerate annually. This provides us with a sort of natural castration and hormone replacement therapy experiment. Many behavioral and morphological traits change in phase with this seasonal change in testis development and testosterone levels. Song, plumage, and the establishment and defense of territory, for example, are all testosterone-dependent spring activities in male songbirds. Male gerbils increase the size of their home range as their testes regenerate early in the breeding season. The growth of antlers in male ungulates heralds the coming rut.

Testis function is correlated with these seasonal phenomena, as is illustrated in Figure 5-10. Castration and replacement therapy experiments have confirmed the causal role of androgens in many of these effects, but the causal connections are not simple one-way influences. While testosterone is a clear causal factor for male copulatory performance, it is also the case that the act of copulation can result in an increase in blood testosterone. So, too, can a mere whiff of the female or even an arbitrary conditioned stimulus that has been paired with her [225]. Similarly, a high testosterone level can be a part of the proximate causation of an animal's picking a fight, but it is also the case that a fight can cause an increase in testosterone.

The causal linkage of sexual performance, aggressiveness, song, mobility, plumage, muscle growth, and any number of other male attributes to the single factor of testosterone is surely to be understood as adapted to their strategic linkage. This relationship is especially clear in species with seasonal breeding. Out of season, songbirds commonly forsake their territoriality and antipathy to neighbors, lose their voice and brilliant

FIGURE 5-10
Seasonality of testis function and related phenomena in the red deer stag

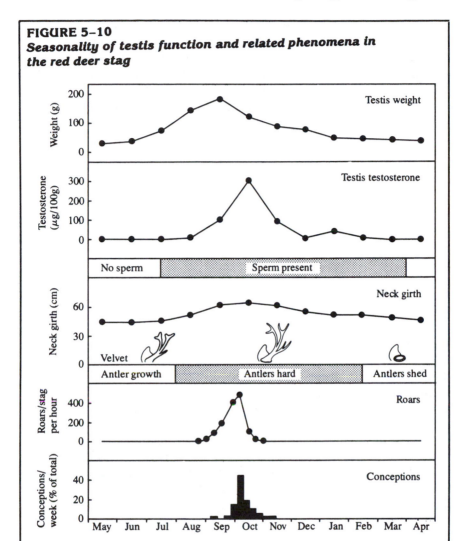

Testes develop in summer in anticipation of the fall "rut" (breeding season). Antlers grow during the spring and become hard during the summer, while neck musculature develops in anticipation of fall battles. Roars are aggressive challenges (see pp. 132-135). During September and October, dominant stags defend harems and subordinate stags attempt to sneak copulations. (After Lincoln, 1971, Figures 1 and 8, and Lincoln, Seasonal aspects of testicular function. In H. Burger and D. deKretser, eds. *The Testis.* New York: Raven Press, 1981, Figure 24 [385, 386].)

plumage, and suspend sperm production; the red flush of the stickleback fades; the gerbil confines his wanderings to one patch of vegetation; the stag discards his antlers and his aggressivity. In season, the value of these androgen-dependent activities and accoutrements is easy to discern. They are quite clearly of service in the competition to fertilize. And it is just as evident that they have their costs. Carrying antlers costs energy, as does territorial defense. Several seasonal stratagems exact a price in greater vulnerability to predators—the conspicuous voice and plumage of the territorial songbird, for example, and the extensive rambles of the male gerbil. And of course there is the great cost inflicted by direct physical strife (recall Figure 5–8).

Testosterone is a cornerstone of the proximate control tactics by which males risk much to gain much when there are fertilizations to be won. Is it any wonder that seasonal breeders switch off the mechanism and the costly strategy it subserves during the quiescent season, when the possibility of reproductive payoffs is absent?

SNEAK FERTILIZERS

There are many species of fish in which brightly colored males defend substrate territories containing a nest scrape, to which they attempt to lure duller females to spawn. Fertilization is typically external, so that the mating pair must release their gametes in close proximity and synchrony, while swimming through the nest together. Exclusive male care of the eggs is then very common among fish, the female immediately departing. Sometimes a curious thing happens in several of these species: An apparent second female joins the spawning pair in a brief ménage à trois [323]. She doesn't seem to be releasing any eggs, and the alert ethologist is likely to question whether she is a female at all. "She" isn't. It has now been demonstrated in a number of species that the third party is a sneaky little male whose mimicry of the female reduces the likelihood of an attack by the territory-holder. In some of these species, sneaking is a juvenile reproductive strategy: Only small, young males employ it, and once they get big enough, they take on male coloration and establish a territory instead. But in sunfish (*Lepomis*) in lakes in Ontario and New York, there are two dramatically distinct kinds of male life histories: Some males breed only as territorials, while the sneaks are sneaks for life [160, 233].

Male bluegill sunfish spawn "colonially" in clusters of some 8 to 150 substrate territories. When third parties are observed in spawning interactions and are netted for examination, they invariably prove to be males.

Three types of males are distinguishable both morphologically and behaviorally. The largest are "parental" males who have a dark orange ventral flush, build nests, and defend territories. Next come "satellite" males who mimic females in both behavior and color pattern. The smallest and palest are "sneakers," who stay near the bottom and make rapid dashes into and out of nests, unlike the satellites who enter nests more slowly like a spawning female. (Both "satellites" and "sneakers" correspond to what we have called sneaks above.)

At this point in the investigation, a remarkable attribute of fish becomes relevant. The ages and growth histories of individuals can be read from their scales, just like reading the growth rings of a tree stump! In one study [232], it was found that parental males ranged in age from seven to eleven years; satellites were mostly four years old; and the little sneakers were mostly two or three. So far the three behaviorial modes appear to correspond to age classes, but the story is more complicated than that: The parental males had grown more rapidly during the first six years of life than the sneakers or satellites. When the lake was seined to sample fish that were *not* attending the spawning site, it was found that the great majority of males under seven years of age were staying away and growing rapidly as the parental males had evidently done according to the growth record of their scales. These stayaways were concluded to be immature males of the parental type, and they were found to have immature testes less than a tenth the size of the large spermatogenic testes of the little sneakers. It seems clear that males follow one of two growth trajectories: "Sneakers" evidently become "female mimics" as they grow older, but they never become territorial "parentals." Even where the female mimics live to the age at which parental males mature, they never grow to territorial size, but stick with the sneak strategy for life [160].

Sneak fertilization as a minority male strategy appears to be widespread in the animal kingdom, and it is often manifest in distinct morphological types. Red deer populations, for example, have long been known to include "hummels," mature males who never grow antlers and avoid fights with those who do. Stag beetles, so called for the immense horns with which mature males fight, prove to parallel their namesakes in male dimorphism too: Many species include both horned and hornless male morphs [205]. Madhav Gadgil has suggested that sexual competition and resultant selection might often exaggerate weaponry over generations until the energetic and other costs of being able to compete were so high as to permit a "drop-out" male type, who shunned combat altogether, to become established. Such males would need an alternative, often "sneaky," route to reproductive success.

MIXED MATING STRATEGIES

Even more common than alternative male morphs are alternative tactics, each of which any individual male is capable of adopting according to circumstances. The choice of tactics is conditional upon what the other males are up to. If too many competitors are hovering around the females' preferred oviposition sites, for example, it may pay to try to intercept a female somewhere else. If big frogs have control of the whole pond, it may pay to wait silently in the mud and pounce on passing females drawn to those deep croaks [281]. In cricket choruses, some males sit silently near the singers, hoping to intercept an occasional female [82]. This sort of "satellite" exploitation of calling or otherwise advertising males by silent partners seems extremely widespread, and it is not always clear whether the alternatives represent a single conditional strategy or two or more different types of individuals (a "polymorphism") pursuing distinct unconditional ("pure") strategies.

In some cases, the alternative tactic is clearly a second choice, forced upon subordinate males who cannot compete. But in other cases, there seems to be an equilibrium in the fitness payoffs of the alternatives. Dung flies, for example, lay their eggs in fresh, steaming cowpats. (Their Latin name *Scatophaga*—dung eaters—is a scurrilous slander. The adult's diet is actually *other* flies who frequent the dung.) Males evidently choose either to guard territories on the cowpat or to lurk in the grass according to the density of other males; there is some evidence that the two tactics are equally successful [481]. But whether or not the alternatives are equally fit, we should like to be able to show that the males' choice of tactics is responsive to their relative costs and benefits [see 62]. A nice demonstration comes from a study of crickets by Bill Cade of Brock University. Cade showed that calling males get more copulations than their silent satellites but are also more often mortally parasitized by warble flies, which are attracted to the singing cricket and lay their eggs in him [82]. Comparing geographic populations, Cade further demonstrated that the proportion of males adopting the satellite approach was greatest where there were the most warble flies. As the cost of singing increases, the safer, low-return satellite behavior looks increasingly attractive! How the males could make such a decision is not immediately obvious, and the likeliest explanation would seem to be different thresholds for switching strategies in the various populations, based on genetic differences resulting from differing histories of selection. Cade is presently breeding crickets in captivity to test for such populational differences, and it now appears that the two behavior patterns may be heritable alternatives maintained as a genetic polymorphism, with the equilibrial proportions varying between populations [83].

FIGHTERS AND SCRAMBLERS

A herd of bison, like a baboon troop or a beachful of elephant seals, consists of a group of reproductive females and their attendant males. Such aggregations are by no means universal in mammals or other animals. A more typical pattern, in fact, is for reproducing females to shun one another and to defend territories containing adequate resources for their personal reproduction. The effect of territoriality is that reproductive females are often overdispersed in suitable habitat—more than merely failing to aggregate, they live farther from one another than a random set of locations. This rather complicates things for a male with polygynous ambitions.

In choosing breeding situations, females are concerned with energetic and other resources, and with the minimization of predation risk. By and large, they don't need to expend much mating effort—as a resource valued by males, they can count on the males to come to them. Where the usual sex difference in parental investment holds, females thus distribute themselves according to the availability of essential reproductive resources. The males then have to distribute themselves according to the availability of the females. This is the usual story in mammals. Where females aggregate, there is probably an ecological rationale—limited breeding sites or antipredator solidarity, for example. The relationship of ecological variables to social behavior (''socioecology'') is a topic to which we shall return in Chapter 7. Here we simply wish to point out that females do not aggregate for the convenience of males. It is up to the males to adapt their reproductive strategies to the females' behavior.

A herd of fertilizable females is a treasure worth fighting over, and the most combative males appear to be those, like the bison, whose females are gregarious. Where reproductive opportunities are more dispersed, males are likely to compete by scrambling after them instead of by fighting for rank. A successfully polygynous bull bison or elephant seal is a lumbering giant, with a high cost of locomotion. He can keep lesser males away from the females, all right—just as long as the females stick together and don't rush off. A successfully polygynous male gerbil, on the other hand, may be tough in a fight too, but he doesn't fight nearly as often or as effortfully. First and foremost, he has to be a good traveler.

This distinction between *contest* and *scramble* competition was originally drawn with respect to foraging strategies, but it nicely fits males ''foraging'' for females too. A big, local resource bonanza—whether food or females—is worth fighting for. Its high value warrants riskier, more dangerous tactics. And combat *is* dangerous. Ethologists once believed that animal fights had evolved to be gentle rituals in which no one was likely to be hurt, but intensive field studies in recent decades have repeatedly found that fatal combat is far from rare, accounting for significant mortality in animals ranging from insects [252, 609] to our nearest relatives, the chimpanzees [216].

The shift from dangerous combat to less destructive scramble competition is largely a matter of the value of the particular clump of resources being contested. But another factor is also relevant. Other things being equal, a male should be less inclined to use dangerous combat as a competitive strategy the more closely related he is to his opponent. W. D. Hamilton, the theorist originally responsible for the inclusive fitness concept, has presented evidence that **nepotistic restraint of male-male competition** is a real phenomenon.

Hamilton [252] studied Brazilian fig wasps. Fruiting fig trees provide immense bonanzas of high-quality food, and many species of insects have evolved to exploit them. Fig wasps constitute a large group that lays its eggs (oviposits) inside the fruit, where the wasps develop to maturity before dispersal. In many species, there are wingless males who mature inside the fig, breed with the females who hatch out there, and die there. The males never see the Brazilian sunlight, leaving it up to the winged females to disperse. As we might suppose from the fact of dense aggregations of fertilizable females, wingless male wasps engage in lethal combat, and they have evolved specialized weapons for the task. (It seems clear that the males are fighting for the females and not for the fig, since several species of wasps commonly inhabit a single fig and yet the mayhem is strictly intraspecific.) That, at least, is the story for wingless males of the genus *Idarnes*. Their wingless counterparts in the genus *Blastophaga,* however, are unarmed pacifists, who scramble rather than fight, scurrying about inside their fig in search of females while ignoring males altogether.

Hamilton ascribed this striking difference in behavior to a difference in the usual level of relatedness among competing males in the two genera. The *Idarnes* female has a long ovipositor with which she drills holes in many figs, leaving one or a few eggs in each. Several females regularly parasitize the same fig, with the result that the males who are hatched out within a single fig and compete there for fertilizations are often utterly unrelated. *Blastophaga* females, on the other hand, struggle into the fig, shed their wings in the process, lay their whole brood there, and die there. Rival males growing up within a single fig are then very likely to be brothers, and their nonviolent ways can be viewed as nepotistic restraint. By itself, this interpretation of fig wasp diversity might seem a rather stretched conjecture, but Hamilton was able to show similar trends, associating fraternal mating rivalry with reduced armament, within beetles, mites, and thrips, three insect groups quite unrelated to fig wasps. In mating rivalries, as elsewhere, it appears that kinship softens conflict.

POSTCOPULATORY COMPETITION

We have argued that males have commonly been selected to maximize insemination frequency, but the battle is not necessarily over once the copulation has been won. Recall the bison's postcopulatory guarding of the still-receptive cow. Copulating with a female does not guarantee a male the role of fertilizer. He must resort to a variety of other tactics to protect his reproductive prospects in species in which females are likely to accept multiple mates.

Students of insect behavior have long been aware that the females of some species mate repeatedly, whereas the females of related species do not. (It is this latter circumstance that makes possible the sterile-male method of pest control.) Male mating strategies are clearly adapted to the different female mating strategies. A good example comes from John Alcock's studies of solitary bees [7]. In species where the female bee mates but once, the premium is upon being the first male on the scene. This can lead to extreme male tactics such as digging young virgin females out of their underground abodes even before they can emerge by themselves, as they will do if unaided. In species where the female mates repeatedly, however, the greatest reproductive success generally accrues to the male that achieves the *last* mating before the female deposits her eggs. In these cases, males are likely to approach females as near as possible to the oviposition time and place and to do what they can to prevent subsequent matings.

Here may lie the most fundamental constraint upon male promiscuity. A male's basic strategy is to achieve as many fertilizations as possible, but he cannot just copulate and depart with the assurance that the mated females will in fact procreate his genes. In many cases a mating will bear fruit only if the male invests in protecting it against sabotage and usurpation by other males. So male-male competition is by no means at an end when a copulation is won. In fact, E. O. Wilson [680] has listed eight distinct types of postcopulatory competition among males:

1. Sperm displacement
2. Induced abortion and reinsemination by the second male
3. Infanticide of the first male's offspring and reinsemination by the second
4. Mating plugs and repellants
5. Prolonged copulation
6. A passive phase in which the male simply remains attached to the female after copulation
7. Guarding of the female without such physical holding
8. Departure of the mated pair from the vicinity of rival males.

The first three of these are devices by which a male eliminates a previous male's progeny and replaces it with his own. In the first of these the males compete to fertilize the same ova. Male salamanders, for example, place a spermatophore on the ground for the female to sit upon and inseminate herself. Competing males plunk their own spermatophores squarely atop the rivals' so as to get there first [19]. Males of various insects have genitalia that seem designed to penetrate the female's sperm-storage organs and flush out their prior contents.

If the female's ova have already been fertilized by a rival male, a would-be father is not without recourse, for he may be able to eradicate the fruits of that prior fertilization. One possible stratagem is induced abortion. This is called the Bruce effect after Hilda Bruce, who discovered that female mice, early in pregnancy, would respond to the odor of a novel male by aborting and returning to estrus [72]. The effect has now been shown in many rodent species, sometimes well into the pregnancy [e.g., 546]. This phenomenon has aroused great interest among sociobiologists, who have remarked the obvious benefit to the abortion-inducing male and have wondered how the response might serve the female's interests [353, 551]. But, as plausible as the Bruce effect's hypothesized role in male-male competition may seem, we are not aware of any conclusive evidence that the effect occurs in natural populations, rather than being an artifact of captivity.

Wilson's third type of postcopulatory competition, infanticide, has been especially well documented in lions and in langurs, which are largely terrestrial monkeys of the Indian subcontinent. In both cases several females and their young are associated in a group. The females are kin—sisters, cousins, mothers and daughters, aunts and nieces—who live out their lives in their natal groups. In the lion pride [51, 548] there is a group of males, often related to one another, though not to the females (but see [477]). Young adult males band together and search for a pride that they can seize for their own by deposing the resident adult males. When such a coup is successful, the usurpers often kill the cubs of the displaced males. The mothers' lactation is thus abruptly terminated, with the result that they come into estrus months sooner than would otherwise be the case. They can then begin the propagation of the new males' genes sooner. For this reason it is clear that infanticide, however ghastly and however damaging to the reproductive success of the lionesses, promotes the fitness of the murderers. The story is much the same in langurs, with the important difference that the killer is a single male [286, 589].

That such a phenomenon occurs is startling in view of the cost to the females. In the case of lions, it seems that the females simply do not have the physical power to resist the huge males and so make the best of a bad situation by getting on with the business of breeding with their new mates. The lionesses are also obliged to put up with male freeloading at mealtimes;

the females do the hunting and the males eat first. Among langurs, there is just the one male, and he does not seem so very much stronger than the females. Nevertheless, the females seem unable effectively to defend their infants from him. They may try to do so, even cooperating in their efforts. But the male is persistent and devious. He may resort to a sudden attack at any time for weeks after taking over the troop. In the end, he is likely to succeed in eliminating all the infants. In multimale troop species such as baboons, infanticidal immigrant males can also be a risk, and defense of their probable offspring may become a major preoccupation of the males who have been around longer [80].

Many biologists resist the notion that these repugnant acts of infanticide can really be evolved adaptations, although their positive fitness consequences for the murderers have been well documented [51, 286]. That natural selection will act to maintain the behavior, by penalizing any male that lets another's infants live and thus delays his own reproduction, seems irrefutable. Such infanticide is really only a particularly gruesome instance of a widespread tendency that we have already remarked—the selective nurture of one's own offspring concurrent with rejection of, and even attacks upon, alien young.

Wilson's last five modes of postcopulatory competition among males are all widespread devices by which the first male defends against usurpers attempting to displace his sperm [479, 480]. "Copulatory plugs" are mechanical devices by which he endeavors to seal off the female's reproductive tract. Often, special substances are added to the male's seminal fluid at ejaculation, causing it to coagulate and thus block access for subsequent males. And in one species of worm, the male uses his "cement glands" not only to seal the female but also, in struggles that have been described as "homosexual rape," to shut down rival males [1]! Finally, there is the bizarre case of a fly, *Johannseniella nitida*, in which the male leaves his genitalia as a plug to assure his paternity. The rest of his body is then eaten by the female, affording his offspring a proteinaceous, auspicious start in life [479]. This is surely a pinnacle of intensive male parental investment!

A GENERAL PATTERN

At this point some readers may protest that this chapter is one-sided. There are, after all, species in which the males do not kill themselves in lustful, promiscuous pursuits. There are species—most birds, for example—in which father and mother share parental duties; and there are cases, especially among fish, where post-zygotic parental care is exclusively a *male* affair. But this one-sided account is neither arbitrary nor unfair—polygyny *is* widespread, and its converse, polyandry, is exceedingly rare. Fights over females are commonplace, whereas a fight between females over a male is a curiosity. There is a bias in nature—a real and persistent asymmetry between the sexes—and the concept of an initial disparity in the parental investment of the two sexes affords an illuminating perspective on that bias. In the discussions that follow, the concept of parental investment will also prove valuable in dealing with the unusual. Wherever we find departures from the two basic sexual strategies outlined in this chapter—intensive nurture by females and the maximization of copulations by males—we also find that the male has been brought back into the parental picture. As his contribution to the nurture of his offspring increases, the distinctions between male and female strategies dwindle, and the related phenomena we have described, such as differential mortality and aggressive competition, are mitigated. Should the male's parental role increase to match the female's, we are apt to observe a stable monogamy. It is rare, however, and especially so in mammals, that the initial disparity in parental investment is ever so completely redressed. Instead, we almost always find some more or less extreme form of the two basic sexual strategies—the nurturant female and the prodigal male.

SUMMARY

Animals have limited resources, time, and energy to expend on reproduction. Females generally invest more in each offspring than do males and therefore have a lower reproductive potential. Males exhibit great variance in reproductive success. Some sire many more young than a single mother could ever produce, while other males die barren. There is great reproductive competition among males, who fight and even die for the chance to inseminate females. The female's greater parental investment is a resource for which the males compete. Females compete too, but more for nutrients than for mates, and less intensely than males. Male competition does not end when a fertilization is won, for another male may still usurp the female's reproductive capacity.

Sex differences in competitiveness and mortality are consequences of the disparity in parental investment and are mitigated wherever that disparity is mitigated.

SUGGESTED READINGS

Ghiselin, M. T. 1974. *The economy of nature and the evolution of sex.* Berkeley: University of California Press.

Thornhill, R. & Alcock, J. 1983. *The evolution of insect mating systems.* Cambridge, Mass.: Harvard University Press.

Trivers, R. L. 1972. Parental investment and sexual selection. In B. Campbell, ed. *Sexual selection and the descent of man 1871-1971.* Chicago: Aldine.

6

Sexual selection: Mate choice

Darwin distinguished between two processes in his original formulation of sexual selection [137]. The first we have just considered at length: competition for mates among the members of one sex, usually males. The second sexual selective process is the exercise of mate choice, usually by females.

CHOOSEY FEMALE, INDISCRIMINATE MALE

Throughout the animal kingdom males generally woo females, rather than the reverse. We saw in Chapter 5 why males are usually the more competitive sex and why they are concerned to maximize matings, while females are not. Michael Ghiselin has aptly labeled this element of the male strategy "the copulatory imperative" [212]. As much concerned with quantity as with quality, males are often rather indiscriminate in courtship, but the female's situation is different.

There is a basic strategic difference between the sexes that demands that the female be more discriminating in her sexual responsiveness. Consider the consequences for each partner of participating in a bad mating, one that is unlikely to produce viable, fertile offspring. It is a bad mating, for example, if the mate is of the wrong species or suffers from a genetically based defect that is likely to hinder survival or reproduction. If a male indulges in such a mating, he wastes some sperm. Sperm are relatively cheap. The loss will not diminish his future reproductive possibilities should a better mate appear tomorrow or next week or month. A possibility, albeit slight, that he will sire fertile offspring and perpetuate his genes is worth the investment of a few sperm, and natural selection should have inclined him to make that investment.

The female is in a different situation. The penalty she incurs from a bad mating is likely to be severe. Her great parental investment in each of her potential progeny necessitates that she choose her mate well. Having mated badly, her future reproductive possibilities are diminished. She may not be in a position to accept a better mate should one present himself subsequently: Months of nurture may be committed to an inviable or infertile hybrid. As we might then expect, females are generally more selective than males in their choice of mates. Bateman was aware that the females are mainly responsible for maintaining reproductive isolation between related species of fruit flies. Males will court females of other species as readily as their own. It is the females' task to check the males' credentials. Among invertebrates as diverse as butterflies and hermit crabs males are apt to court an astonishing variety of objects, indeed almost anything that bears some resemblance to a female.

In the vertebrates, with their more complex capabilities for perceiving and processing information, males are seldom quite so indiscriminately eager. Nevertheless, the task of fine discrimination and the prevention of bad matings continues to reside mainly with the females, just as we would expect from considering the differential penalties that such bad matings impose upon the two sexes. Naive male guppies, for example, are willing to court females of the wrong species, but even naive females will not accept the wrong males [382]. The males learn to confine their attentions to the appropriate females only by being rejected when they err. Females cannot afford to learn by trial and error, for the errors would be too costly. We could cite further examples of the greater discriminatoriness of the female sex of various birds and mammals too, but it will perhaps suffice to note that the principle also holds up in our own species. Men are much more susceptible than women to misdirected sexual behavior, a subject that will be discussed further in Chapter 11.

"Choosey female, indiscriminate male" is of course a bit of a caricature, although it reflects a real asymmetry. Individual sperm are indeed cheap, but whole ejaculates may not be [152], especially when they constitute a nutrient-rich "nuptial gift," as we shall see. And where males make a major parental contribution, as in most birds, a mismating can be as much a disaster for males as for females. In such cases, males have to be a little choosey too [289]. Males also have to be somewhat selective when there is a significant time cost of mating. In isopods, for example, an order of invertebrates including sowbugs, males often guard females from other males for several days before mating, actually carrying the female about until she is ready to copulate. The males are then selective, preferring to guard larger females who will produce larger numbers of eggs [413, 567]. But it is precisely because the female is so often the scarce resource, valued and fought over by males, that she is frequently in a position to be more choosey and to demand that her suitors assume at least some of the costs of the reproductive enterprise.

FEMALE CHOICE: MAKING THE MALES PAY

The black-tipped hangingfly is a slender predatory insect that hunts in the dense low foliage of temperate forests. Its prey consists of houseflies, aphids, daddy longlegs, and other small animals of the forest floor; and according to Randy Thornhill of the University of New Mexico [605], female hangingflies prey upon arthropods of various sizes in proportion to their abundance. Not so the males. When a male hangingfly catches a dead arthropod, he feeds on it only briefly and then does one of two things. A small prey item he discards to look for another. But if the prey item is a certain threshold size, the male everts a pair of abdominal glands and lets a special chemical waft away on the breeze, the advertisement of a successful hunter.

The target of the male's chemical signal is of course a female, and when she approaches, the male passes her the prey item (without relinquishing his hold on it!) and attempts to mate while she feeds (Figure 6-1). If the prey item is inadequate, she breaks off copulation before any sperm are transferred and departs. If it's a big juicy fly, on the other hand, she will permit copulation to continue for the full twenty minutes necessary for complete sperm transfer. Once that goal has been attained, the *male* becomes increasingly likely to break off copulation and snatch back his gift,

FIGURE 6-1
Hangingflies, **Hylobittacus apicalis.**

During copulation, the female is feeding on a prey item brought by the male. (Photograph by R. Thornhill.)

which may still be good enough to win him another female! Meanwhile, his mate ignores advertising males for about four hours and goes off to lay about three eggs. Her egg production is to some degree a function of the size of the nuptial gift. Males are quite clearly buying fertilizations.

The female hangingfly seems to have taken control of the mating system, and she lets the male serve her. When things are going well, she hardly has to hunt at all. Hunting is dangerous for these animals, for as they move through the foliage in search of prey, they frequently run afoul of their major source of mortality—web-building spiders. And it is males, who travel about twice as far as females, who mostly get snagged.

So females with their precious ova are not quite such helpless victims of roving males as might be imagined. They have their ways of making males pay. Thornhill [606] has reviewed a fascinating range of such transactions among the insects. In some species the male must provide a proteinaceous spermatophore, which the female assimilates, if his sperm is to be accepted. The spermatophore may weigh up to a quarter of his body weight. In some flies, grasshoppers, and other insects, females feed upon glandular secretions from the male during copulation; if he stops secreting, she stops mating. Finally, it is not rare for the male himself to be consumed by his mate.

Nuptial feeding is even more conspicuous in birds. Ornithologists have long been aware that males feed females in courtship and during nesting, but until recently little attention was paid to the energetic value of the gift. Courtship feeding turns out to be no arbitrary ritual [536]. Terns, for example, are colonially nesting birds who plunge-dive for fish and shrimp. In a fine study of common terns by Ian Nisbet of the Massachusetts Audubon Society [461], it was found that the rate at which males feed their mates is maximal just when the energetic cost of egg production is maximal. Even more interesting was the fact that pairs varied in their nuptial feeding rates, and that the rate at which the male fed his mate was predictive of the weight of her clutch.

In Chapter 5 we discussed some cases in which males can use two or more alternative mating strategies (pp. 103–104). In such a case, the females may prefer the male strategy that is most profitable for them, discriminating against the alternatives. This is very clearly the situation in scorpionflies (*Panorpa*), who are close relatives of the hangingflies and, like them, have been studied by Randy Thornhill [607, 608]. Like hangingflies, *Panorpa* suitors acquire a dead arthropod, emit a pheromone to attract a female, and present it to her as a "nuptial gift." But when they cannot get a suitable prey item, male scorpionflies switch tactics. They secrete a salivary mass and advertise for a female in the same way they would with a prey. Male salivary glands are enormous—perhaps a quarter of the insect's total weight—and their secretion is nutritious. So a male with no other nuptial gift to offer gives the female of himself. And there is a third avenue—rape. With neither a nuptial gift nor a secretion of pheromone, the male may simply try to knock a passing female out of the air, immobilize her with specialized claspers and clamps, and inseminate her forcefully. Males prefer the arthropod ploy—if you give them an arthropod, they'll use it and forsake the alternatives. Without a prey, they'll use saliva if they can afford it. Rape seems to be a last resort. That the males would prefer to use a prey item rather than their own saliva is not surprising, since salivary secretion is debilitating. But it is less obvious why both advertising strategies are preferred over rape. It appears that this priority is female-enforced. Females flee from and resist males who attempt to use force, and they manage to escape about 85 percent of the time. They have a further line of defense too. In the laboratory, Thornhill found that insemination was successful in all "honest" resource-transfer copulations, but was somehow blocked as often as not if the female had been raped. By discriminating against rapists, then, female scorpionflies selectively maintain a style of male courtship that offers them some material gain and some choice in who will inseminate them.

In a few cases, the resources that male insects provide to females are so substantial that the male has evidently become a limiting resource for female reproduction, rather than the reverse. In the Mormon cricket (actually a katydid), for example, the male provides a proteinaceous spermatophore equal to 20 percent of his body weight. Unlike other katydid species with smaller spermatophores, female Mormon crickets fight over males, and males are coy and choosey [237]. We shall return to the subject of "sex-role reversal" in Chapter 7.

RESOURCE-DEFENSE POLYGYNY

A male can achieve some degree of polygynous monopoly by aggressively guarding several females. It's an exhausting strategy, and as we saw in the bison, a total monopoly is often not attainable, but female defense by a dominant male can certainly pay off in a better-than-average number of fertilizations.

But what if the females won't stick together? Simply being the toughest male around may not be much help. As we have seen, males may switch to scramble competition when the females are not defensible. There is also a third possibility: Though the male may be unable to herd and control the females, he can set up a territory in some place that females are sure to come to, and he can exclude other males. A steadily producing food source that will regularly attract mobile females may be worth defending. Hummingbirds, for example, feed on nectar, and a flowering bush may be guarded by a male who excludes other males while permitting females to feed in exchange for copulations [691]. This is rather like the hangingfly's nuptial gift, except that the nectar source is stationary, and the male has no other claim to it than his ability to exclude others. In many insects, amphibians, and fish, ideal sites for oviposition or spawning are limited and are defended by large, relatively successful males who are able to keep their rivals at bay and fertilize the eggs of multiple females. Finally, where there is post-zygotic care, as in all birds and mammals, females often establish their separate breeding sites close enough together for males to attempt to defend exclusive territories encompassing two or more breeding females. It is this latter sort of territorial polygyny that we should like to examine in a little more depth.

THE POLYGYNY THRESHOLD

Most migratory songbirds—robins and warblers and finches and so forth—are predominantly monogamous. A male establishes a territory, attracts a single mate, and shares parental duties with her over the nesting cycle. But a minority of species—the ubiquitous red-winged blackbird of North America is one example—are regularly polygynous. Individual males are commonly able to attract two or three or four or more females to nest within their territories. Why? It would seem, from the arguments developed earlier, that males of *any* species might strive to accumulate multiple mates. If there is always an advantage to polygyny for the male, and yet it does not always occur, then the circumstances under which it does occur must be those in which there is also some advantage for the female. This, at least, is the argument advanced by Gordon Orians in an attempt to explain the diversity of mating systems [473]. He elaborated upon the earlier ideas of Verner and Willson [642], who coined the term "polygyny threshold."

This concept is best explained with reference to migratory songbirds.

FIGURE 6-2
The Orians-Verner-Willson "polygyny threshold"

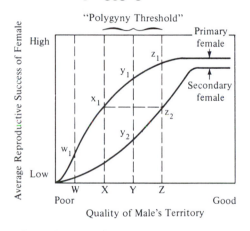

Male territories vary in quality in such a way that females tend to enjoy greater reproductive success on a "good" territory than on a "poor" one. Males arrive first and set up territories. Imagine, for example, that four males have territories of quality W, X, Y, and Z. Females then arrive and evaluate the territories. The first female should choose the territory of quality Z and thus enjoy reproductive success z_1. The second should opt for Y and y_1. But the third female can do as well by being the second mate on the Z territory or monogamous on the X territory: $z_2 = x_1$ and the difference in territory qualities has reached the "polygyny threshold." The territory of quality W is so poor that females would do better to be bigamously mated on either Z or Y. The male there may remain a bachelor, while the males on Z and Y each have two mates and the male on X has one. (After Verner & Willson, 1966 [642] Copyright © 1966 by the Ecological Society of America; Orians, 1969 [473].)

Males arrive upon the breeding grounds in the spring before the females, and begin to sort out territories among themselves. Some males get choice territories, while others settle for poor ones. When the females arrive, each settles on the best available territory for nesting; in so doing, she incidentally picks a mate (though to what extent females may be influenced by the male himself is still unresolved; see p. 128). The consequences of variation in the quality of the territories are illustrated in Figure 6-2. Where the difference between the qualities of available territories exceeds the polygyny threshold, as shown in the figure, a female will do better as a second mate on a good territory than as a first mate on a bad one. What makes a territory good or bad varies with the species in question and with the ecological circumstances. It might be the availability of suitable nest sites or materials, of food for the nestlings, or of good conditions for the detection of approaching predators. What is crucial is that territory quality varies in some way that the female can assess.

The two curves in Figure 6-2 represent the cases of monogamy and bigamy. We assume that a female will fledge fewer young when bigamous than when monogamous on a territory of the same quality because bigamous females must share some resources, most notably the parental contributions of the male. (Hence, in species where the male contributes relatively little, the curves should be relatively close together.) The curves reach an asymptote because we assume that there is a ceiling on reproductive success beyond which improvements in territory quality are irrelevant.

Other curves could be added to Figure 6-2 to represent the cases of three females on a territory, four, and so forth. Territory qualities might then be such that the first female to arrive settles with male 1 and the second with male 2. The third becomes the second mate of male 1, the fourth joins male 3, the fifth becomes the third mate of male 1, and so forth. Each successive female picks the best available circumstance in terms of expected reproductive success. This picture of a succession of female arrivals and choices is consistent with Verner's observations of long-billed marsh wrens and with Orians's and Willson's of blackbirds.

According to the polygyny threshold model, monogamy should result when the curves are relatively flat (and/or relatively far apart) over the range of existing territories. More plainly, monogamy is expected where territory quality does not vary greatly, so that each arriving female will do better to accept the best remaining bachelor territory than to accept second female status on a slightly better one. This suggests that those species of migratory songbirds that are regularly monogamous are not irrevocably so but practice monogamy in response to particular ecological circumstances. This suggestion gains some plausibility from observations of occasional polygyny in a large number of species that are usually monogamous.

Is polygyny in fact prevalent in those species where the variance in male territory quality is especially large? Measurement of territory quality is difficult, but there is some relevant evidence at a broad level of comparison. Ecological data suggest that marshes, which contain lush patches distributed irregularly according to water depth and other variables, should be habitats with especially great diversity in territory quality. According to Verner and Willson [642], only 14 of the 291 species of songbirds nesting in North America are regularly polygynous and only 18 are marsh nesters or marsh feeders during nesting, yet there are 8 species of polygynous marsh users. Polygyny is therefore far more prevalent among marsh users (44 percent) than among other songbirds (2 percent). This correlation is very impressive. However, in Europe, ornithologists have not found polygyny to be especially prevalent among marsh nesters. The polygyny threshold argument, that polygyny will occur where there is great variance in territory quality, may still be applicable to the polygynous birds of Europe, with the availability of suitable nest sites being the crucial and highly variable aspect of territory quality [238, 686].

If females are really assessing the relative promise of available breeding situations and settling accordingly, and if they are at all accurate in these assessments, then several predictions would seem to follow. The same males who are the first to get a mate should also be the first to get a second and a third (at least if females are reacting to overall territory quality and not to specific nest sites). Females settling for successively less preferred breeding situations should raise successively fewer young. And perhaps the most interesting prediction of the polygyny threshold model is that those opting for secondary female status should do just about equally well reproductively as those who opted for monogamous status *at the same time*; in Figure 6-2, $x_1 = z_2$. All these predictions have been neatly corroborated in an exceptionally detailed study of lark buntings conducted by Wanda Pleszczynska [208, 503, 504].

In the grassy Great Plains of North America the dramatic black-and-white male lark buntings make sure their presence is known, while the females are inconspicuous. (After marshes, grasslands seem to be the habitat where songbirds are most often polygynous.) A rare virtue of Pleszczynska's study was that she was able to identify a simple measure of territory quality that proved to be highly predictive of both female preference and reproductive success. That measure was shade, for it seems that a shady nest site on the treeless plains is a precious commodity. Having observed that shade was related to female choice and to fitness, Pleszczynska verified that the relationships were causal by adding plastic rosettes of leaves to alfalfa plants in some territories; the effect was both to increase fledging success and to enable the lucky males on those territories to attract *three* females, something that never occurred in the absence of experimental intervention! Male lark buntings, like so many other species, get their territories sorted out before the arrival of the females. By intensive observation during the two-week period of initial female arrivals, Pleszczynska was then able to confirm each of her predictions (Figure 6–3).

In those highly polygynous songbird species in which the male does not participate in parental care, our emphasis may be somewhat misplaced when we speak of females opting for a polygynous or monogamous breeding situation. In some populations of red-winged blackbirds, for example, males are never observed to feed young. The female's choice may simply be a matter of the best available nest site, with the male's mating status and territorial boundaries largely irrelevant to her. In fact, the local territorial male may not even be the sire of all her young, although he probably sires most of them [65]. But where the male contributes to the young, the female should be more concerned with the distinction between being monogamously mated or polygynously mated.

FIGURE 6–3
Polygyny in the lark bunting

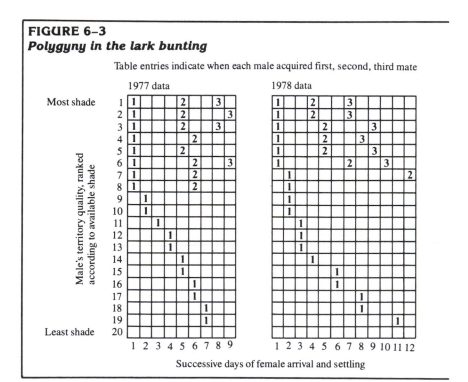

Table entries indicate when each male acquired first, second, third mate

Male's territory quality, ranked according to available shade

Successive days of female arrival and settling

In several species of polygynous songbirds the male helps his first female feed her brood but leaves any subsequent mates to fend for themselves [e.g., 485]. Even so, a female may choose to become mate number 2 on a fine territory while territorial bachelors are still available. An example is the bobolink (*Dolichonyx oryzivorus*, Figure 6–4), a regularly polygynous bird of North American pastures and cropfields that is in many ways reminiscent of the lark bunting. In a careful long-term study, Stephen Martin found that the unaided second female had decidedly less reproductive success than the assisted first female [414]. Second females laid 4.8 eggs per clutch and successfully fledged young from 49 percent of their eggs; the comparable figures for first females were 5.5 and 64 percent, respectively. The lower success rate of the second females was clearly due to their lesser ability to feed the nestlings without male assistance. That females continue to choose this second-female status over mating with remaining bachelors suggests that their reduced payoff must still exceed what they could hope for on the poorer territories, even with male help.

Females behave in accordance with the polygyny threshold model, successively occupying territories according to their quality, and choosing polygynous status on the best territories in preference to monogamy on the worst.

Moreover, reproductive success is maximal for those females choosing first, and declines with successively less preferred breeding situations. Monogamous and bigamous females settling at about the same time do about equally well.

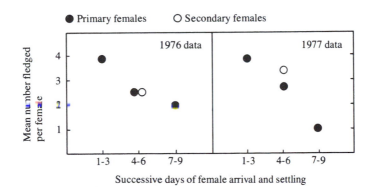

(After Pleszczynska & Hansell, 1980, Tables 2 and 3 © 1980 by the University of Chicago [504]; Garson, Pleszczynska & Holm 1981, Table 1 [208].)

FIGURE 6–4

The strikingly patterned male bobolink at right directs a crouching, wing-spreading display at the female at left immediately before copulating with her. (Photograph by S. G. Martin.)

POLYGYNY AND FITNESS IN MAMMALS

There is some reason to doubt that the polygyny threshold model can be extended to mammals without major modifications. The yellow-bellied marmot is a heavy, terrestrial squirrel that lives in subalpine meadows in western North America. In suitable habitat, marmots may be found in concentrations ("colonies"), but the species is only moderately sociable. According to a study by Downhower and Armitage [164], adult males are territorial and mutually antagonistic. The territory that a male defends encompasses one to four breeding females. As we would expect, a male's reproductive output, measured by the number of yearlings produced, increases with harem size. However, female reproductive success declines sharply as harem size increases (Figure 6–5). Downhower and Armitage interpreted their results as contrary to Orians's notion that polygyny occurs because of its advantages to females. Monogamy is clearly the best mating system from the point of view of the female yellow-bellied marmot, yet polygyny exists. The usual harem size of two is a compromise, they suggested, between the still more polygynous inclinations of the male and the female's preference for monogamy.

The argument has been criticized by Elliott, who doubts whether the large-harem female suffers reduced fitness [177]. The loss of fitness may not be shared equally among all females. There is a dominance hierarchy among the females, and the senior female is not necessarily harmed by the presence of the others. So it may be only the young females that do worse in big harems than they would in a monogamous situation. Their present loss may be compensated by improved survival prospects, resulting from the predator-detection advantages the larger harems enjoy in comparison to more isolated smaller groups. Elliott still favors the notion that polygyny

FIGURE 6–5
Harem size and reproductive success in the yellow-bellied marmot, **Marmota flaviventris**

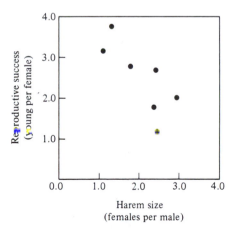

Females in larger harems produce fewer surviving young. (After Downhower & Armitage, 1971 [164].)

must serve the female in order for it to have evolved. Nevertheless, the polygyny threshold theory is decidedly less compelling in mammals than in migratory songbirds. A terrestrial rodent like the marmot does not have the mobility to visit all available male territories and thus make an informed choice. Moreover, in some harem-forming mammals male coercion clearly limits female choice. We have already described the male Hamadryas baboon that herds his females and punishes those who stray, and there are a number of hoofed mammals that behave similarly. As Downhower and Armitage quite rightly stress, the optimal reproductive strategies of a male and female are not perfectly harmonious, and there is no reason to suppose that either mate will be able to achieve its own optimum in conflict with the other. The concept of a conflict of interests between mates will be developed further in the next chapter.

The available human data on polygyny and reproductive success are reminiscent of the marmot findings: Polygyny costs the female and benefits the male. Among the Temne people of Sierra Leone, for example, anthropologist Vernon Dorjahn found 54 percent of 246 married men to be polygynous [161]. Men gained in fitness from increasing degrees of polygyny, while women lost (Figure 6-6). The decline in female fertility may appear to be small, but it is statistically reliable and by no means trivial. Figure 6-6 shows that the reproductive loss in polygynous women has two separable components. The rate of live births declines, but only after the man has four or more wives. Infant mortality becomes a factor sooner. It is higher at all levels of polygyny than in monogamous marriages—41 percent of children born to bigamous households died, an identical proportion in the still more polygynous families, but only 25 percent in monogamous households.

What are we to make of such data? It is tempting to suppose that the heightened infant mortality in polygynous Temne families is somehow due to the reduced possibility for male parental investment in each offspring, but this is speculation. The phenomenon is all the more striking in view of the fact that it is the more affluent men who have multiple wives, and they are presumably best able to contribute to their children's welfare.

But is it at all legitimate to apply to such human data the kind of strategic considerations that we have discussed in relation to other animals? There are substantial pitfalls. The actual extent of female choice and male coercion is difficult to assess, and there are more interested parties in a marriage contract than just the mates. Moreover, people have expressible intentions that often do not jibe with their "best interests" in terms of biological fitness. And their actual behavior may accord with neither! We shall discuss this further in Chapter 11, but for the moment an example will suffice. It would seem from the Temne study and other similar data that polygyny is

FIGURE 6-6
Polygyny and reproductive success in Temne people

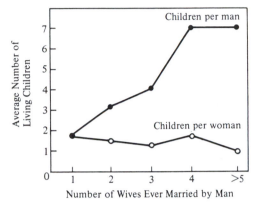

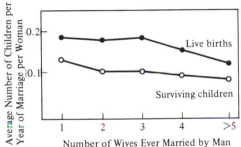

(a) Men gain in fitness with increasing polygyny, while women lose. (b) Live births per wife decline in highly polygynous households. Infant survival is reduced by any degree of polygyny. (Data from Dorjahn, 1958 [161].)

contrary to a woman's reproductive interests. We might suppose that she should dislike the practice, but we could be quite wrong. Among the Ibo and Yoruba of Nigeria, for example, it is (or was in recent times) slightly humiliating to be the sole wife, and lonely too [173, 650]. Monogamous women commonly encouraged or goaded their husbands to marry again. It seems that a form of Elliott's argument applies here: The reproductive success of a senior wife is apparently not compromised by polygyny, which might even serve her interests by bringing in a helper [309, 628]. But we should not be greatly surprised to discover, in any particular case such as this, that fitness is decidedly not maximized. People limit their reproduction in a multitude of ways, both deliberate and incidental to other goals. Any species that includes substantial numbers of childless couples, exclusive homosexuals, and religious celibates is sure to present some complications for the evolutionist! How the theory of natural selection for reproduction maximization remains relevant to the human animal is the subject of our concluding chapters.

FEMALE CHOICE: SHOPPING FOR GENES

If the females refuse to stay in a group, and the resources that females value are too scattered or too ephemeral to defend, what is an aspiring polygynist to do? He might abandon his ambitions and settle down with one female, reallocating reproductive effort to parental investment; male parental care and monogamy are topics to which we shall return. However, this often doesn't seem like much of a solution. In a grazing antelope, for example, or a browsing grouse, neither a mother nor her precocious young has much need of a male hanging about. He can't feed them—the food doesn't lend itself to gathering—and he's probably likelier to attract a predator than to successfully defend his family against one. Everyone would be happier if he would just leave [697].

In situations such as this, where the male provides the female with nothing other than a gamete, the female's best strategy seems clear. She should pick the best gamete she can find, the set of genes that will best contribute to the fitness of *her* genes when united with them in an offspring. And this female strategy suggests what a male must do: propagandize. Without so much as a nuptial gift, the male's plight is like an old song lyric: "All I have to offer you is me." It's up to the singer to make that offer appear attractive, *more* attractive in fact than the same offer being sung by every other male in the vicinity.

Whether females actually assess male genetic quality in choosing mates has troubled field biologists and theorists for decades. In territorial, polygynous birds such as the red-winged blackbird, many workers have tried to discover whether female choice is affected by both the territory and the male or just by the territory alone. This is not an easy question to settle. The best males will generally hold the best territories, so one must somehow eliminate the correlation between the two factors if one is to separate their effects. Experimental approaches are not necessarily successful, since the animals may respond by changing positions or boundaries and reestablishing the correlation. There are some complex arguments in favor of the hypothesis that male quality is not altogether ignored [655, 701], but the ef-

FIGURE 6-7
Effect of age on mating success in male black grouse

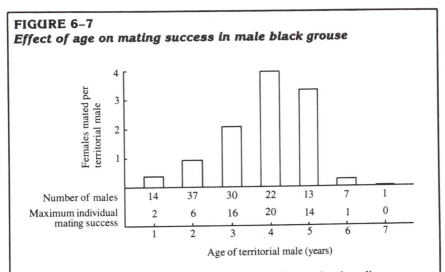

Number of males	14	37	30	22	13	7	1
Maximum individual mating success	2	6	16	20	14	1	0
Age of territorial male (years)	1	2	3	4	5	6	7

On a lek observed intensively for fifteen consecutive spring breeding seasons, four-year-old cocks attained the greatest mating success. Most males first held a territory in their second spring, and had disappeared by their fifth. In an average year, 9.6 males held territories, but the single most successful male got 54 percent of the matings. Only 3.6 percent went to non-territory-holders. (After De Vos, 1983, Appendices 2 and 3 [146].)

fect of male quality upon female choice in resource-defense systems is still unresolved.

Where the male offers nothing but himself—that is to say his genes— the situation seems clearer. And the clearest such mating system is that referred to by the Swedish label **lek.** Males congregate at a fixed site, the lek, and display en masse while defending a small area each from the others. Lek sites commonly persist for many years. Females visit the lek, peruse the selection of strenuously posturing males, and choose in whose territory they will tarry and be mated. Leks are used only for mating. No feeding occurs there, and the females depart quickly once serviced. These communal display grounds may be conspicuous at great distances as a result of the spectacular displays of the assembled males. Lek systems occur sporadically in several bird families and in a few mammals, fish, and insects.

In lek-breeding grouse [146], it has been demonstrated that the preferred males are the oldest. A young cock must attend the lek for years with little or no sexual reward before he can move up to the status of a successful breeder (Figure 6-7). Each time one of the more central cocks disappears, the younger, more subordinate, more peripheral cocks jockey about to move in a notch toward the charmed inner circle. (In other lek-breeding birds, such as manakins, the most successful males do not necessarily occupy territories near the geometric center of the lek.) What the hens seem to be selecting for, then, is longevity with vigor. To reach the top a cock must eat well, dodge predators, stay healthy, win fights. And that is certainly not a bad set of attributes to pick on if one were aiming for an across-the-board evaluation of genotypic quality.

BUT IS MALE QUALITY HERITABLE?

There seems to be no question that the males on a lek offer females nothing of value other than genes. And it has been abundantly verified that females are choosey—and highly consistent in their choices. Nonetheless, many biologists hesitate to accept the conclusion that females have evolved to effectively select the best male genotype. The argument against the gene-shopping interpretation of mate selection is a theoretical one. The lek mating system is one where the variance in male reproductive success is large [e.g., 272, 347, 673]; visiting females evidently select males upon consistent criteria, so that a few preferred males get a huge proportion of all copulations. Now, if *every* male is the offspring of a genetically superior father, then surely inferior genotypes must be rapidly eliminated by selection. Recall that any trait that has been subject to selection is likely to have a low heritability (p. 33). How, then, can fitness itself be heritable? It would seem that the males on a lek are unlikely to vary much in their genetic fitness potential. According to this argument, variation in fitness must, for the most part, be environmentally induced, especially in highly polygamous mating systems, and if that is so, then females cannot use phenotypic cues to shop for good genes.

This objection has been raised by many leading theorists, but it can be countered both theoretically and empirically. The idea that fitness variations must have only a minor heritable component is based on population genetical models with additive fitness effects of the genes at each of several loci. Such models are indubitably oversimplified. As soon as one considers that particular *combinations* of genes may have nonadditive fitness benefits, important genetic differences between males on the lek become feasible. And possibly even more important is the organism's overall level of heterozygosity across *all* genetic loci. Considerable variability in individual degrees of heterozygosity is to be expected, and there is evidence that high levels of heterozygosity contribute to fitness; if a female wishes to maximize heterozygosity in her offspring, she should mate with the most heterozygous male available [61]. By choosing the oldest, most vigorous, or most dominant male on a lek, then, females may simply be managing the breeding system to maximize offspring heterozygosity.

A recent experiment confirms that selective mating can contribute to offspring fitness. Linda Partridge of Edinburgh University [483] assigned female fruit flies to one of two conditions. The flies either mated in a group situation or in no-choice randomly constituted pairs. The group-mated females thus had the chance to mate with "better-than-average" males, whether as a result of the males competing or of the females choosing. If male quality is heritable, this group-mating procedure should produce better-quality offspring on average than the no-choice condition. The critical test pitted the progeny of the two sorts of matings in a competitive situation against opponents of a standard strain. The result was that the progeny of group matings were better competitors than those from no-choice matings. As Partridge concluded, "fitness may indeed be heritable." And that is an encouraging result, for if there were no such thing as heritable variation in male quality, then female choice, at the lek and elsewhere, would seem a meaningless charade.

RUNAWAY SEXUAL SELECTION

When Darwin first theorized that certain sexually dimorphic characters might be "sexually selected," he drew a distinction between those characters of service in competition between individuals of the same sex (usually male) and those which served to attract members of the opposite sex (usually female). He remarked, however, that the same character—such as horns or other male armaments—might be effective in both contexts, and this is probably the general case. Effects of certain male attributes upon male-male competitive success and upon female choice have been demonstrated empirically, but an intriguing question remains unresolved: Are any of the traits that we observe utterly arbitrary products of a "runaway process" of preferential mate choice? As Sir Ronald Fisher [191] pointed out, preferential mate choice has a peculiar capacity to be self-reinforcing. If some females find some arbitrary characteristic attractive—orange hair, say —then a rare orange-haired male will be on average a little fitter by virtue of this little attractiveness edge, all else equal. Well, as soon as there is a selective advantage to orange hair, there is immediately instituted a selective advantage for those who *prefer* it, since such females are likelier to have orange-haired sons who will enjoy the fitness benefits of . . . and so on. A sort of bandwagon effect ensues; nobody wants to be left out: Once most females prefer orange hair, the sons of any female who does *not* conform are likely to be losers. Fisher felt that a runaway process of this sort was necessary to account for such "extravagant" display features as the cock pheasant's tail, and he verified that such a process was feasible with population genetical models.

Fisher's formal analysis has been widely accepted and elaborated upon, but many biologists question the importance of runaway sexual selection in nature. If there are "real" differences in genetic quality between the males that court a female, then fancying orange hair would seem rather frivolous. "Extravagant" display characteristics are furthermore not just arbitrary differences like orange versus green hair—they are positively costly! The flight and foraging abilities of males who drag great plumes behind them are compromised. Mightn't there be something more to this than runaway sexual selection? In 1975, Amotz Zahavi, a voluble Israeli biologist, suggested that the costliness of extravagant characteristics is precisely their point: They are "handicaps" that are displayed to advertise their bearer's capacity to endure them [706].

Zahavi argues his perspective in an informal, anthropomorphic style. A male with a huge tail, he suggests, is transmitting a message: "Look at the burden I bear! Can you doubt that I must be exceptionally nimble in avoiding predators, and skilled in gathering food, if I can afford to divert resources into such an absurd impediment?" Reaction to this argument has been mostly critical—even hostile—but that reaction seems to be due more to its style than to its content. Zahavi disdains formal models, and his flamboyant translations of the messages in animal displays are easy targets for nit-picking. Several population geneticists have responded by constructing models in which a "handicap gene"—a mutant allele imposing a handicap and hence a fitness cost on its bearer—cannot spread by sexual selection. One leading theorist has set the handicap theory on the shelf with this declaration: "I see little point in further discussion until it has been shown to work in at least one plausible genetic model" [424, p. 174]. It is not altogether clear, however, that the genetic modelers approached Zahavi's argument in the right spirit. The "handicap" is probably not best conceived as an inevitable fitness penalty. It is a burdensome trait that is expressed only in those individuals who can "afford" it [707]: "Look at the size of *my* antlers. I must be very vigorous to be able to support them." When the handicap principle is modeled in this way, then it is indeed plausible [181].

HONEST ADVERTISING

The point that Zahavi has been making is in fact more general than we have yet indicated. It is that signal systems tend to evolve toward "truth in advertising." Suppose, for example, that females are concerned to choose the male of the best genetic quality. Any phenotypic indicator of male quality that they use to make that choice should be a valid one, or else females will be selected to ignore it. What is to keep a second-rate male from "faking" high quality? Females must choose on the basis of characteristics that second-rate males are *incapable* of faking, and that would seem to mean characteristics that cannot be produced cheaply. This is the sense in which females should have evolved to prefer males with "handicaps." It is not that the trait imposes a fitness cost on its bearer, but that it *would* impose a fitness cost on a *lesser* bearer. The trait should thus be developed only to the extent of the individual's ability to bear it, and where this is true it will be an honest indicator of quality.

Honest advertising is to be expected outside the context of sexual selection too [e.g., 505]. Harris's sparrows, for example, exhibit variable winter plumages—the higher a bird ranks in the dominance hierarchy of the winter feeding flock, the larger is a black area of breast and head feathers. Lower-ranking birds give way to approaching dark birds, and the latter eat better and survive the winter better. The hierarchy seems to be maintained with little overt aggression. Why, asked Sievert Rohwer of the University of Washington [533], don't low-ranking birds "lie"? Why don't they grow more black feathers and thereby improve their competitive standing? To find out, he dyed some low-ranking birds so that they appeared to be dominants, and you may have guessed the outcome. The dye job didn't do them any good. The impostors were unable to make good on the bluff when they were aggressively challenged. The dominance hierarchy is generally maintained without *much* overt aggression, but the relative rank of birds of similar status is occasionally tested. And when advertising is then revealed to be false, the aggression persists and intensifies. Honesty seems to be the best policy for a Harris's sparrow.

Mate attraction displays ought to be similarly unfakable. Female toads, for example, have been shown to be attracted to the deep croaks of the largest males. The pitch of a croak is determined by the dimensions of the sound producer in such a way that small males are simply incapable of mimicking big males' voices [139]. An almost identical circumstance has been demonstrated in the red deer [105]. The hinds are attracted to the loudest, longest roars, and the volume and duration of a stag's roar are limited by his chest capacity (Figure 6–8).

It is again the lek mating system where the effects of female choice ought to be clearest—undiluted by male defense of either females themselves or the resources that females value. When males strut and puff in front of a critical audience of potential inseminees, we must assume that it is the females who have selected, over evolutionary time, for the strutting and puffing. Why? It seems unlikely that females find malicious fun in making males look ridiculous! In a variant of honest-advertising theory that he calls the "war propaganda" model, Gerry Borgia [61] has argued that females insist that males on leks—and indeed territorial males in general—put on

FIGURE 6–8
Red deer: stag with hinds.

(Photograph by T. H. Clutton-Brock.)

conspicuous displays in order to test the veracity of their claims to dominance. The point is that displays attract the attention of rival males as well as females, so that any male who has not established his dominance rank fairly through a history of male-male interactions will be challenged and found out.

A clear example of just this sort of enforcement of honest advertising is provided by some outstanding studies of brown-headed cowbirds by Meredith West and Andrew King [660]. In these brood-parasitic birds, females are induced to solicit copulation by male song. Dominant males sing more potent songs than subordinates, and one is immediately led to wonder why anyone would sing a relatively ineffective song. The answer has been provided by experiment. It is his competitors who ensure that a male's song reflects his true status. Isolate a male and his song will change over time to a dominant, potent one. But return him to a competitive situation and the other males will attack him whenever he sings, although they will not attack an equally strange male if he sings a subordinate, ineffective song. Like the Harris's sparrow, the cowbird who advertises falsely is persecuted mercilessly.

An idea similar to the "war propaganda" model was advanced for elephant seals by Cathy Cox and Burney LeBoeuf, although in this case the female herself loudly draws attention to any male who may be bluffing high status [117, 118]. When a bull seal mounts an estrous female to attempt copulation, she makes such a fuss that other bulls are attracted to challenge him, and it is then the most dominant male who generally ends up with the copulation. Cox and LeBoeuf called this "female incitation of male-male competition" and the parallel to Borgia's "war propaganda" model should be clear: Although in the one case it is the female and in the other the male who raises the attention-getting ruckus, in both cases it is ultimately the females who are really insisting that the males' status be confirmed.

Whatever one may think of the concept of "handicaps," then, Zahavi has drawn attention to a fundamental aspect of animal social behavior. Individuals are always striving to manipulate one another to their own fitness advantage, and propaganda should therefore be tested before it is believed. Fakable information is no information at all. Communication systems are likely to evolve in the direction of honest advertising, not because honesty is a virtue, but because chronic lies and uninformative signals will come to be devalued and ignored.

SUMMARY

Females are generally more selective in mating than are males, as is to be expected since a mismating can cost a female much wasted nurture and lost reproductive potential.

Females sometimes demand nuptial gifts from males in exchange for fertilizations, and they sometimes choose mates on the basis of territory quality. In some cases, such as lek systems, females apparently select a mate for the quality of his genes.

Females selecting mates, and indeed animals responding to any sort of social signals, should evolve to attend only to those aspects of signals that cannot easily be faked and are therefore truly informative.

SUGGESTED READINGS

Borgia, G. 1979. Sexual selection and the evolution of mating systems. In M. S. Blum & N. A. Blum, eds. *Sexual selection and reproductive competition in insects.* New York: Academic Press.

Bradbury, J. W. 1981. The evolution of leks. In R. D. Alexander & D. W. Tinkle, eds. *Natual selection and social behavior.* New York: Chiron Press.

Orians, G. H. 1969. On the evolution of mating systems in birds and mammals. *American Naturalist,* 103: 589–603.

7

Comparative reproductive strategies

In the foregoing pages, we have made many allusions to species differences in sociosexual phenomena, but we have said little about the significance of this diversity. In this chapter, we hope to show that the view of organisms as fitness-promoting "strategists" can illuminate much more than the most general aspects of the male-female relationship. Organisms vary in their reproductive strategies, and a functional perspective is essential to an understanding of that variation.

It is really only within the last few years that science has acquired a collection of reasonably accurate descriptions of the characteristic social

systems of a number of animal species. An important generalization
emerges from an examination of these data: **Taxonomy is a poor predictor
of an animal's social behavior**. Suppose, for example, that we identify an
unfamiliar animal as a bird. We can then confidently make a number of
predictions about its mode of existence. The creature in question should be
feathered rather than hairy and should maintain a body temperature
somewhat higher than our own. It will have hatched from a yolk-rich egg
that was incubated externally. Its vision will be excellent and its chemical
senses relatively poor. It will probably fly. There are even more basic attributes of which we can be quite sure. It will have to eat something. It will
be incapable of digesting cellulose. These and many more features are
reliably to be found in any bird because birds have descended from a common ancestor. Indeed, if more than a few of these predictions turn out to be
wrong, we conclude that the animal is not really a bird but fooled us by its
superficial resemblance to one! Predictions, assumptions, and deductions
of this type are part and parcel of the logical framework of comparative
biology, which works very well. But having identified our animal as a bird,
we are then at a loss to predict on that basis whether it will be gregarious or
solitary in its habits, whether it will establish a territory and, if so, whom it
will exclude. There is no typical avian social organization.

Social systems are immensely variable. Within every major animal
taxon we are likely to find species that are highly gregarious and others that
shun contact. There are monogamous "pair-bonding" species among birds,
mammals, fish, insects, and even isopods. Leks can be found among particular species of antelope, grouse, bats, and fruit flies. In fact, there are
striking contrasts in social systems within every large mammalian order,
within families, genera, and even species. (Although it is usually accurate
and certainly convenient to speak of species-characteristic social organization—and we shall do it—even this can amount to excessive generalization.)
This diversity demands explanation.

COMPARATIVE SOCIOECOLOGY

Behavioral differences between species can be explained in terms of distinct
causal processes, but it is often most interesting to address the question
functionally. Someone who asks why crows are more inclined to flock than
owls, for example, is unlikely to be satisfied with the explanation that owls
react with fear or aggression to one another, however true an account that
may be of an immediate mechanism of spacing. What the questioner probably wants to know is why it is to the individual crow's advantage to join a
flock and not to the individual owl's.

We have already discussed some examples of functional explanations that trace species differences in social organization to differences in ecology. The tendency for marsh-nesting birds to exhibit polygyny (p. 121) is an example of the sort of correlational generalizations which have been uncovered by comparative research. This approach has been dubbed "comparative socioecology" by John Crook, one of its leading practitioners [123].

Consider, as an example of the comparative socioecological approach, the social systems of African antelope as described by P. J. Jarman [314]. The seventy-four African species of antelope represent some twenty-eight genera in nine subfamilies of the family Bovidae (cattle and antelope). All seventy-four share certain basic features that are nearly universal among the ungulates (hoofed mammals): Females bear a single, large, precocious offspring each year, for example, and the diets are almost exclusively vegetarian. But the different species vary in numerous other respects—body size, group size, mating system, seasonality of movements, and composition of diet, to name a few. These variables are interrelated in a complex way, outlined in Table 7-1, in which the species are divided into five classes according to the correlated dimensions of body weight, group size, and dietary selectivity.

Jarman proposes that an antelope species' size is a major determinant of the kind of diet it can use efficiently, both because of the inverse relationship between metabolic rate and body size, and because of the dexterity with which the small species can select plant parts. The small class A antelope takes entire plant parts such as flowers while foraging; an animal following in its wake will find slimmer pickings, will expend more search time, and will fall behind. On the other hand, the class D grazer does not remove whole items but reduces the plant in successive bites, leaving the food dispersion the same. Followers can therefore feed without falling behind as long as the area has not been too heavily grazed. This suggests that it is feasible for the class D antelope to feed socially but not for class A.

The smaller species—selective feeders upon a variety of plants—benefit from intimate knowledge of a limited home area. They can find highly nutritious foods the year round in their small, relatively dense, diverse habitat. It is therefore both desirable and practical for them to defend small territories; their antisocial tendencies are also suited to an antipredator strategy of remaining inconspicuous in dense cover. The class D wildebeest, by contrast, cannot remain on a small territory. With irregular rainfall, there are major seasonal and spatial variations in the availability of choice grasses. While good grass exists, however, it is likely to be adequate to feed thousands, so there is nothing to be gained from territorial defense.

What all this implies is that body size, diet, and predation pressure in relation to habitat features are naturally correlated determinants of whether

TABLE 7–1
The socioecology of African antelope

Social class	Exemplary species	Body weight	Group size	Principal habitats	Feeding style	Typical mating system and mobility	Anti-predator tactics
A	Dikdik, duiker, klipspringer, steinbok, grysbok	3–60 kg.	1 or 2	Forest, thicket	Many plant species; selective for parts; relatively high-energy foods, such as fruits, buds	Stable pair on territory	Hide in dense cover
B	Reedbuck, gerenuk, oribi	20–80 kg.	2 to 12	Brush, riverine grassland	Browse or graze with selectivity for plant parts	Stable harem group on territory	Hide, flee

→ increasing body size →

→ increasing group size →

→ decreasing cover →

→ decreasing selectivity

→ increasing effective polygyny →

→ increasing conspicuousness

					for hi-energy foods →		& defense →
C	Waterbuck, kob, impala, gazelle	20–250 kg.	2 to 100	Riverine woodland, dry grassland	Graze or browse with moderate selectivity	Large home range; male territoriality in mating season	Flee, hide in herd
D	Hartebeest, wildebeest	90–270 kg.	4 to 150 (1000's in migratory herds.)	Grassland	Graze with selectivity for stage of growth	Temporary harems or territories within mobile herds	Hide in herd, flee, stand and defend
E	Eland, buffalo	300–900 kg.	10 to 1000	Ubiquitous, especially grassland	Graze relatively unselectively	Male-dominance hierarchy within herd carries mating rights	Mass defense and attack on predators

Source: After P. J. Jarman, 1974 [314].

antelope are likely to wander in herds or to defend established territories. The different mating systems can then be seen to result from the males' attempts to maximize their reproduction within the constraints imposed by varying patterns of female aggregation and mobility. The class A male is likely to attach himself monogamously to a territorial female. In classes B and C the females are more gregarious and may be less predictable about turning up at the same places every day; the male may therefore defend a harem or hold a territory in an area of such attractive resources that females are likely to frequent it. In the case of a class D migratory herd the male cannot hold a piece of turf with any reasonable expectation that females will come his way; better to follow the herd and to try to control some females within it. In class E species the huge males, permanently associated with the female herd, vie for dominance status and mating rights like bull bison (p. 90).

Many details of antelope society remain to be explained, but the broadest distinctions, such as that between stable monogamy and hierarchical mating competition among males, seem clearly to reflect adaptations to ecological differences. Just as in the case of Saharan gerbils (p. 95) it is illuminating to look first at female dispersal as a food-resource-related strategy and then to consider the strategies of the males, for whom females are a prime resource. In fact, this principle has been usefully applied in socioecological analyses of several mammalian groups [e.g., 63, 696].

PRIMATE SOCIOECOLOGY

As our nearest relatives, the primates merit special interest. Anyone concerned with the origins of human society must study the monkeys and apes, if only to be assured of their irrelevance! Yet scientists already knew a great deal about the lives of birds, bees, fish, and flowers before they cast so much as a glance at nonhuman primates. Study of primate social systems in the wild finally got under way in the 1930s when an American scientist, C. Ray Carpenter, began collecting observations of several species in Central and South America and in Asia. He soon found evidence that individual primate species had distinct and characteristic social organizations: the monogamy of gibbons, the group territoriality of howlers, and so on. But it took another twenty years for his promising start to be taken up by a number of other fieldworkers. By the 1960s, anthropologists, zoologists, psychologists, sociologists, and some who called themselves primatologists were all following bands of monkeys, with notebooks, binoculars, and tape recorders at the ready. At last a rich picture of the diversity of primate societies began to emerge. Today we have at least some descriptions and data concerning the behavior of a majority of the 180 or so living species,

TABLE 7-2
Existing members of the order Primates

Suborder Superfamily	Family	Number of genera	Approximate number of species	Exemplary common names
Prosimii	6	16	32	loris, potto, galago, bushbaby, lemur, indri, sifaka, aye-aye, tarsier
Anthropoidea Ceboidea (New World monkeys)	Callitrichidae	5	35	marmoset, tamarin
	Cebidae	11	29	night monkey, titi, saki, uakari, howler monkey, capuchin, squirrel, spider and woolly monkey
Cercopithecoidea (Old World monkeys)	Cercopithecidae	8	50	macaque, mangabey, baboon, mandrill, gelada, guenon, vervet, patas, talapoin
	Colobidae	6	24	langur, proboscis monkey, colobus
Hominoidea	Hylobatidae	2	7	gibbon, siamang
	Hominidae	4	5	orangutan, chimpanzee, gorilla, human
		52	182	

Source: After A. Jolly, 1972. *The evolution of primate behavior.* London: Macmillan; and J. R. Napier & P. H. Napier, 1967. *A handbook of living primates.* London: Academic Press.

about two dozen of which have been the objects of painstaking studies of several years' duration.

The existing members of the order Primates are summarized in Table 7-2. As we have stressed for other groups, taxonomy is imperfectly related to social structure, so any category of social types is likely to include animals of diverse taxa. Among the several monogamous species, for example, there are a few prosimians, a number of New World monkeys, at least one Old World monkey, and all the Hylobatidae (gibbons and siamang). A predominantly solitary mode of existence characterizes a number of small nocturnal prosimians and an ape, the orangutan. Both major suborders of monkeys, the Ceboids (New World) and Cercopithecoids (Old World), contain many species with a one-male, multifemale troop structure and many others with a multimale, multifemale troop structure.

The first important attempt to make socioecological sense out of this diversity was made in 1966 by two experienced observers of African monkeys, John Crook and Steve Gartlan [125]. They categorized primate species into five social "grades." Their scheme was rather like Jarman's antelope story, and that is no accident. There are real parallels between the

FIGURE 7-1

Red colobus monkeys, *Colobus badius*. (Photograph by T. H. Clutton-Brock.)

FIGURE 7-2

Black-and-white colobus monkeys, *Colobus guereza*. (Photograph by J. F. Oates.)

primate and antelope radiations, and Jarman deliberately modeled his effort on Crook and Gartlan's. Their "grade I" primates, for example, paralleled Jarman's "class A" antelope in several respects—both are relatively small and lacking in sexual dimorphism, occupy dense habitat, consume relatively high-energy diets, hide as an antipredator strategy, forage singly or in pairs, and are often monogamous.

The Crook/Gartlan analysis left many social variations unexplained, and others have elaborated upon it as more data have been accumulated. It now appears, however, that no such global scheme will give us a satisfactory account of the adaptive significance of all primate societies. Further progress has been achieved primarily by finer analyses of the ecology of particular species [106]. Tim Clutton-Brock, for example, has explained in terms of feeding ecology the contrasting social organizations of the red colobus monkey (Figure 7-1), which lives in multimale troops of forty or more, and the black-and-white colobus (Figure 7-2), which lives in one-male troops of some five to ten in the very same forests as the red colobus

[104]. Both are arboreal and eat a mixture of leaves and fruits, but the black-and-white feeds in a much narrower range of tree species, and when none of these is fruiting, it subsists on their leaves. A small troop of black-and-white colobus can therefore fill its needs on a small and defendable home range. The red colobus, however, needs fruit in its diet the year round, so it feeds in a greater variety of tree species and requires a larger home range. This larger area can support a correspondingly larger troop, since the average productivity of colobus food per acre of forest over the whole year is similar for the two monkey species; it is the temporal and spatial patterns of food availability that are important. Within-species variations in troop size and structure can sometimes be traced to these variations too [e.g., 587].

Not every detail of primate societies need be explicable in terms of socioecological adaptation, however. Thomas Struhsaker has suggested, for example, that the phylogenetic heritage of a monkey species imposes constraints upon the ecological adaptability of its social system [586]. Large multimale troops, for example, may indeed be an adaptation to terrestrial life as many authors have argued, but in the primarily arboreal genus *Cercopithecus* certain rather terrestrial species have retained the one-male organization of their tree-dwelling relatives. Conversely, mangabeys are largely arboreal and yet retain the multimale troops typical of their nearest relatives, the terrestrial baboons. From such examples Struhsaker drew the conclusion that a monkey's taxonomic affinities are at least as good a predictor of its social system as is its ecological niche. This illustrates what E. O. Wilson calls "phylogenetic inertia" [680], which is the retention, among evolving species, of characteristics of their taxa of origin. Clearly, its ancestral attributes are major determinants of the way a species can adapt evolutionarily to a novel ecological circumstance.

We would hardly expect that ecology could tell the whole story. Ecological factors indubitably influence the costs and benefits of every sort of action and hence are selection pressures that must influence the evolution of social behavior. The threat of big cats in open country, for example, exerts a steady pressure in favor of gregariousness in baboons. But the social milieu within the troop is *itself* part of the environment, and it too imposes costs and benefits upon individual behavior [106, 221]. This is especially clear with regard to reproductive strategies.

Consider again the male mating competition among Hausfater's yellow baboons (pp. 84–88). Dominant males reproduce more successfully than subordinates. The upshot of this cannot be simply a selection of "dominant genes" over "subordinate genes" or all the alleged subordinate genes would soon be eliminated and everyone would be a dominant. The subordinates are not baboons of a different kind but baboons in a different social role or life stage. As Hausfater points out, there are year-to-year changes in dominance hierarchies, so that most males who live a full life may get their turn at high status and reproduction [265]. But that position must still be won, so selection is constantly reinforcing the capacity and inclination to play the game of ascending a dominance hierarchy. This requires brawn and brain, and males deficient in either are likely to miss out. These considerations do not tell us how and why this mating system evolved, but they offer some insight into how it perpetuates itself: The extant social practices of the baboon troop are the critical context within which an individual's behavior succeeds or fails. Such self-perpetuating aspects of social systems must be a major part of phylogenetic inertia.

REPRODUCTIVE ISOLATING MECHANISMS

There is another reason, besides phylogenetic inertia, why we should not expect every detail of species-typical social practices to have an ecological rationale.

Reproductive isolating mechanisms are differences between species that make hybridization unlikely. The term refers particularly to differences that have evolved precisely in order to achieve that function—in other words, to mechanisms that have evolved specifically because of selection against hybridization. The "isolation" is that between the gene pools of different species. Indeed, the definition of a species hangs on this isolation: A species is a set of animals that (virtually) confine sexual union to mates within that same set. (It must not be forgotten, however, that it is individual animals that are concerned with mate selection. The term "reproductive isolating mechanism" is perhaps unfortunate in suggesting an adaptive significance for the gene pool rather than for the individual strategist.)

A well-studied mechanism for achieving reproductive isolation exists among fireflies. There are twenty-eight species of fireflies of the genus *Photinus* in the United States and Canada. On a summer night males fly about, flashing. Females on the ground answer males of their own species by flashing back. A series of reciprocal flashes then guides the male to the vicinity of the female, where courtship may proceed. The female recognizes a male of her own species—and a human observer can soon do so too—by particular patterns of flashing. One species displays a succession of points of light by flying steadily and flashing briefly at regular intervals. Another flashes for a little longer while moving and thus displays a longitudinal strip of light. Another zigzags, and another flashes in bursts. The effect of all this is reliable species recognition. Zoologist Jim Lloyd coaxed males to the ground with a flashlight by duplicating the flash patterns of females of the right species [391].

In the only case in which Lloyd found that captive members of different species would actually copulate, the two species did not naturally overlap in range but had been brought together artificially. If such reproductive isolating mechanisms have evolved in order to prevent hybridization, then we have no reason to expect effective isolating mechanisms between closely related species whose geographical ranges do not overlap. The leading theory of the evolution of reproductive isolating mechanisms supposes first that two populations of a single species become geographically isolated—by a glacier, for example, or a body of water—that they evolve slightly different adaptations in the face of slightly different environmental circumstances in their two locales, and that they later come into contact again. At this point, the populations may interbreed and wash out their divergences. However, they may have diverged sufficiently so that the hybrids produced by cross-populational matings are less fit than either parental strain, by having compromised the special adaptations of both. If this occurs, there arises a powerful selection pressure against indulging in cross-populational matings and producing unfit hybrids, and this selection pressure will favor any device that helps individuals confine their matings to their own strain. Once this exclusivity is achieved, the two populations are reproductively isolated and have become two species.

An expected result of this process is that closely related species that are sympatric (geographically overlapping) will differ in some feature crucial to courtship and mating to an extent greater than might be anticipated on the basis of their other similarities. Perusal of any bird guide will show that this expectation is often fulfilled. The parulid warbler genus *Dendroica,* for example, contains twenty-two species breeding in North America, all of similar size and habits but with a dazzling array of distinct breeding plumages. Outside the breeding season the plumages are much more alike. The most similar species, moreover, such as the yellow-throated warbler and Grace's warbler, are not sympatric. In other taxa, such as the tyrant flycatchers, sympatric species may look very much alike but sing utterly different songs.

Any efficient reproductive isolating mechanism should enable an animal to recognize an inappropriate mating prospect reasonably early—before wasting a lot of reproductive effort in producing an unfit hybrid. The animal that squanders time and energy in the pursuit of unproductive matings is at a selective disadvantage in reproductive competition with any conspecific better able to confine its amorous attentions to the right animals. This selective imperative is important to both sexes, but it weighs more heavily upon the sex with the greater parental investment in each offspring, and that is usually the female. If she is to commit herself to an intensive parental effort, her choice of a male should be relatively precise and foolproof. The male is less concerned that he mate only with the best available female. With her by all means! But also with anyone else that is willing, including even those of doubtful specific identity. We have already mentioned greater female choosiness with respect to mate quality, and the issue is all the more crucial where species identity is in question. So it is that species recognition and reproductive isolating mechanisms reside largely in the female: It is she that chooses whether to answer the male firefly's flash. It is she that has more to lose in the event of a mismatch.

A particularly well-analyzed case is that of the isolating mechanisms operating among a group of South American fishes, the guppy and its close relatives, studied by Robin Liley [382]. Four species of these little fishes, members of a single genus, occur together in the drainage trenches of Georgetown, Guyana. Hybrids are sufficiently sterile or inviable to assure a near-total isolation of the four species. The males display distinctive species-specific patterns and courtship postures, and the females respond only to males of their own species. As expected, the viviparous females, risking a substantial proportion of their reproductive potential at every mating, are the more discriminating sex. In laboratory experiments Liley has shown that even naive virgin females recognize their conspecific males. Males, by contrast, at first court indiscriminately, but they learn by experience to confine their attentions mainly to conspecifics, though never so single-mindedly as the females.

The strikingly different courtship patterns of these fish virtually eliminate the time and energy wastage of cross-species courtship, as we suggested that an efficient reproductive isolating mechanism should do. This aspect of isolating mechanisms complicates the scientist's task of interpreting specific diversity in reproductive control mechanisms. We have argued that observed species differences can be explained socioecologically in terms of differing ecologies, mating systems, and parental strategies. But the necessity for mechanisms assuring reproductive isolation suggests that some species differences will never be explicable in socioecological terms: Their adaptive significance lies in the very fact of differentness. The flash patterns of a particular firefly species, for example, are probably not adapted to the peculiarities of that species' way of life; they differ in the service of discriminability.

A COMPARISON OF BIRDS AND MAMMALS

We have stressed that taxa are not good predictors of social organization and that any classification of social systems is likely to cut across taxonomic groups. This fact is not yet universally appreciated. It is an oversimplification even to treat social systems as fixed attributes of species, ignoring such complications as habitat-related changes in group size or occasional polygyny in primarily monogamous species. So we have stressed situational and ecological determinants of social structure, but it is easy to overstate their importance. Taxonomy is not totally worthless as a predictor of social behavior. As we noted with respect to primate genera varying in arboreality, basic attributes of the biology of a taxonomic group may constrain evolutionary possibilities and amount to preadaptations for particular social systems. This point becomes clearer from a broad comparison of the mating and social systems in two vertebrate classes, mammals and birds.

There are fundamental differences between the reproductive practices of these two classes. Birds lay eggs, while mammals (with the exception of the Monotremata: the platypus and echidna) are live-bearing. Either way, the young must somehow be nourished both before and after they can ingest and digest food. Mammals deal with early nurture by means of internal gestation and (except in the marsupials) a placental food-delivery system. Birds cope with the same task by packaging the embryo with a yolky food supply and letting it get on with development outside the mother's body. For later nurture, mammals offer a special food produced only by the female—milk. Birds generally cope with the same problems by feeding nestlings natural foods that can be collected by adults of either sex. (And where they do manufacture a special baby food, the "crop milk" of doves and pigeons, it is made by both parents.)

These practices dictate a major parental role for the female mammal. Not only must she bear her developing young through a pregnancy, but there is a postnatal dependency on her milk too. She is therefore committed to a major parental investment in each offspring, an investment that a male cannot substitute. In birds the story is quite different. By getting the egg out of her system early, the female bird creates a situation in which her mate can play a major parental role much earlier in the developmental chronology of the offspring. It is therefore more practicable for birds to partition the parental role relatively equitably between the sexes than it would be for mammals. Since the mammalian male can offer relatively little effective parenting, he instead channels his reproductive effort into the pursuit of matings, a circumstance that favors greater polygyny, less stable mateships, and more intense male-male competition for mates.

To summarize, the crucial contrast between birds and mammals is this: A male bird can make a useful parental contribution rather soon after mating by incubating the eggs and then by feeding and guarding the young, whereas a male mammal can usually make little contribution to early nurture and is probably wasting time and mating opportunities by hanging uselessly about. With these considerations in mind, let us look at the different mating systems prevalent among these two groups.

Among the birds, David Lack estimated that 92 percent of the eighty-six-hundred-odd species are primarily monogamous, 2 percent polygynous, 6 percent promiscuous (as, for example, the lek maters), and less than 0.5 percent polyandrous [356]. (We say "primarily monogamous" because occasional polygyny appears to be widespread among such species.) No such precise estimates are available for mammals, since the mating systems of the majority of the four-thousand-odd mammalian species are quite unknown. Any such treatment by percentage of species would emphasize the order Rodentia, which comprises almost three thousand species, at the expense of other groups. In any case, it seems clear that polygyny and intense male-male competition are far more common in mammals than in birds, which is what we would expect from the differences in parental behavior. The prevalence of polygynous mating systems holds up within each of the better-known orders—rodents, insectivores, pinnipeds, ungulates, primates. A possible exception is the carnivores, in which monogamy may be more prevalent. Monogamy also occurs, as we have seen, in some ungulates and primates, and is infrequent but widespread in other mammalian orders. Stable polyandrous associations characterize a few human societies (Chapter 11, pp. 286–288) and may occur exceptionally in social carnivores such as wolves and hyenas, but there are no grounds for suspecting them elsewhere. Hence, the mammals are predominantly polygynous, sometimes monogamous, and virtually never polyandrous, whereas birds are sometimes polygynous, predominantly monogamous, and occasionally polyandrous. There is indeed an overall difference in the distribution of mating systems in these two vertebrate classes, a difference consistent with the difference in parental investment in offspring.

ON ESTIMATING SPECIES-TYPICAL DEGREES OF POLYGAMY

The variations among African antelope species (Table 7–1) illustrate the fact that species differ in their characteristic degrees of polygamy, ranging from monogamy (e.g., klipspringer) through small harems (e.g., gerenuk) to dominance hierarchies (e.g., eland) and lek systems (e.g., kob) in which successful males inseminate large numbers of females. It will be useful for comparative purposes to treat this variation as part of a dimension ranging from polyandry (unknown in antelope) through monogamy to polygyny. Sexual selection theory, as discussed in Chapters 5 and 6, suggests that the position of a species on this dimension should be predictive of the magnitude of various sex differences, including differences in mortality and in morphological structures used in intrasexual reproductive competition. How exactly to define this polygamy dimension proves problematic.

It is common to define polyandry, monogamy, and polygyny rather vaguely in terms of relatively stable social bonds between an individual and one or more mates. This approach seems inadequate for a discussion of selection pressures acting on the two sexes, since the intensity of sexual selection is a matter of the within-sex variance in reproductive success (see p. 82). Some authors have suggested that the definition must hinge on the relative numbers of members of each sex who contribute gametes to zygotes, but this approach also ignores variance: If ten males inseminate a hundred females, the system is by this definition equally polygynous regardless of whether each male gets ten females or one gets ninety-one and the others one each. We think that the **degree of effective polygamy** of a mating system is often best conceptualized as a **ratio of male fitness variance over female fitness variance**. The higher the ratio, the more polygynous the system; a ratio of one means effective monogamy; and a ratio less than one indicates effective polyandry. In the example in Figure 5-1 (p. 80), the degree of polygyny in Bateman's fruit flies is 4.6, and in the Xavante Indian example (Figure 5-6, p. 89) it is 3.1. This measure of the extent to which a mating system is effectively polygynous or polyandrous breaks down when the sex ratio departs greatly from 1:1. It also does not discriminate simultaneous polygamy, as in the maintenance of a harem, from successive polygamy, in which mates are repeatedly changed. (Confusingly, the latter system, which is fast becoming the norm in the North American population of *Homo sapiens,* has been called both "successive monogamy" and "successive polygamy.") But at least a ratio of fitness variances provides us with a conceptual scale.

This is all rather academic, for in comparative studies, practical measures of the degree of effective polygamy are generally much cruder. There are not many data sets from which within-sex variances in fitness can be estimated. For most species, the best we can do is to estimate the number of mature females per male in reproductive groups (the "socionomic sex ratio" [179]) or to estimate the number of females that seem to be controlled by a single dominant male ("harem size").

POLYGYNY AND SEXUAL SIZE DIMORPHISM

More than a century ago Darwin [137] proposed that sexual dimorphism in size and ornamentation was the result of sexual selection and hence male competition for mating opportunities. This implies that in a comparison among species a greater degree of polygyny will be associated with a greater degree of sexual dimorphism. This association is well documented in birds [553], at least with respect to a dichotomous categorization: In monogamous species the sexes are not too different in size, whereas in effectively

FIGURE 7-3

The relationship between mean harem size and sexual dimorphism in size in pinnipeds, ungulates, and primates

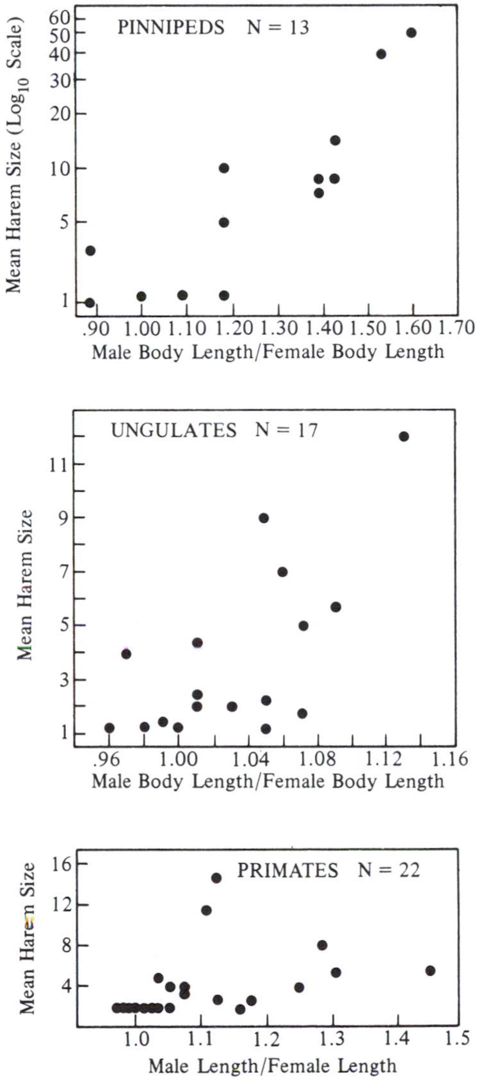

(After Alexander et al., 1978, Figures 15-1, 15-3, 15-5 [12].)

polygynous species such as lek-breeders the males tend to be much larger. In mammals, a correlation between the degree of sexual dimorphism in size and the degree of polygyny has recently been documented [12] by comparisons within the orders Pinnipedia (seals), Artiodactyla (even-toed ungulates), and Primates (Figure 7–3).

In some cases, two presumed correlates of the intensity of sexual selection can be shown to be related to one another, even though the effective degree of polygamy is not estimable from available data. Among grouse, for example, most species are lek-breeders, but only a few have been studied in sufficient detail to allow direct estimates of the degree of polygyny. Lacking data on reproductive success and its variance, however, we can still test a prediction of sexual selection theory: If body size dimorphism and sex differences in mortality (cf. pp. 92–97) are both selective consequences of sexual selection, we might expect to find a correlation between size dimorphism and population sex ratios (which reflect differential mortality). Figure 7–4 shows that these variables are indeed strongly related, a fact that provides indirect support for the hypothesis that the intensity of male-male competition is the primary determinant of sexual dimorphism in this particular group of birds.

We should not imagine, however, that we can "read" the historical in-

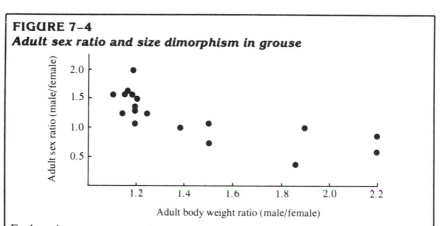

FIGURE 7–4
Adult sex ratio and size dimorphism in grouse

Each point represents a different species of grouse. In those species in which males are only slightly larger than females, adult males are more numerous; this probably reflects greater mortality of females during nesting, a common situation among birds. Where the males are much larger than the females, however, the lower sex ratio implies that male mortality has increased relative to female mortality; these are the lek-breeders with the most polygynous effective breeding systems. (After Wittenberger, 1978, Figure 2 [687].)

tensity of intramale sexual selection from size dimorphism alone. There are at least three reasons why the relationship between sexual size dimorphism and degree of effective polygyny will not always hold.

In the first place, large male size is indeed likely to be favored when access to females is settled by *fighting,* but competition between males is not always of a sort that favors large size. The Weddell's seal, for example, is the one rather polygynous species of pinniped in Figure 7–3 in which males are nevertheless smaller than females. This species also has the peculiarity of mating in the water, where male-male competitive skill is presumably less a matter of size than of agility [103]. Much the same story seems to apply to a completely different group of animals, the turtles. In terrestrial species, males fight over females and are the larger sex. In swimmers, the males are almost invariably smaller than the females, who evidently choose their mates in response to elaborate aquatic courtship displays [47]. Among snakes, females are usually larger than males. Competition for females is often a slithering free-for-all, but males seldom fight. In that minority of species where they *do* fight, however, the males then tend to be larger than the females [564]. Exactly the same cross-species relationship between sexual dimorphism and male combat has also been found in the amphibians [565]. These facts certainly seem to implicate sexual selection in the evolution of sexual dimorphism, but we have no basis for predicting what form of mating competition will be associated with the more effectively polygamous mating systems.

In the second place, the extent of sexual size dimorphism will be constrained by ecological demands outside the reproductive context, and these demands will vary between species [362, 516]. Terrestrial monkeys, for example, are more sexually dimorphic than arboreal monkeys, and the terrestrial-arboreal dimension accounts for some of the residual variance in dimorphism that is not predictable from the socionomic sex ratio. One interpretation of this difference is that the tendency for intense male-male competition to lead to exaggerated body size is more strongly countered in arboreal species because of more steeply rising costs of being oversized [107].

The third reason that the dimorphism-polygyny relationship is imperfect is that sex differences may be selectively favored outside the context of sexual selection. Males and females may be specialized differently either to avoid competition with one another or to perform complementary roles. The most convincing example of the former possibility concerns several species of woodpeckers on Caribbean islands, where there is a substantial sex difference in the size of the bill and tongue, related to differences in feeding style (probing versus gleaning); monogamous birds sharing a territory thus avoid competing with their mates for the same food items [552].

Complementarity of roles occurs in systems where one sex broods, incubates, lactates, or the like, while the other hunts or defends the nest site against conspecifics or predators.

SEX-ROLE REVERSALS IN PARENTAL CARE

Here and there in the very diverse ramifications of evolutionary history it happens that the burden of parental care falls mainly or entirely upon the male rather than the female. The reasons for this development are not well understood in every case, and they are certainly not uniform. The phenomenon is relatively infrequent in comparison with predominantly female care, but it is widespread. It occurs in insects and reptiles, frogs, fish, and birds, but not in any mammal.

These scattered instances of a reversal of the usual sex-typical parental roles deserve a special scrutiny, for it is often the unusual case that provides us the clearest insight. Let us briefly review a few such cases, one each from the insects, fishes, and birds.

Water bugs

The giant water bug *Abedus herberti* is a resident of creeks in Arizona, preying upon creatures smaller than its own one-inch length. The peculiarity of reproductive behavior in these bugs and some related species is that the females lay their eggs upon the males' backs [577].

The male water bug has a hundred eggs or so individually glued to his back, covering the entire available area. There they remain for an incubation period of twenty-one to twenty-three days. During that time the attentive father performs several caretaking behaviors that help to protect the eggs from desiccation, suffocation, and fungal infection. This paternal care is confined to the incubation period. Once the eggs hatch, the little nymphs linger only a few minutes on their father's back before embarking upon their own independent predatory careers. At this stage the brooding male will not attack and eat the new nymphs even if he is hungry, but he has no compunctions about such cannibalism when he has not been brooding eggs and available nymphs are therefore unlikely to be his own offspring.

In courtship the female water bug assumes a relatively aggressive role, approaching the male repeatedly and sparring with him. He may then initiate copulatory behavior. If he does not, the female displays to him. If he is already egg-laden, however, she will reject him as a suitable mating partner. In mating, the male controls the sequence of behavior by signals initiating and terminating both copulation and oviposition by the female.

FIGURE 7–5

A sea horse, *Hippocampus hudsonius*. (Photograph by Runk/Schoenberger for Grant Heilman.)

Furthermore, he steers the female to lay in any vacant slot in his dense array of eggs. Once the female has laid her eggs, her parental role is complete, and she aids the father in his prolonged brooding duties not a whit.

Sea horses and pipefishes

Sea horses (Figure 7–5) and pipefishes make up a family of rather odd-looking fishes, the Syngnathidae. They have extremely slender bodies and peculiar heads that have reminded whimsical zoologists of horses. But surely their most remarkable feature is internal gestation by the male! Male sea horses and pipefishes come equipped with a brooding pouch, a slit down the belly into which the female lays her eggs. Males then gestate the young for about three weeks, utilizing a placentalike structure to nourish the fry. When gestation is complete, the young are expelled with some force in little clouds of miniature adults from the male's belly.

The German zoologist Kurt Fiedler has studied the courtship and mating behavior of several species in laboratory aquaria [189]. In pipefishes, and to a lesser extent in sea horses, the female is the aggressor in courtship. Brighter than the male in her special courtship colors, she approaches him and displays to him, eventually inducing him to follow her. Even the act of copulation entails a sex reversal, for the female inserts her genital papilla into the male's pouch, transferring anywhere from a very few eggs to a few hundred, according to the species.

Phalaropes

The phalaropes are a family of little shorebirds that look like sandpipers but differ from them in their lobed toes and their fondness for swimming. There are just three species. The red phalarope and the northern phalarope nest in the Arctic, the world round. Wilson's phalarope (Figure 7-6) nests on the prairies of central and western North America. In all three species the females are bigger and more aggressive, and display spring plumages that are bolder, brighter versions of the males'.

The following observations on the breeding biology of Wilson's phalarope near Edmonton, Alberta [275, 276], may be taken as illustrative of the behavior of phalaropes generally. In the spring females arrive first on the breeding grounds, in contrast to most migratory birds, and the males soon follow. The birds breed in little aggregations of some two to eight pairs on prairie sloughs and ponds, nesting in the marshy vegetation. They pair off much like ducks, but with the sex roles reversed: Each female picks out a male and jealously accompanies him, threatening away other females as the couple swims about together feeding on tiny creatures on the water's surface. Males may threaten one another too, especially later in the breeding cycle, but they are never so aggressive as the females. Once eggs are laid, parental care is the exclusive province of the male, who drives off the female [282]. He incubates without female assistance, and when the eggs hatch, he alone broods the precocious chicks and leads them about.

These sex-role reversals provide us with a special opportunity to test sexual selection theory. Recall Bateman's principle (Chapter 5), according to which the sex that invests more in the young becomes a *resource* for which

FIGURE 7-6

A pair of Wilson's phalaropes, *Steganopus tricolor,* on a prairie slough. The slightly larger and brighter female is on the right. (Photograph by J. & D. Bartlett.)

members of the less parental sex compete. We have argued, following Robert Trivers [619], that sex differences in aggressive competition, in courtship behavior, and in mortality all follow from the parental investment disparity. The less nurturant sex is, of course, generally the male, but the theory should work both ways. *Whichever* is the sex with greater parental investment will be the sex that is courted, that competes less, and that survives better. If the theory is wrong about the ultimate causation of the prevalence of male courtship of females, for example, then we should expect that male courtship of females might persist even in these cases of a reversal of parental roles.

Insofar as these phenomena have been investigated, the predictions of parental investment theory meet the challenge well. In each of the animals we have just considered—water bugs, pipefishes and sea horses, and phalaropes—the reversal of sex roles occurs not merely in parental behavior but in the courtship phase as well. Females do indeed compete for and court coy, reluctant males. This striking behavioral sex reversal may have morphological correlates too: The female phalaropes and sea horses are bigger and brighter than their males (though, as we saw in turtles and snakes, larger females are not so very unusual even in species where sex roles are not reversed).

Male parental care is *not* associated with sex-role reversal in substrate-spawning fish such as sunfish (p. 102) or sticklebacks, however. In three-spined sticklebacks, for example, a species so much studied as to be called the "laboratory rat of ethology," brightly colored males vie for territories, build nests, court females, stimulate them to spawn, guard and aerate the eggs, and tend the fry [296]. Female reproductive effort is concentrated upon the production of egg biomass. Even though males perform the greater part of the parental duties, they are the suitors too. This state of affairs is not so paradoxical as it may at first appear. The male stickleback remains an avid polygynist: The more females he can persuade to spawn in his nest, the greater will be his reproductive success. His work is somewhat increased by an increased clutch, especially the work of aeration, but twice as many eggs certainly do not require anything like twice the effort. His parental capacity is therefore not so limited as a water bug's, a sea horse's, or a phalarope's. Recall that Trivers's definition of parental investment (p. 77) specified investment in one offspring at the expense of parental capacity to invest in another. By that definition, the male stickleback's care of eggs is *not an investment,* or at least increments in egg number do not lead to a linear increment in investment. A useful distinction in this context is between "whole-brood" investment that has a fixed cost regardless of the number of young and "individual" investment that increases with brood size [506]. Male parental investment in the stickleback is predominantly of the whole-brood variety (guarding), while female investment is of the individual type (egg production). Hence the male stickleback still has a higher reproductive ceiling than the female, his nurturant role notwithstanding, and it follows that females are still the limiting resource for males.

POLYANDROUS BIRDS

The reversal of courtship sex roles in animals such as the phalaropes suggests that the parental investment of the male actually surpasses that of the female, so that males have become a limiting resource for female reproduction rather than the reverse. If this is so, then we might expect female phalaropes to be at least occasionally polyandrous, which they are [e.g., 549]. A female leaves her mate to care for the eggs and may go on to pair with another male in the same season.

This sort of successive polyandry may be quite common in the order to which phalaropes belong (Charadriiformes, which include shorebirds, gulls, and several other families, all with rather aquatic habits and precocious young). Among several species of sandpipers, for example, it is not unusual for females to produce two clutches of four eggs each, with each parent caring separately for one. In these species, females sometimes remate between clutches, with the apparent intent of leaving the second male too and perhaps squeezing a third clutch into the short Arctic summer [498, 499].

This "double-clutching" by female sandpipers probably derives from a primary adaptation for rapid clutch replacement. Nest predation is extremely high in these ground-nesting birds, so that even in monogamous pairs, females are concerned to recoup the energy they have expended in eggs and thus to be able to replace a clutch as soon as it is lost. This situation seems to have led to the prevalence of male incubation and brooding in shorebirds. Parents do not feed the young in this group but merely guard and brood them, and it would appear that one parent is not much less effective than two. Hence it becomes possible for each parent to tend a brood separately. It is easy to suppose that this is a situation from which successive polyandry might arise. And it has.

A regularly polyandrous species is the spotted sandpiper, a common North American bird with a distinctive tail-wagging signature. Spotted sandpipers have been studied in Minnesota by Lew Oring and colleagues [179, 419]. As in the phalaropes, there is a reversal of sex roles with females substantially bigger than males (though no brighter—there is no plumage dimorphism in the sandpipers). Females set up territories and court males, setting up each mate with a clutch of four eggs before trying to attract another. If, as often happens, the nest is destroyed by a predator, the female will return to provide a replacement clutch, and if she does not succeed in attracting another male, she will help a little with the incubation. Her senior mate may resist addition of a second, attacking a new male, but if so the larger female soon puts him in his place. Females sometimes fight severely over a newly arrived "surplus" male. Males cannot fledge more than a single brood of four in a season, but a successful female may fledge young by two or more males; the seasonal record for a female is nine fledglings. Females concentrate their effort on foraging and egg production, and it seems most probable that this system has evolved from monogamy by the route outlined in the last paragraph.

Another well-studied polyandrous Charadriiform bird is the American Jaçana, *Jaçana spinosa,* a tropical bird whose most striking physical feature is its prodigiously long toes. Jaçanas frequent ponds, walking atop lily pads and other floating vegetation, picking off invertebrate tidbits. Their breeding behavior has been studied in Costa Rica over several years by the American zoologist Don Jenni [316, 317].

Jaçanas do not seem to have any cessation of breeding: Clutches of eggs are apt to be laid in any month. The males maintain relatively small territories on a pond, and the females, almost twice as big as the males, maintain large territories that may contain several male territories. One pond, studied in seven different years, always had three to five female territories and seven to eleven male territories. There were generally two males to a breeding female, an average of 2.2, and a maximum of four. (This is not due to any shortage of females: Nonbreeding adults of both sexes are always available to take over any vacated territory and thus join the breeding population.) The female jaçana deposits a clutch of four eggs with each of her males, who tends them just like the male spotted sandpiper. The jaçana, more than the spotted sandpiper, seems a true example of simultaneous rather than successive polyandry. The birds are not migratory, but reside on their territory the year round. The female continues to interact with all her males, copulating with them regularly and helping them in their territorial defense. The mates are not merely attached to a common territory but have a personalized bond: If a bird of either sex has its territory taken from it, the newcomer is not immediately accepted by the evicted bird's mate.

In the polyandrous birds we have discussed hitherto, one female's males are spatially separated, each with his own brood. Other species are *cooperatively polyandrous,* with a female and her mates attending a single nest. This system has been described in two species of hawks [185], and in the Tasmanian native hen *Tribonyx mortierii,* which is actually a flightless rail [526]. In the latter species, a female usually takes only one mate, but it is not unusual for her to accept two brothers; on rare occasions three or even four males may mate with a single female. The group, which may rarely contain two adult females as well, then cooperates in the incubation and rearing of a single brood. In the hawks, it is not known whether the males are brothers. There is not the striking sex-role reversal characteristic of other polyandrous birds in these cooperatively polyandrous species. Males are not the primary parental caretakers, and females do not court. The whole system is very like that found in some jays and several other birds: A flock (with certain kinship links) cooperates in the rearing of one female's young. The only difference, but an important one, is that the female may mate with more than one male.

The great majority of birds are monogamous and share parental duties. From this basis, some tendency for greater male incubation and posthatching care is not a surprising development, since it may free the female to recover from the energetic costs of egg production and perhaps permit her to begin to develop another clutch somewhat sooner. If already committed to monogamy, the male thus benefits himself as well as the female. Although the male assumes much of the burden of incubation and posthatching parental care, he may continue to play the "masculine role" with respect to territory establishment and courtship. A reversal of these premating roles is to be expected when the male's parental investment surpasses that of the female overall, so that males become a resource for which females compete, and then polyandry becomes a possibility too. Given the predominance of monogamy and shared parental duties in the birds, it may seem odd that polyandry is so much rarer than polygyny. But we can think of at least two reasons why it should still be unusual for males to make as large a parental investment as females.

In the first place, it remains the case that the male can increase his reproductive output by an adulterous mating, while the female cannot. Although monogamous, a mated male should be prepared to seize any sexual opportunity, whereas a mated female will derive no benefit from so doing. Hence, even a mated male may continue to expend some mating effort. The second reason why a male might not abandon himself quite so fully as a female to a career of intensive parental investment is the possibility of cuckoldry: Males generally must be more susceptible than females to the danger of being tricked into wasted investment in young that are not their own.

Female jaçanas mate with all their males repeatedly. Jenni has seen a female copulate with three males within twenty minutes. While laying for a particular male, however, and shortly before, she stays mostly in his territory and copulates mostly with him. The female lays one egg per day in his nest for four days, and she waits at least a week before beginning another clutch elsewhere. The male therefore has some grounds for confidence that the young he raises are his own. However, there remains a real possibility that some or all of the chicks are not his. The spotted sandpiper story is the same. A clutch is laid over four consecutive days, and there is generally an interval of several days before the female lays for another male. Nevertheless, paternity mixups seem possible. One female, for example, had laid the first egg for a second mate when her first male suddenly needed a new clutch and she returned directly to him, thus laying for different males on successive days.

The uncertainty about paternity that these antics must engender would seem to favor any male strategy that would increase the chances of inseminating a female before she lays in another male's nest. In other words, there is a constant selection pressure for the male to redistribute his reproductive effort into the pursuit of matings and away from parental investment. This selective situation is a barrier to the evolution of polyandry, and that seems to be a major reason why polyandry remains so much rarer than polygyny.

ANTICUCKOLDRY TACTICS

The males of internally fertilizing species have a special problem to which we have alluded several times: Paternity is difficult to verify. A male has been *cuckolded* (as biologists use the word) if he rears another male's young as if they were his own. A deserted female of a species with biparental care would clearly do well to persuade another male to help her raise her young. Just as clearly, there should be powerful selection against male acquiescence or gullibility in this situation [619]. Like the duped victim of a brood parasite, the cuckolded male expends his considerable parental investment with no chance of any procreative payoff whatever.

"Cuckoldry" as we have just defined it can occur only where the male makes a parental contribution. But competition among males means that mates are to be defended against rivals even in the absence of parental care; recall the postcopulatory guarding by bison bulls (p. 90) and the precopulatory guarding in isopods (p. 115). When there is to be parental care of the young, however, we might expect that the premium on barring the door to rival sperm should be greater still.

Several recent studies of monogamous birds indicate that males are often prepared to expend a great deal of time and energy in guarding their mates during their fertile period. Bank swallows, for example, nest colonially in burrows in earthen banks, usually near water. Large colonies may contain several hundred pairs of birds living at close quarters. The typical clutch consists of four eggs laid on consecutive days. During that time and for about three or four days before the first egg is laid, the male flies with the female on every departure from the nest burrow, virtually never letting her out of his sight [37]. Swallows are spectacular aerobats, hawking flying insects on the wing, often low over water. On a spring day near a bank swallow colony, one can observe many swooping pairs, and it is marvelous to see how closely one individual can follow the rapid turns of another. Sometimes the aerial chases involve three birds. It turns out that where there is a single follower, it is generally a male, the leader's mate; two followers are generally two males, the mate and someone else. That someone else is evidently intent on rape. A male guards his mate for about a week during her fertile period, but once the clutch is complete and there is no longer a risk of cuckoldry, he will leave her unescorted and harass other females who are still fertile. It has been shown that a male will indeed try to force copulation upon such a female if he can evade her mate and knock her out of the air. It has furthermore been shown that the rapists are selective, pursuing with great zeal those females who are still laying (especially on those rare occasions when such females are for some reason unaccompanied) and ignoring females who are not [37].

"Forced extrapair copulation," that is to say rape of mated females, seems to be something of a chronic social problem in monogamous, colonially nesting birds. A recent review by Douglas Gladstone cited observations of such behavior in various species of swallows, gulls, herons, albatrosses, and others, eighteen species in all [213]. It was noted that all these were cases in which mated males sometimes tried to force themselves on their neighbor's unattended mates, and furthermore that the females always resisted. And all eighteen were species in which males make a substantial parental contribution, establishing the nesting territory and sharing in nest building, incubating, and brooding the young. Gladstone therefore suggested that mated females resist rape because their mates' contributions are a valuable resource that might be withdrawn in the event of cuckoldry. This hypothesis seems plausible, and yet it begs the question of how an absent male is to know. If the neighboring male were to offer an inducement, a fish perhaps, we might not be surprised to find a female acquiescing to copulate. But we are not aware of any such observations in monogamous birds.

Extramural fertilizations are certainly a possibility. In one study of red-winged blackbirds [65], some of the males on several small marshes were vasectomized, an operation that renders the males infertile without affecting their behavior. Many of the females nesting on the vasectomized males' territories continued to lay fertile clutches, though the probability of a fertile egg declined as the distance to the nearest fertile male's territory increased. Even the vasectomy of every territorial male on a marsh did not reduce egg fertility to zero. The inescapable conclusion of this study is that red-wing females do not typically confine themselves to a single mate. The data suggest that most of the eggs laid on a male's territory are fertilized by him but that a substantial proportion are not.

Now, red-winged blackbirds are not monogamous, and there is very little male care of the nestlings. Indeed, we should not *expect* males to care for young where there is a high risk of cuckoldry; selection should favor reallocation of male effort into other pursuits. There is a bit of a chicken-and-the-egg problem here: Paternity is unreliable, so males don't tend the nest but instead strive to be polygynists, so paternity is unreliable. But whatever factor is given causal priority, some degree of correlation does seem likely: High probability of paternity and intensive male care of young ought to go hand in hand. And when we put it that way, we have a proposition that can be tested comparatively. The closely related yellow-headed blackbird nests on the same marshes as the red-wing with a similar pattern of multiple nesting females within a male's territory, but males *do* feed the young. We therefore predict that yellow-head males have more paternity confidence than red-wings, so that a similar vasectomy experiment with yellow-heads should lead to a higher degree of infertility on the vasectomized males' territories.

We cannot discuss the guarding of paternity in birds without some consideration of ducks, who are generally monogamous and yet exhibit no conspicuous "paternal" investment. The drakes of many species guard their mates assiduously during the laying of a large clutch, but keep away and perhaps desert altogether by the time the brood is hatched. Rape is extremely prevalent among ducks [429], and it appears that male effort is directed primarily at keeping rivals away from one's mate. By subsequently deserting, males evidently survive better than females, who suffer predation on the nest. The result is that males outnumber females and hence that pressure from bachelors necessitates the mate defense already described. What seems to be responsible for this vicious circle is, first, that males are emancipated from parental care (like sandpipers, the precocious ducklings can be adequately cared for by a single parent), and second, that female dispersion, seasonality, and a prolonged laying period of several weeks' duration together make it impossible for a male to monopolize two or more mates.

PATERNITY "CONFIDENCE"

Several authors have argued, as we have, that a high probability ("confidence") of paternity is a necessary condition for paternal care to be selectively favored. However, it has recently been suggested by John Maynard Smith that this confidence-of-paternity argument contains a fallacy [424]. This critque has been quite widely accepted [e.g., 234, 569, 688], and a little digression to defend the confidence-of-paternity argument therefore seems in order.

Maynard Smith argues that a male may either guard young or desert them to seek another mate. The suggestion to be criticized is then that uncertain paternity disinclines him to guard.

> The gain from desertion is the chance of additional mating. But, if paternity is uncertain, the value of additional matings is reduced by exactly the same factor as the value of guarding. It may pay a male to desert if this greatly increases his chance of remating, but confidence of paternity has little to do with it [424, p. 178].

It seems to us that this critique contains a fallacy of its own. It treats reduced paternity confidence as a fitness loss to all males, but ignores the available fitness gains for those males who cuckold others. Imagine a male with a tidbit that he can either feed to the hungry chick in his nest or offer to the female next door in exchange for sexual favors. According to Maynard Smith, the species-characteristic level of confidence of paternity reflects both the probability that the importuning nestling is indeed its putative father's very own, *and* the probability that the extrapair copulation will bear fruit. If confidence of paternity is low, both of these probabilities are alleged to decline together. But these two quantities are not the same, nor are they likely to be positively related. In fact, just the opposite is true: The *lower* the species-typical level of "paternity confidence," then the *more* extrapair fertilizations there are going to be. Therefore, chronically low paternity confidence both *reduces* the expected fitness return from feeding putative offspring and *increases* the expected fitness return from mating effort outside the pair bond. We conclude that any evolutionary change that lowers confidence of paternity—a loss of overt signs of the timing of ovulation might be an example, or an increase in the capacity for prolonged sperm storage by the female—any such change will tend to select for some shift in the male's reproductive effort budget away from parental effort and toward mating effort.

Forced copulation attempts by males other than the mate are clearly a hazard in many monogamous birds and in other animals too. Whether there is extrapair copulation in the absence of force we do not know. Females in polygynous species are known to sometimes solicit copulation from males other than their ostensible mates. In encounters between baboon troops, for example, females have been observed to approach and copulate with males of the other troop! The reproductive consequences of this extramural sexual activity remain to be elucidated by paternity determinations. But there is no question that the males of many species behave *as if* the risk of cuckoldry were a real and imminent cost. Among starlings, for example, nest attendance is shared between the parents during incubation, but males insist on guarding their mates rather than their nests until the last egg is laid. They pay a real penalty for doing so [508]: While the nest is vacated, other starlings quite frequently "dump" extra eggs in it too. This is just one of numerous documented cases of *intra*specific nest parasitism [702]. It appears certain that the penalty of such parasitism could be reduced were the male and female to time-share nest attendance. That the male will have no part of it suggests that leaving his mate unguarded must also be costly [508].

Guarding is one matter. *Reaction* to being cuckolded is another. Do males ever "divorce" on evidence of infidelity? Obviously, human males do precisely that, and this is a topic to which we shall return in Chapter 11. Whether there is analogous behavior in other species is an open question. In an early study of mountain bluebirds, a stuffed male placed near the nest evoked attack from the resident male, and the attacks were directed not just at the intruder but also at the resident female, who was allegedly driven off and replaced [27]. The author suggested that circumstantial evidence of infidelity provided sufficient grounds for male-imposed "divorce." However, more elaborate versions of this experiment, using both the mountain bluebird and the related eastern bluebird, have not confirmed the "divorce" reaction, and there is reason to doubt the original interpretation [222, 507]. There is no question that the resident male responds aggressively to the "intruder," but even these attacks are not necessarily to be inter-

FIGURE 7-7
Anticuckoldry behavior in bluebirds?

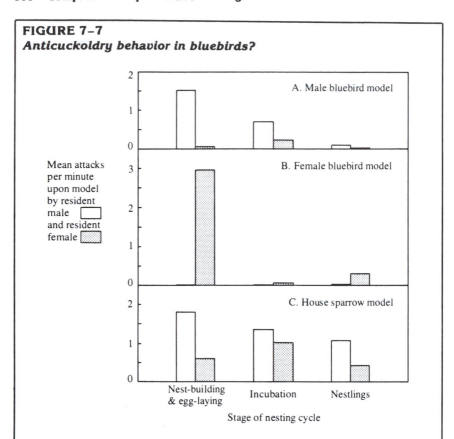

A stuffed model of a bird placed near the nest hole of a pair of breeding Eastern bluebirds elicits aggression from the resident pair.

Male aggression to a male bluebird model is maximal early in the nesting cycle (A). This effect (here controlled for effects of repeated exposure by testing each pair only once) was first demonstrated in the related mountain bluebird by David Barash, who interpreted stage-specific aggression as anticuckoldry (defense of paternity) behavior.

This interpretation must be suspect, however, since males exhibit a similar stage-specific aggressivity to a house sparrow model (C). House sparrows are nest hole competitors.

It is females, not males, who show different patterns of aggression to conspecific rivals (B) versus sparrows (C). The female bluebirds attack other females during that stage when they are susceptible to intraspecific brood parasitism. (After Gowaty, 1981, Table 2 [222].)

preted as anticuckoldry behavior, since bluebirds attack stuffed sparrows much as they do bluebirds (Figure 7-7). Sparrows and strange bluebirds are both potential competitors for the scarce resource of a suitable nest hole, so the behavior may in both cases be mainly or entirely a nest-site-defense reaction.

ANTIPHILANDERY TACTICS

When fertilization is internal, only males have to worry about cuckoldry, but females have a problem too. The world is full of suitors and promises are cheap. How is a female to ensure that the male of her choice will provide for her rather than absconding?

One option is to demand a direct transfer of resources as a condition of mating. That is what female hangingflies (p. 115) and scorpionflies (p. 117) do. The male is obliged to offer the female food in fairly direct proportion to the number of fertilizations he is granted. Of course, these are cases in which post-zygotic parental care is minimal, and a significant proportion of the energetic cost of the female's reproductive effort can be paid by the single feeding. Where there will be prolonged, energetically expensive parental care, it is a little trickier for the female to find ways to control the situation. A meal is nice, but it pays only a trivial proportion of her costs. She can, of course, give up on getting resources from her suitors altogether and simply pick a mate for his genetic quality; that seems to be what is happening at leks (pp. 129-130). If possible, however, she should prefer to receive something of value, over and above a high-quality sperm. Thus, in some polygynous, territorial species of birds, males build nests as courtship offerings to females, who evidently choose their mates at least partly on the basis of the nest quality.

If the female is seeking significant post-zygotic contributions from her mate, then her problem is most severe. How is she to protect herself against a deceiver with parental commitments elsewhere? The answer seems to be "coyness": Make the males court you long and hard, and those with competing commitments will not be prepared to stay the course. In a species with prolonged bonding and biparental care, a long courtship period can serve the interests of both parties, protecting the male against cuckoldry and the female against philanderers [619].

The ring dove is a pale, slim African member of the pigeon family, whose reproductive cycle has been exceptionally well studied. The proximate causal factors in the breeding biology of this monogamous species have been the subject of years of careful experimentation by the late Daniel S. Lehrman and his many students at Rutgers University [376, 101]. Ring doves (Figure 7–8) are ideal experimental subjects because they will carry out their reproductive cycle on demand in laboratory cages. What these studies reveal is that both the physical environment and stimulation from the mate have complex effects upon each bird's reproductive physiology.

Both birds need to be exposed to days of about fourteen hours' duration if they are to become sexually active. Given adequate daylength and a cage with food, water, a suitable nest site such as a saucer, and some nest material, a male who is placed in the cage with a female will almost immediately begin to court her with persistent bowing and cooing. After some hours of this he begins to spend a good deal of time in an oblique posture, head down and tail up, near the vicinity of the future nest site, still cooing. The female joins him in this activity, which bird-watchers call "showing the nest site."

Neither bird is ready to engage in any reproductive activities until it has been on the receiving end of some stimulation from the other [180, 398]. A female kept alone rarely lays eggs. A male similarly isolated will neither build a nest nor incubate eggs presented to him. Neither bird will engage in the stereotyped behavior of showing the nest site until exposed to the mate

FIGURE 7–8

Ring doves, *Streptopelia risoria*. A mated pair and a squab. (Photograph by R. Silver.)

for a while, and the whole process will break down if there is not an adequate nest site and nest material. It is the sight and sound of the courting male that stimulates the female's reproductive system. It takes at least five days from the time of initial pairing, and often longer, before the first egg is laid. The second follows some forty hours later, and there are always just two. If courted sufficiently, a female will lay her two eggs even if the two birds are barred from contact and copulation by a wire screen, though the eggs, of course, are infertile.

Over the first few days after pair formation, the female spends an increasing amount of time sitting in the nest depression, vibrating the tips of her wings. The male does most of the carrying of nest material to the nest site, where the female sits and turns about and tucks the material under herself. Copulation may occur as early as the first day. It recurs frequently until about day 4 or 5, declining about a day before the first egg is laid [399, 415]. After the first copulation, the male immediately begins to attend the female more closely than before, a behavior pattern that has been interpreted as mate-guarding [405]. Even more interesting is the fact that the male is considerably less hospitable when the female to whom he is introduced has recently been cohabiting with another male [709]. Such a female has already been courted, and is therefore in a different physiological state from one that has not. It seems that males detect such females, for they direct more aggression and less courtship toward them. Perhaps they would reduce cuckoldry risk in a more natural situation by thus avoiding liaisons with females that have had the prior attentions of a rival male.

Once the clutch is complete, a week or so after pairing, steady incubation begins in earnest, with the female sitting on the eggs for about eighteen hours a day and the male spelling her for a six-hour stint at midday. After fourteen days of incubation the eggs hatch, and for several days the parents continue to sit on the squabs and brood them. The family of doves and pigeons is unusual among birds in that they produce a specialized baby food. This is "crop milk," a regurgitated substance that consists of epithelial cells sloughed off from the inner walls of the crop, a side chamber of the esophagus. These walls thicken and change while the parents are incubating the eggs so that the whole crop increases in weight from about three quarters of a gram to some three grams before it begins to regress at about the end of the parental feeding phase [76]. The young doves first leave the nest about ten or twelve days after hatching, and they are fully fledged by about two weeks. They continue to beg from their parents, which become gradually less responsive to their appeals. By the end of three weeks posthatching the young birds are fully grown, and the parents cease to feed them altogether.

Clearly, the biparental enterprise is an elaborate one, and the reproductive strategic interests of the parents are similar. Each party should be concerned that the other live up to its commitments. Yet despite this equitable situation, it is the male who is initially obliged to engage in strenuous courtship for several hours. The fact that the female requires such stimulation seems best interpreted as an antiphilandery tactic. In this regard, the biparental ring dove may be contrasted with a uniparental lek-breeder. There males court strenuously too, but it is not because the female requires prolonged exposure to such courtship before she will ovulate. On the contrary, female visits to the lek are often extemely brief. The female has no strategic need of antiphilandery insurance, since there is no paternal investment—indeed, she is likely to *prefer* the most promiscuously successful male. It follows that she has no physiological need for long hours of courtship stimulation.

The monogamous female's interest in testing her male's commitment suggests a somewhat different perspective on male "guarding" than the one offered on pp. 90-91. Having her mate in attendance may sometimes be as much in the female's interest as the male's, both because of antiphilandery considerations and because of his providing defense against molesters and vigilance against predators. It has been suggested [404] that the reason the female ring dove often solicits copulation for several days before the first egg is fertilized is in order to disguise her fertile period from her mate and thus force him to guard her longer.

Lest the reader doubt that philanderers are a real problem for nonhuman females, we conclude this section with a description of deceitful "polyterritorial" males. Recall first, for purposes of comparison, the bobolink (p. 123). Males help at the nests of only their first mates, consigning second females to hard work and lower fitness. But male bobolinks at least appear to be honest advertisers: The choice of second-female status seems to be an "informed choice," the female opting for what is in fact a better breeding situation than the low-quality bachelor territories still available. In some other species, by contrast, it appears that males attract second females by hiding their mated status. Among pied flycatchers, for example, it is not unusual for a male to maintain two territories that are geographically separate and to attract one mate to each. Once the first female is incubating, the male can afford to leave her unattended long enough to court a second female on territory 2 and convince her of his good intentions. Once established, however, she is abandoned to care for the brood alone, while the male returns to help at his first nest [6]. In this case, it does not appear that the polygyny threshold has been surpassed and an informed choice is made. There are bachelors still offering better bargains. Rather, it seems that the female has been deceived.

BIPARENTAL CARE IN MAMMALS

A major male parental role is much rarer among mammals than among birds. That very rarity makes the parental behavior of mammals a subject of special interest. Why does it occur when it does? And what, if anything, have the scattered instances of paternal investment among the mammals to tell us about the evolutionary history of the human animal?

Published reviews of "paternal" care in mammals contain long lists of cases where male-young interactions have been described, whether in captivity or in the wild. These reviews can easily be misinterpreted as evidence of widespread paternal investment in mammals, for captive males of many species have been observed to groom and retrieve young, to cover them over with nesting material, and in general to show elements of normal maternal behavior. However, captivity distorts behavior, and these observations tell us little about behavior in natural settings. Even in such an incorrigibly antisocial species as the golden hamster, for example, a caged male who is presented with a pup is liable to retrieve it, to lick it, even to assume the mother's special nursing posture over it. But he is also liable to kill the pup. His pseudomaternal behavior is no sign of true paternal solicitude. It is merely one example of a widespread capacity in animals for the expression of behavior appropriate to the opposite sex. We would be quite wrong to assume that behavior that can be exhibited by both sexes is functional in both sexes. Female mammals can and do make the movements characteristic of male copulation, and males sometimes exhibit lordosis. After an injection of male hormones an adult female white-crowned sparrow will sing the same song that is sung in nature by the male alone [325]. The neural organization underlying behavior typical of one sex often lies dormant in the brain of the other sex too, and can be made to manifest itself by the right stimulus. For this reason we must be skeptical of any observations of paternal care other than those gathered in the natural habitat.

In fact, where male mammals lend a hand parentally, they seldom or never do so simply by participating in "maternal" activities [333]. Instead, biparental care in mammals generally involves some division of labor, which is hardly surprising since only the mother can nurse. In many carnivores, particularly canids (dogs and their relatives), there is a prolonged association of male and female that continues after the young are born. A nursing female may depend upon the male to hunt while she stays with the pups [364]. He regurgitates food to his mate and later to both mother and pups. Such a division of labor is found in foxes; in African wild dogs; and, to a lesser extent, in jackals and wolves. It appears that the diet of portable meat makes paternal investment a more practical option for carnivores than for most herbivores. (Biparental birds also divide labor; in many species only the female incubates, depending upon the male to feed her. Again, the food is mostly of an animal nature, and highly portable.)

There is a trade-off between mating effort and parental care, so that we may expect biparental care to be best developed in monogamous mammals and absent in highly polygynous species. This expectation accords reasonably well with available evidence. The promiscuous elephant seal, for example, is as nonparental as can be imagined. He is oblivious to pups. In fact, he not infrequently kills them accidentally in his mighty, clumsy quest for matings [371]. A monogamous male marmoset monkey, by contrast, makes a substantial parental contribution, carrying the babies, whom he passes to their mother for nursing [304]. But although the paternal investment seems to be a reasonably good indication that polygyny will not be extreme, it is not clear that we can turn the prediction around: Monogamous species do not necessarily exhibit conspicuous paternal investment. Among monogamous primates, for example, male marmosets and siamangs carry the babies, but male gibbons seem to do surprisingly little for their offspring. Jarman's class A antelope species (Table 7-1, p. 140) provide another example: There seems to be rather little a male ungulate *can* do for his young, who are born large and precocious, and are almost exclusively herbivorous when weaned. A father can neither brood his offspring nor feed them or their mother. However, these monogamous males may be making important indirect contributions to their young as, for example, antipredator sentinels and territory defenders. In the monogamous elephant shrews, a male evidently interacts little with his mate and young, but he invests much effort in the maintenance of runways that are used by the whole family [519].

WHY DON'T MALES LACTATE?

In suggesting that there is sometimes rather little a mammalian father can do for his offspring, we have taken as a given that mothers lactate. But why don't fathers [128]? The crop milk of doves and pigeons, after all, is produced by both sexes. It seems a fair question why male lactation has never evolved in any monogamous mammal, since it might alleviate the sex difference in parental investment, and hence contribute to the fitness of both mates by allowing the female to recoup lost energy and breed again sooner.

The least interesting possibility is that "Why don't male mammals lactate?" is a question like "Why don't pigs have wings?" That is, maybe it's just too great a modification and hence not a real evolutionary option at all.

In order to evaluate this possible answer, we need to consider the physiology and ontogeny of mammalian lactation and of the differences between the sexes.

Development of the functional mammary gland in the normal mammalian female proceeds through three main stages. The first is prenatal, at which time the nipples and the system of ducts leading to them are differentiated. Development is then arrested until the approach of puberty, at which point further growth of mammary structures resumes in surges associated with early ovarian cycles. Finally, there is further development of the milk-secreting cells and of the alveoli into which they secrete during pregnancy and early lactation. The hormonal induction of these developments is exceedingly complex, and varies among species. At all three stages—prenatal, circumpubertal, and pregnant—certain hormones from the ovaries, the pituitary and the adrenal glands are all causal factors for further mammary development. Perhaps the most important chemical stimuli are ovarian hormones at puberty, and pituitary hormones, especially prolactin, near delivery.

At first glance, it would appear that nursing males are indeed like winged pigs—too many steps removed from reality to ever appear. But this is too hasty a judgment. The essential pituitary and adrenal hormones are available in males as well as in females, and surges in prolactin secretion can be induced by nipple stimulation in both men and women [337]. Obviously, puberty is very different in females and males, but both involve surges of gonadal steroid hormones (Figure 10-5, p. 255). The ovarian and testicular hormones are chemically very similar (Figure 10-6, p. 256) and they are to some degree functionally substitutable, as we shall see in Chapter 10, where we discuss sexual differentiation in some detail. Some degree of breast development is in fact a common clinical syndrome in otherwise normal pubertal boys [623]. So it seems to be mainly at the earliest prenatal phase that sexual differentiation precluding male lactation occurs. And just how much sexual differentiation there is at that stage is highly variable. In rats and mice, males don't even develop nipples. In primates, dogs, and some others, fetal males develop ducts and nipples that are indistinguishable from their female counterparts; differentiation involves subtle differences in chemical response thresholds.

In any event, male lactation is very rare, but it is not a complete nonstarter. There have been scattered clinical cases reported in the medical literature, sometimes following trauma but sometimes more "spontaneous" [229, 291]. And reproductive biologist R. V. Short tells us that fertile stud billy goats sometimes produce a good supply of milk (albeit with a rather special billy goat flavor!). So it does not seem that we can explain the total absence of male lactation as a normal phenomenon in any mammalian species by dismissing it as a physiological and phylogenetic impossibility. Functional explanations for its absence instead seem called for.

It is by no means clear that the advent of male lactation could enhance the fitness of a monogamous pair [128]. Consider a pair of coyotes, wolves, or other canids (dog family), for example. There is an annual period of heat after which the female is pregnant for two months and then nurses her pups for another two, feeding them only milk for about the first month. During nursing, the father helps provision the female and pups, and even after the pups are fully weaned they remain dependent upon parental hunting for at least two more months. So would it be useful for the male to pitch in with lactation? Probably not. The size of the litter that can be reared successfully is apparently limited, not by maternal lactational capacity per se, but by the family's ability to defend a hunting territory and to extract sustenance from it. Pup mortality is maximal in the first winter, long after weaning; it is then that the environmentally imposed ceiling on the parents' reproductive capacity really comes into play. So we hypothesize that monogamous male mammals do not lactate because they have evolved other, complementary forms of paternal investment and would not profit by reallocating effort into nursing.

This must remain a tentative hypothesis until a good deal more is known about the energetics and nutritional aspects of mammalian reproduction in the wild. Next to nothing is known, for example, of the extent to which pregnancy and lactation deplete maternal calcium stores, nor of the time course of maternal repletion. We do not know whether covert maternal depletion of substances like calcium influences birth intervals or the quantity or quality of subsequent progeny. Laboratory studies indicate that such depletion can be substantial (see p. 211). In fact, lactational effort seems to involve a sort of whole-body mobilization: Not only do female mammals store more calcium and more iron than males in anticipation of reproduction, but they undergo enormous growth of kidneys, heart, intestines, and especially liver during pregnancy and lactation [669]. These somatic commitments apparently facilitate the elevated energetic flow necessary for reproduction, but at what cost to the mother's own somatic maintenance we do not know. We should like to know all these things and much more before we could feel that we fully understood maternal reproductive effort budgets or the rationale for the particular paternal contributions that we observe. Theoretical concepts like "parental investment," "honest advertising," and "antiphilandery strategies" will have to be integrated with a lot of hard facts to produce complete and satisfying accounts of reproductive strategies. But considerable insight into the diversity of such strategies in nature has already been achieved by the socioecological approach.

SUMMARY

The different mating and social systems of related animal species can often be traced to ecological differences. Food distribution, for example, may dictate the optimal patterns of movements and gregariousness for females, and the males' reproductive tactics will then vary accordingly.

Broad taxonomic constraints are also relevant in understanding comparative social systems. Mammalian females are specialized for parental care—gestation and lactation—that males cannot provide; males therefore channel their reproductive effort primarily into competition for fertilizations. In birds, by contrast, males can provide parental care as readily as females and are much more likely to channel their reproductive effort in that direction, resulting in a high incidence of monogamy and even occasional polyandry. Polyandrous mating systems are much rarer than polygynous ones, however. The risk of cuckoldry and misdirected parental care biases males toward mating effort. When males do offer parental assistance, then females must protect themselves against philandery.

SUGGESTED READINGS

Clutton-Brock, T. H. & Harvey, P. H. 1977. Primate ecology and social organization. *Journal of Zoology,* 183: 1–39.

Emlen, S. T. & Oring, L. W. 1977. Ecology, sexual selection, and the evolution of mating systems. *Science,* 197: 215–223.

Ridley, M. 1978. Paternal care. *Animal Behaviour,* 26: 904–932.

8

Life history strategy

Pacific salmon begin their lives as small round eggs buried in the gravel of a fast-flowing stream. After a month or two, they hatch as inch-long translucent larvae attached to large yolk sacs. When the yolk is spent, the salmon fry emerge from the gravel bed and begin to prey upon even tinier animals. After a variable time period—almost immediately in chum salmon, as much as two years in some sockeye—the young salmon, now called "smolt," descend the system of freshwater lakes and rivers to the Pacific Ocean.

Some major physiological changes take place before the little fish enter the ocean. They acquire a coat of prominent scales. Their swim bladders enlarge, making them more buoyant. Their tails lengthen and become deeply forked. Traveling mostly at night, when the risk of predation by birds is reduced, the smolt travel downstream in the spring. Before the building of the Grand Coulee Dam on the Columbia River, some juvenile chinook and sockeye used to travel almost two thousand kilometers to the sea [458].

Once the fish are in the ocean, growth begins in earnest. Traveling over thousands of kilometers of ocean, often following species-typical routes by navigational programs as yet undiscovered, the salmon grow for one or more years and then in the last year before maturity put on fat to pay the enormous energetic cost of the return migration up the natal streams to the spawning grounds, a trip that in long, fast rivers like the Columbia can take months.

At the spawning ground, the female scoops out a little nest depression with her tail, while her mate engages in aggressive interactions in an attempt to keep other males away. During the final migration, the male has metamorphosed, to a greater or lesser degree according to species, into a fighting machine. In particular, his jaw has been so transformed into a hooked weapon that he may be incapable of feeding.

The female spawns but once, laying some three to five thousand eggs, and dies, energetically spent. The male may succeed in spawning with successive females, but he too ends his life in this single intensive reproductive effort. All resources are spent in the migration, in the territorial battles for the best spawning sites, and finally in the great material expenditure of the spawn itself. Having spawned, the dying adults have no further use at all, unless it is to decompose and fertilize the waters in which their young will hatch.

What we have just described is the characteristic "life history" of a salmon. Some kinds of organisms, such as Pacific salmon, breed only once, others many times. Some must grow to a certain size before sexual maturity, while others mature more as a function of age than of size. Fish commonly continue to grow after sexual maturity is attained; birds generally do not. Clearly, these variations in life history characteristics must be interpretable in terms of their adaptive significance. The basic problem is how an organism is to allocate its efforts over its lifetime in order to maximize fitness. When *should* an organism first reproduce if it is to maximize its lifetime fitness? With what investment of effort? And how should that investment be partitioned? When should available energy be devoted to the organism's own growth and when to reproduction? The offspring may be few or many, much or little cared for, fed or merely guarded; and parental investment may or may not be equally shared among them. A duck may lay a clutch of over a dozen eggs; a petrel invariably lays but one. A vole or a hamster may deliver her own first litter just two months after she was herself conceived; an elephant will take a decade to reach sexual maturity and then gestate her embryo for twenty-two months. A pair of condors may feed a nestling for two years; many songbirds send their young on their way after two weeks. A species-typical life history is a reproductive strategy, and this chapter is about the adaptive significance of the tactics.

SUICIDAL REPRODUCTION

The Pacific salmon is probably the best-known example of an organism with a *semelparous* life history. **Semelparity**, or "kamikaze reproduction," is the extreme strategy of expending all one's resources and dying in a single reproductive effort. The alternative, more familiar strategy of repeated reproduction is called **iteroparity**. The distinction can be a little fuzzy when reproductive activity is somewhat protracted, so that the boundary between single and multiple bouts of reproduction becomes arbitrary. It is usual to confine the term "semelparity" to the more explosive cases. The real point, however, is that reproductive expenditure is irreversible in semelparous organisms, whereas iteroparous species return to their prereproductive states. Many insects, for example, metamorphose into a mature form that is incapable of feeding, and then expend in reproduction the energy they accumulated as larvae; we consider such species semelparous. Iteroparous insects, such as hangingflies (p. 115), are also energetically depleted by oviposition, but they can feed, recover spent resources, and lay again repeatedly.

Semelparous life histories are common among plants. Most obvious are the many annual species that flower and die in a season. Other semelparous species grow for many years (more than a century in some bamboos!) before a spectacular, suicidal flowering. Semelparity is also common in invertebrate animals. Few vertebrates, however, are so obligatorily semelparous as the Pacific salmon, although certain other fish species exhibit similar life histories.

The most extremely semelparous life history that has yet been described in a mammal is that of male marsupials of the genus *Antechinus*. These small, predominantly nocturnal, insectivorous animals occupy highly seasonal habitats in Australia, where they produce a single litter after a brief mating season in the winter or early spring. The best-studied species is *A. stuartii,* but several other species seem to exhibit the same remarkable life history [372]. Pregnancy in *A. stuartii* lasts about a month, and all births in a population occur within a two-week period. The newborn are nursed for at least three months, the first of which is spent permanently attached to the nipples. But the fathers never see their offspring: While the females are pregnant, all the males drop dead!

Just why the males die, all within three weeks of the onset of mating, has been the subject of considerable research. In the field, the immediate cause may be predation or a fatal fight, but animals captured and removed to a safe laboratory at the time of natural mortality die anyway. Anemia, gastric hemorrhages, elevated adrenal hormone levels, and suppression of the immune system have all been implicated. Male *Antechinus stuartii*, at eleven and one half months of age, self-destruct. This physiological collapse follows closely upon dramatic behavioral changes during the brief mating season, including great increases in aggression and in ranging, a switch to round-the-clock activity, and repeated copulations of several hours' duration. If they are captured and removed to the laboratory well before the mating season begins and are isolated from sexual and aggressive encounters, then males fade more slowly, sometimes surviving for a few months beyond their natural life span, but none lasts until the next breeding season. It seems that the male *A. stuartii* is truly obligatorily semelparous. Females, on the other hand, at least occasionally live for a second or even a third breeding season. Indeed, there is no known mammalian species in which females are incapable of raising more than a single litter.

COLE'S PARADOX

Semelparity seems a bizarre reproductive strategy, but further reflection may lead us to enquire why it is not more common than it is [109]. The reason for asking this question is shown in Figure 8-1. For the sake of simplicity, the example deals with an asexual female of a species with a single discrete generation (or "cohort") per year; the logic of the argument is not prejudiced by these simplifications. The surprising fact portrayed in Figure 8-1 is that the semelparous female needs to produce only one more offspring than the iteroparous female in order to match her fitness, even with the extreme and unrealistic condition that the iteroparous female is reproducing at no cost whatever to her reproductive potential! The same is true with any brood size: The semelparous female need produce just one more young (or two more, if reproduction is sexual) than an immortal female who breeds annually, and she will be just as successful in terms of her genetic representation in future populations.

Can this be true? If so, why don't more species go in for kamikaze reproduction? This is "Cole's paradox." Surely the females of many species could increase their clutch or litter size by throwing all their accumulated energy into a single effort. The logic of Figure 8-1 suggests that they would profit by doing just that.

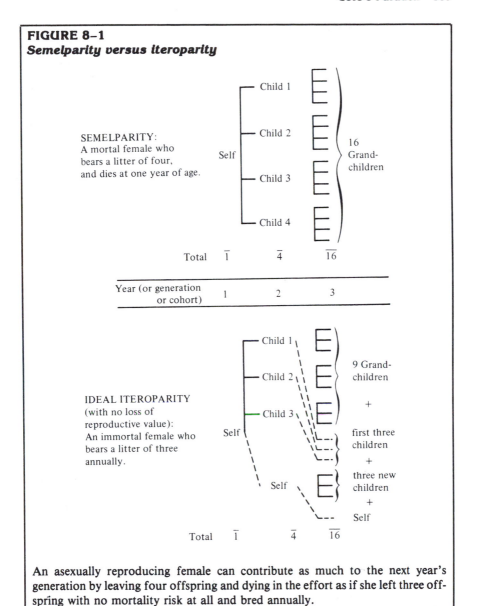

FIGURE 8–1
Semelparity versus iteroparity

SEMELPARITY:
A mortal female who
bears a litter of four,
and dies at one year of age.

Self

Child 1

Child 2

Child 3

Child 4

16
Grand-
children

Total   1̄   4̄   1̄6̄

Year (or generation
or cohort)   1   2   3

IDEAL ITEROPARITY
(with no loss of
reproductive value):
An immortal female who
bears a litter of three
annually.

Self

Child 1

Child 2

Child 3

Self

9 Grand-
children

+

first three
children

+

three new
children

+

Self

Total   1̄   4̄   1̄6̄

An asexually reproducing female can contribute as much to the next year's
generation by leaving four offspring and dying in the effort as if she left three offspring with no mortality risk at all and bred annually.

The situation portrayed in Figure 8–1 is of course deceptively oversimplified. It is not so much the artificial features of asexuality and discrete
generations that are misleading; rather, it is the assumption of a freely expanding population and the treatment of the (already mature) mother as the

equivalent of one child. What, instead, if the population is limited (as it always is, to some degree)? What if each offspring's chance of surviving to become part of the next breeding generation is far from perfect [580]? If the child's chances of surviving to breed are only one in ten, for example, the immortal iteroparous female in our example would be represented in the second generation by an expected 1.3 individuals (herself plus 3 offspring times 0.1 each); the mortal semelparous female would have to produce 13 offspring of the same viability to match her. Now our example is becoming more realistic, for juveniles are commonly more susceptible to mortality, especially as a result of predation, than are adults.

We can make our example more realistic still by supposing that the surviving mother can do something useful for her offspring and can thus increase their survival prospects by her own survival. Then the semelparous female's young will be less viable than those of the iteroparous female, and her output will have to be even greater if it is to match the output possible with iteroparity. Adding realistic complications to our model makes the apparent advantage of semelparity shrink. And indeed a number of sophisticated life history models indicate that organisms are often well advised to "hedge their bets" by distributing reproduction in time, with relatively little effort at any one time [581].

The interesting question for functional theory, then, is why certain species are semelparous while others are iteroparous. If both strategies are to be understood as adaptive results of the process of natural selection, it follows that they must be adapted to different circumstances. For this reason we must again appeal to ecological diversity in order to explain species differences, although taxonomic constraints ("phylogenetic inertia"; see p. 146) may also have to be invoked.

The ecological peculiarity of the Pacific salmon and some other semelparous fishes is readily apparent: It is the enormous cost of migrating to spawn, in terms of both energy and predation risk. The Pacific salmon matures in the ocean but must struggle up a fast-flowing, freshwater river system to its spawning grounds. The expense of that trip is incurred only once, after years of growth at sea, and it seems sensible that the fish does not hold back resources for a risky return trip and repeat journeys in subsequent seasons. In fact, comparisons of several species of the family Salmonidae indicate that those species which make the most costly migrations tend to be semelparous, while iteroparity characterizes species with less arduous migrations [39; see also 547].

Why *Antechinus* should be semelparous can also be surmised, though perhaps more tentatively. The sharp seasonality of the environment and the rather long gestation and lactation make it essentially impossible for females to squeeze two litters into a season. The lactation coincides with a spring explosion of insect food, whereas a second litter would have to be born in the hot, dry summer. Thus, only one litter is attempted, and males would have to survive a full year to breed again. The year-to-year survival prospects of a small, insectivorous mammal facing a dry summer are probably rather poor, with the result that selection should promote maximal expenditure of reproductive effort in the first breeding season [372]. That this argument should apply to males more than to females may be attributed, first, to the generally greater male mating effort and mortality characteristic of polygynous mammals, and, second, to the fact that females must already be adapted to last several months longer than males in order to wean their litters successfully, and therefore have less "reproductive downtime" to survive if they are to breed again. Even so, most females do not make it to a second season either. It is furthermore of interest that certain quite unrelated but ecologically similar mammalian species (tenrecoid insectivores) exhibit a life history approaching the semelparity of *Antechinus* [176], a fact that lends some support to the ecological interpretation above.

THE EVOLUTION OF LOW REPRODUCTIVE RATES

Iteroparous organisms conserve themselves to breed again. They may conserve themselves so well, in fact, that they last for many seasons, carefully nurturing a few offspring at a time or even one. Such life histories appear at first glance to invalidate the basic theoretical framework of this book. Natural selection is allegedly a process that inexorably promotes the maximization of individual reproduction. What, then, are we to make of slow-breeding creatures like elephants and people? Several years may pass between births in such animals. How can this languid reproductive pace persist against a steady selection pressure favoring the population's best reproducers?

Elephants and people are just two extreme examples of low reproductive rates. As extreme cases, they seem particularly paradoxical, but the issue just posed remains the same for any animal: Why not reproduce more? With the question rephrased in this way, the reader may recognize it as one we have asked before. In Chapter 2 we raised precisely this question with respect to the species-typical clutch sizes of birds, and we outlined David Lack's answer to the question, as exemplified by the case of the European swift. The simple answer was that the birds were doing the best they could. Try to raise an unusually large brood and all will starve.

Many seabirds have very low reproductive rates and expend a great deal of time and energy on each offspring. The tubenose order (Procellariiformes: albatrosses, petrels, and shearwaters), for example, includes some of the world's most numerically successful species of birds. They exploit an immense, consistently productive habitat, the oceans, yet they lay just one egg [20]. Tubenoses then endure much hardship carrying out an exquisitely timed parental program in order to raise their single chick. About thirty-five thousand pairs of grey-faced petrels (*Pterodroma macroptera*) breed in burrows on tiny, volcanic Whale Island off northeast New Zealand [297]. Pairs establish their burrow sites and then leave the island. After two months the female returns and immediately lays a single egg averaging 15.5 percent of her own body weight. The male returns about four days after the female and takes over the incubation of the egg while she goes to feed. But her feeding grounds are 650 kilometers away, and so the male sits on the egg alone for seventeen days! Then the female does a similar seventeen-day stint, and the male does yet another, until the egg finally hatches after a fifty-five-day incubation period. The timing is such that there is another male-to-female switch at just about the time of hatching; indeed, there must be, for the female, just returned from feeding herself, can then regurgitate the hatchling's first meal. Had she been sitting on the egg for several days before hatching, she would have nothing to offer: During each seventeen-day incubation spell a parent starves down from about 670 to 530 grams.

The hatchling petrel is guarded continuously for a couple of days and then is visited and fed about once every four nights. It is no small feat for the parent birds to collect a meal hundreds of kilometers away, but they work diligently at the task for four months before the chick is at last fledged and leaves the burrow. An average feed is 97 gm. Sometimes, by chance, both parents arrive home with a meal on the same night; one 245 gm chick thereby gained another 225 gm overnight! Productive though the oceans are, petrels do not have an easy life: With the few suitable breeding sites lying at vast distances from the best fishing grounds, these birds have been obliged to evolve an elaborate parental strategy in order to exploit the ocean's bounty.

Such prolonged and intensive labor to rear a single chick is surprising only because we are familiar with the large broods that some other species are able to rear so much more quickly. But in birds with large broods, such as the great tit (Figure 2-1, p. 26), most individuals die young, and even those who breed seldom last more than a couple of years. Tubenoses, by contrast, are remarkably long-lived: The life expectancy of the fulmar, a relative of the grey-faced petrel, has recently been estimated at 33.9 years for the male and 35.5 for the female [470]! And, of course, it is lifetime reproductive success that matters.

Averaged over many generations, animal populations tend to be fairly stable. What this implies is that the average lifetime reproductive output of each successful parent must be just two breeding offspring, one son and one daughter. Adult female cod each release millions of eggs, but their average output is nevertheless just two more cod that actually survive to breed in their turn. The average tubenose with her single egg per year also has a lifetime reproductive output of two surviving young. Anything over two is better than average: A parental strategy that produces just three breeding offspring wins a growing share of the gene pool. It follows that the evolution of slow reproduction is not after all so surprising: A lowering of the reproductive rate will evolve by natural selection whenever circumstances dictate that going a little slower will increase the lifetime production of surviving offspring.

REPRODUCTIVE EFFORT AND DEATH

"Reproductive effort" is ideally conceptualized as an investment of "reproductive value" or expected fitness units, as we explained on p. 41. This ideal is hard to make practical, however, with the result that life history studies generally resort to energetic definitions of reproductive effort, such as the proportion of available maternal energy invested in gonads or in eggs. This sort of measure ought to be at least monotonically related to the ideal currency: The more a mother puts into a clutch of eggs, for example, the lower is her residual reproductive value (expected future reproduction) likely to be, as a result of depletion and elevated mortality risk.

The expenditure of reproductive effort is debilitating, for females and males, semelparous and iteroparous. A serious question for the evolutionary theorist is why creatures deteriorate and eventually die, even if they are secure from predators, food shortages, and all other conspicuous extrinsic mortality causes. A widely accepted answer, called the Medawar-Williams theory of senescence after the two biologists most responsible for its development, is that attributes can enhance fitness and hence be favorably selected by contributing to reproductive success early in the life cycle even though the same attributes have degenerative consequences later [250]. Such attributes are perhaps clearest in semelparous species. An example is the structural modification of the salmon's jaw, which contributes to success in aggressive competition in the short term but is part of the metamorphosis that guarantees the fish's imminent demise. A slightly more subtle example is the elevation of adrenal corticosteroids in the male *Antechinus*. This hormonal change contributes to short-term coping with the demands of intense reproductive competition, in part by its role in the mobilization of energy; the same hormonal events contribute to mortality by suppressing the immune system and opening the door to a variety of infections [372].

It is not just semelparous organisms to which the Medawar-Williams theory presumably applies. Recall the negative effects of testicular hormones upon the life span of men and tomcats (p. 100). Organisms are not

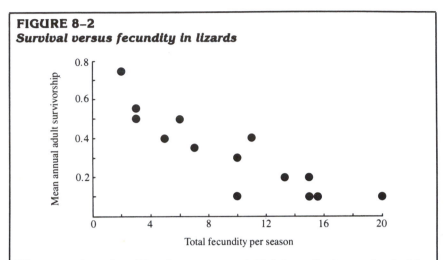

FIGURE 8–2
Survival versus fecundity in lizards

When several species of lizards are compared, high fecundity is associated with a low probability of surviving to breed again. One interpretation of this negative relationship is that expenditure of reproductive effort leads to mortality; conversely, poor survival prospects may select for elevated effort. In either case, there is an evident trade-off between fecundity and survival. (Adapted from Tinkle, 1969, Figure 1 [615] © 1969 by the University of Chicago.)

designed to live forever. It is not life span that natural selection tends to maximize but fitness. Reproduction "competes" with survival for an organism's limited resources. More precisely, we can conceive of the organism's budgeting energy or other crucial resources between "reproductive effort" and "somatic effort" [11, 674], the latter of which may be divided into growth and maintenance. Hence, senescence sometimes reflects a degeneration of organs as a result of the withdrawal of material resources from somatic repair and their diversion to reproduction [84, 330].

The proposition that there is some sort of a trade-off between reproduction and survival can be examined interspecifically. The fecundity and adult survival of various species of lizards, for example, are negatively correlated (Figure 8–2), as we might expect. Those species that expend the greatest effort in egg production evidently suffer the greatest mortality.

Comparisons within species are perhaps even more to the point. Thus, it has been reported that female house martins who rear two consecutive broods in a single summer are likelier to die in the subsequent winter than are those females who rear only one brood [73]. It is far from obvious, however, whether we should expect naturally occurring variations in reproduction to be predictive of mortality in this way, since animals may

adjust their reproductive expenditures to their individual resource situations. We made the same point in discussing clutch size variations: Unusually large clutches are produced by the birds best able to cope with them and therefore need not be predictive of reduced fledging success (p. 26). The same argument would seem to apply to adult mortality. And indeed in a study of song sparrows [576], in contrast to the house martin results, it was found that breeding females who survived to the following year had raised significantly *more* young (3.82 fledglings) than those who died over winter (2.98). Similarly, in magpies, birds who reared large clutches outsurvived birds who reared smaller ones, a result that has been attributed to the fact that both clutch size and parental survival are positively related to territory quality [274].

What seems called for is an experimental approach. We have seen (p. 26) that the effects of clutch size upon fledging success can be separated from individual differences in fitness potential by having the experimenter control clutch size instead of letting the birds do so. The effects of clutch size upon adult mortality can be investigated similarly. We are aware of two such studies. In one, adding offspring and thus increasing the parents' work load reduced the return rate (presumed to reflect the survival) of pied flycatchers, although the effect was significant only in fathers [21]. In a similar study of tree swallows, however, no such effects could be demonstrated, despite evidence that the mothers indeed expended greater effort in order to rear the enlarged broods [144]. That the expenditure of mating effort leads to increased mortality has been more clearly demonstrated in experimental studies of fruit flies. Mated females oviposit and then die sooner than females denied the opportunity to mate [421]. A similar effect can be shown in males too: Those who never mate live much longer than males given access to one virgin female a day, and males given access to eight virgin females a day die sooner still [484]!

TO GROW OR BREED?

We have just seen that reproduction and survival are to some degree "competitors." What should be still clearer is the competition between reproduction and growth, since the issue of allocating energy is even more straightforward. Suppose I have some calories over and above maintenance requirements. Shall I convert them to eggs or shall I use them to grow larger? The decision will affect eventual fitness, and the fitter choice is not obvious.

The conflict between growth and reproduction is an empirically observable fact [84]. Among barnacles of a particular size, for example, those that reproduce expend matter and energy and grow less well after the reproductive effort than those that do not [29, 121]. Hence, the reproducers are most apt to be squeezed out by their neighbors and to lose the chance to reproduce again. Or recall the alternative life histories of male sunfish (pp. 102–103): Males who reproduce early as "sneak fertilizers" cease to grow, whereas those who defer reproduction grow much larger. If we rely on natural variations, however, we run into the same problem that arose with regard to the mortality-reproduction trade-off. Growth and reproduction might in some cases be *positively* correlated, as for example where both are maximal in those individuals with the best food supply. A more careful analysis of the growth-reproduction trade-off again demands an experimental approach.

One possibility is to experimentally eliminate reproductive function and to look at the growth consequences. The trouble with this approach is that most of the ways that we can think of to preclude reproduction will have behavioral and physiological consequences other than those intended. If we isolate an animal, for example, in order to prevent sexual access, the result of this frustration is less likely to be growth than an increase in motor activity that would normally function to improve the prospects of finding a mate.

The ideal experimental approach for studying this trade-off would entail manipulation of the very extrinsic variables that normally exert proximate control over growth versus reproduction in nature. An unusual opportunity for just such experimental control is provided by hermit crabs, and the requisite experiments have been conducted by Mark Bertness of Brown University [50].

The hermit crabs include both marine and terrestrial species, all of which live in discarded snail shells. Without such shells, these soft-bodied animals are quickly preyed upon. It is not surprising, then, that finding a suitable shell is a top priority in a hermit crab's life. Nor is it surprising that an old shell will not be discarded unless a fresh replacement is available. The trouble with living in vacant gastropod shells is that they are restrictive. A hermit crab cannot lug about a shell that is too large, but shells without a lot of space are eventually outgrown, and then a larger replacement has to be found. In many populations of hermit crabs, shells of preferred size are a scarce resource that is aggressively contested. When a crab cannot find a shell a little larger than the one that is beginning to pinch, there is nothing for it but to stop growing. At that point, surplus energy must be spent as reproduction.

Bertness first demonstrated, across a range of populations of the hermit crab *Clibanarius albidigitus,* that local populations with the best supplies of adequate shells grew larger and reproduced less often than shell-limited populations. He then did experiments, in both laboratory and field, introducing large numbers of suitable shells to some populations, while

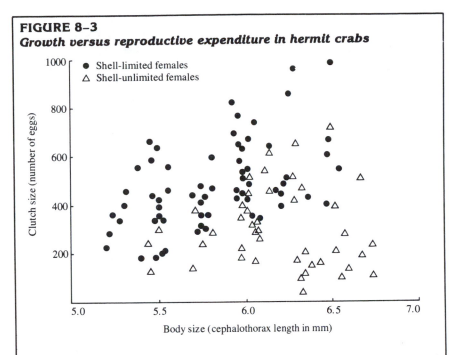

FIGURE 8–3
Growth versus reproductive expenditure in hermit crabs

Hermit crabs (*Clibanarius albidigitus*) whose growth is constrained by limited availability of shells channel more of their resources into reproduction than do shell-unlimited females. After 90 days under experimental shell-availability regimes, shell-limited females had grown from 5.5 to 5.8 mm while shell-unlimited females grew from 5.5 to 6.2 mm. Seventy-six percent of shell-limited females were carrying eggs compared with 45 percent of shell-unlimited females, indicating more frequent reproduction in the shell-limited. The figure shows that shell-limited females also produced larger clutches in each reproduction episode. (Adapted from Bertness, 1981, Figure 4 © 1981 by the University of Chicago [50].)

leaving other, initially similar, populations shell-limited. The effect was clear: Unlimited shell availability led to later age of first reproduction, smaller clutches, and more rapid growth (Figure 8–3). Hermit crabs are able to allocate resources to either growth or reproduction according to the availability of suitable shells to grow into.

WHEN TO MATURE?

Complex multicellular organisms begin life as single cells, grow and differentiate, sometimes for years, undergoing various changes in structure and behavior, until at last they mature and reproduce. How soon should an organism mature? This is in part a question of how large an organism should grow before first reproduction. Bertness's hermit crabs delayed maturity when available shells made growth an attractive option. The optimal solution to the problem of how large to grow before reproducing will depend on a number of factors. Is predation risk reduced by large size, so that I should grow as quickly as possible and defer reproduction until a threshold size is attained? And is my fecundity going to be a function of body size? Many species of fish can reproduce at a wide range of body sizes, with egg production being a rather direct function of female size. The fitness of a male, by contrast, may be less size-dependent, and this seems to be the reason why it is common among fish for males to mature earlier and at a smaller body size than females [e.g., 653]. If the male defends a territory, however, size may again be crucial; recall again the sunfish (p. 103) in which territorial males grow until seven years of age before becoming sexually mature, whereas little sneak-fertilizing males mature at two. (The phenomenon of a sex difference in the age of first reproduction is called **sexual bimaturism**, and it may be either males or females who mature first.)

In birds (and to a lesser extent in mammals), mature adults do not continue to grow, and bodily dimensions generally have a narrow species-characteristic range. Nevertheless, the attainment of adult proportions does not necessarily lead to immediate sexual maturity. This seems perplexing. Other things being equal, selection should favor maturing as early as possible; the most obvious reason is that there is always some risk of death before first reproduction, and the risk can only increase if the organism marks time.

A fulmar petrel attains full adult size in its first year of life, but the average age of first nesting is 9.2 years [470]. Once the birds finally get started, they nest annually, often for 20 years or more. This sort of delay of maturation is typical of long-lived birds with low reproductive rates, and although we are strongly inclined to suppose that the delay must be adaptive, it is not always obvious why. There *is* evidence, for fulmars (Figure 8-4) and many other species, that older, more experienced birds are more successful in fledging young than are individuals breeding for the first few times [459, 538]. It is generally believed that feeding skills and other slowly accumulated bits of wisdom give older birds the advantage, but it is really by no means clear why it should take nine years for a fulmar to acquire enough experience to make a first nesting attempt worthwhile. Studies of the ontogeny of foraging and other skills in species with delayed maturity would be illuminating.

Many authors have suggested that maturity will be delayed whenever younger animals cannot compete with older animals and would only damage their eventual fitness by trying. The main support for this argument

FIGURE 8–4
Breeding experience and fledging success in fulmars

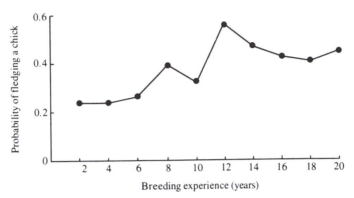

Fulmar petrels first breed at about nine years of age, and then nest annually, sometimes for many years. The probability that the single chick will be successfully fledged improves gradually with parental experience. The data represent the reproductive success, averaged over two-year periods, for thirty-three birds (fourteen females and nineteen males), each of whom was observed during twenty consecutive breeding seasons. The improvement in reproductive success over seasons is significant for each sex considered separately. (After Ollason & Dunnet, 1978, Figure 1 [470] by permission of British Ecological Society.)

comes from an observed correlation between the degree of polygyny of a species' mating system (cf. p. 152) and the extent to which males mature at a later age than females. Among monogamous species of grouse, for example, the sexes mature at the same age, whereas in polygynous species, male maturity is delayed [673]. Since polygyny elevates male-male competition, it seems sensible to interpret delayed maturity as a response to that competition. But again, it is not always obvious *why* it should take several years for a male to reach his maximum competitive ability.

In songbirds, there is a good deal of evidence of the competitive disadvantage of younger males, who regularly have to settle for poorer territories and fewer mates (Figure 8–5). A common solution is for males to become reproductively mature and to attempt to nest at one year of age, but to delay taking on the bright plumage of maturity until they are two. It appears that breeding in immature or femalelike plumage is a compromise that reduces some of the costs of competition with older males, but may also make the bird less attractive and less efficient at territorial defense against his agemates [511, 534].

Age effects upon reproductive success are well known in females too

FIGURE 8–5
Reproductive advantage of older male indigo buntings

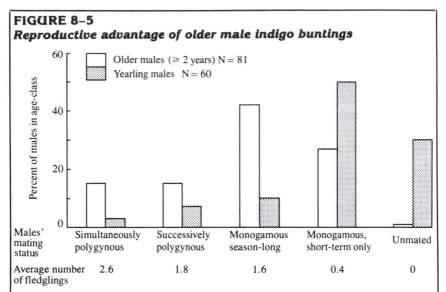

Males' mating status	Simultaneously polygynous	Successively polygynous	Monogamous season-long	Monogamous, short-term only	Unmated
Average number of fledglings	2.6	1.8	1.6	0.4	0

Many more yearlings than older males remained unmated for part or all of the breeding season, although all males held territories. The reproductive advantage of older males apparently derives from their holding better territories. (After Carey & Nolan, 1979, Table 4 [85].)

FIGURE 8–6
Age and reproductive success in female elephant seals

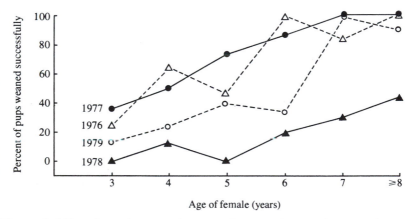

The probability of weaning a pup increases for several years after maturity. The competitive advantage of older mothers, who monopolize the best sites on the beach, is greatest when female and pup density is high. (After Reiter, Panken & LeBoeuf, 1981, Table 1 [522].)

(e.g., Figure 8-4), but even a considerable reproductive disadvantage in youth does not necessarily lead to deferred maturity. In elephant seals, for example, females breed annually from about three years of age up to a maximum of about fourteen, and their probability of successfully weaning their pup continues to increase for several years after maturity (Figure 8-6). The youngest females, competitively consigned to the edge of the rookery where storms and other mishaps often cost them their pups, are seldom successful. Some are even killed by "over-eager male suitors" [522]! It is not altogether clear why they do not delay puberty (though the smallest females evidently *do* postpone first reproduction until four years old or even five). Perhaps some essential experience is gained in the first breeding attempt.

REPRODUCTIVE VALUE AND THE SCHEDULING OF EFFORT

A spawning salmon conspicuously expends its very life in reproductive effort. Evolutionary theory suggests that all organisms are in effect doing the same. Genetic posterity should be the ultimate function of all organismic adaptations, thanks to natural selection. What, then, is an offspring worth to a parent? We must expect the offspring to be valued according to its prospects for contributing to the transmission of parental genes, which usually means reproducing in its turn. And indeed we conceptualize "reproductive value" in units of expected future reproduction (p. 41).

A mature animal's reproductive value diminishes with time, and this may be so even if she has not yet reproduced: A childless woman of thirty has a lower residual reproductive value than a childless twenty-year-old because her expected number of future offspring is lower. What may be less obvious is that reproductive value actually increases during early development. Some infants will die before they reach maturity. Therefore, each infant will, on average, produce fewer offspring than can be expected from each average pubertal individual. The probability of reproducing, and hence the reproductive value, increases as an animal approaches maturity, simply by virtue of survival.

What this implies is that immature offspring become more valuable from the parental point of view as they approach maturity. We might therefore predict that a parent with several offspring will favor the eldest, but this prediction leaves out of account the possibility that the offspring's needs differ. If older children can manage to some degree on their own, favoring younger children may be the optimal parental strategy despite the older child's greater reproductive value. We can predict, however, that parents should risk more for young that are nearing mature independence than for *equally helpless* young that have further to go. This prediction has been verified in both songbirds and in fish. Nesting birds will stand their ground longer and display more vigorously at an approaching predator when they have hatchlings in the nest than when they have eggs, and they will accept still greater risk as the nestlings near independence [16, 26, 231] (Figure 8–7). Similarly, a male stickleback will incur greater risks to defend older eggs [509].

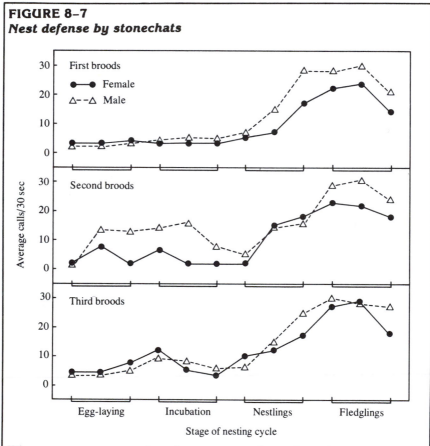

FIGURE 8–7
Nest defense by stonechats

When a person approaches the nest, stonechats (*Saxicola torquata*) emit predator-distraction calls ("chacks") at a higher rate as the reproductive value of the offspring increases. (By the end of the fledgling period, young are mobile and evidently less in need of parental protection.) (After Greig-Smith, 1980, Figure 6 [231].)

The decision mechanisms determining acceptable parental risk are in fact even subtler than the preceding account implies. The stickleback is sensitive not only to the age of his eggs but to their number, incurring greater risk to guard a large brood than a small one [509]. And in the case of songbirds, parents evidently assess the particular risks that a potential predator represents. In the white-crowned sparrow, for example, snakes, hawks, and jays are responded to by parents more or less intensely at different stages of the nesting cycle, according to whether the particular predator is a threat to eggs, to nestlings, or to adults; within this complicated pattern, however, the increasing value placed upon older offspring is still discernible [487].

Besides being sensitive to the offspring's reproductive value, parents may be expected to be sensitive to their own. If one has *no* future reproductive prospects, for example, then one ought to throw caution to the winds in the present effort, and that in a sense is what a semelparous organism such as the male *Antechinus stuartii* does in one fatal burst of mating effort.

Many songbirds are seasonal breeders who can squeeze two or more nesting cycles into a summer. Overwinter mortality of adults is commonly high, so we can say that a parent's reproductive value drops rapidly as the breeding season draws to an end and the prospects for a successful renesting disappear. This circumstance should select for greater reproductive effort later in the season, and there is at least one study of parental energy expenditure that supports this prediction. Recall that house martins sometimes rear a second brood with resultant elevated mortality risk for the mothers (p. 188). Second broods average 2.92 nestlings, and are significantly smaller than first broods of 3.52. We might, therefore, suppose that the energetic expenditure is lower for second broods, but we would be wrong. The smaller later brood in fact demands about 30 percent *more* parental energy expenditure. This has been shown by the use of an ingenious method for measuring metabolism in unrestrained animals, the "doubly labelled water" technique [242]. The method involves injection into the bird of a known quantity of water composed of radioactive isotopes of both hydrogen and oxygen. A sample of bodily fluid is later collected and is tested for retention of the isotopes, from which a measure of total metabolism over the interval between injection and sampling can be computed. The logic behind the calculation is that the loss of both isotopes is indicative of body water turnover, but any loss of radioactive oxygen in excess of loss of radioactive hydrogen is indicative of CO_2 production, hence of total respiration and metabolism. Without this method, the house martin researchers would have had no inkling of the great elevation of energetic costs of second broods, an elevation that is probably attributable to a seasonal change in the particular insect foods available to the parents [242].

The measure of tolerable risk in the form of reaction to approaching predators is a still clearer index of parental "effort" than is energy expenditure. There is some indication in the stonechat study (Figure 8–7) that this sort of risk-taking increases not only with the age of the brood but with successive broods over the season as well: The birds do not emit predator-distraction calls during the incubation stage with first broods, but do so with subsequent broods. Other interpretations of the data in Figure 8–7 are possible [231], and further study of seasonal patterns of risk-taking is needed.

The seasonal increase in reproductive effort is a fairly straightforward and elegant prediction, although it has seldom been tested. A somewhat messier prediction is that reproductive effort ought to increase with age once the animal begins to senesce (see p. 188) and hence has a diminishing reproductive value. The prediction is messy in that there is a certain circularity to the hypothesized selective forces. Low reproductive value allegedly selects for increased expenditure of effort, but it is partly that expenditure that *causes* senescence and the reduction of reproductive value

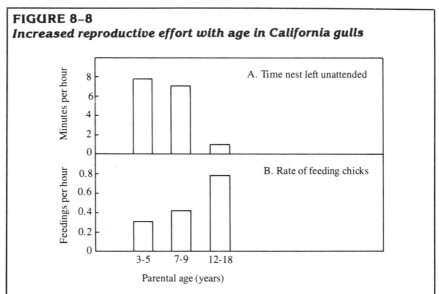

FIGURE 8–8
Increased reproductive effort with age in California gulls

Older parents leave the nest unattended less than younger birds, and feed the chicks at a higher rate. Moreover, older birds maintain their maximal feeding rate for longer than younger birds, suggesting that the difference reflects a greater expenditure of effort and not just a greater foraging efficiency. (After Pugesek, 1981, Figure 1 [514]. Copyright 1981 by the American Association for the Advancement of Science.)

[330, 580]. The most convincing demonstration of increased reproductive effort in aging animals comes from a study of California gulls [514]. The birds first breed at about three years of age, and the study compared young (three to five years old), middle-aged (seven to nine), and older (twelve to eighteen) birds. The older birds left their chicks unattended less than the other groups, but nevertheless managed to feed them more frequently and continued to feed them longer (Figure 8-8). They also engaged in more territorial defense. The greater parental solicitude seems really to reflect a greater expenditure of effort rather than just that older birds are more effective parents, for if they were simply better at parenting, why should they continue to feed their young for longer? Unfortunately, this study did not include the approach-of-a-standard-predator experiment, as has been done in studies of within-season effects (Figure 8-7). It would be interesting to know if older birds will incur greater risk to defend chicks than will younger parents.

PROFLIGACY OR CAREFUL NURTURE? *r-* VERSUS *K*-STRATEGY

The diversity of life histories boggles the mind. Some general, organizing set of principles would clearly be an aid to comprehension. The most influential general scheme that has yet been proposed is that of *r-* **versus** *K*-**strategists**. This concept, which we owe to Robert MacArthur [406], refers to a hypothetical continuum between two extreme types of strategy. The terms *r* and *K* derive from two quantities in the mathematics of theoretical population biology, but we needn't mathematicize in order to understand them.

The quantity *r* is a measure of the **intrinsic rate of increase** of the size of a population that has been freed from resource limitations, expressed, for example, as the increment in total population size per thousand adults per year. Clearly, population growth at rate *r* is an abstraction and cannot be sustained indefinitely. Sooner or later, a growing population must come up against a resource limitation and cease to expand. The maximum population that a species can maintain in a particular habitat is its **carrying capacity** for that habitat, symbolized by *K*.

Now, an *r*-selected species, or an *r*-strategist, is a creature that has been adapted by natural selection to maximize *r* because circumstances in the evolutionary history of the species have repeatedly permitted rapid population growth. In any population that regularly encounters conditions in which a spurt in numbers is possible, selection will operate to maximize the capacity for rapid, prolific breeding because the quickest breeders will win the largest share of the expanding gene pool during each population explosion. A K-selected species, or K-strategist, on the other hand, is one whose adaptations are more concerned with survival and reproduction when the population size is at or near K, the maximum that the habitat can support. Selection should have operated to maximize these abilities in species in which population size generally remains stable near K. In fact, the adaptations of such K-selected organisms should tend to maximize K by promoting the efficient use of limited resources.

Basically, then, the difference between r- and K-strategy is this: The r-strategist is a profligate speed-breeder, whereas the K-strategist carefully husbands scarce resources to raise relatively few offspring. The r-strategist has some chance of landing several young in the breeding population. The K-strategist faces a no-vacancies situation in which only the most intensively nurtured offspring will stand a chance. The tactics of these alternative strategies are outlined in Table 8-1. Various measurable aspects of reproduction and life histories tend to be correlated with one another. A relatively large number of offspring tends to go with a relatively short life, relatively rapid development of the young, and so forth [497]. The combination of these life history tactics is subsumed by the single label r-strategist.

The dimension of r- versus K-strategy is of course relative. It only makes sense to call a particular species K-selected in comparison with a more r-selected creature. Frogs that use temporary puddles of water that will soon dry up, for example, are r-strategists in comparison with other frog species that inhabit more stable bodies of water. In fact, habitat stability is one key to parental strategies: r-selected species are often "pioneers," invading transient habitats and exploiting a temporary abundance of resources. Rodents, birds, and other animals characteristic of transient ("successional") vegetational communities, such as the brushy habitat that appears after a forest fire and is eventually replaced by new trees, tend to be r-strategists; in comparison, the relatively K-selected species are typical of more stable ("climax") communities like the forest itself.

Voles, for example, are the mammalian r-strategists par excellence [342]. These tiny herbivorous rodents have a tremendous capacity to expand their numbers in transient habitats. Large litters are produced after a gestation period of less than three weeks, and when food is ample, the young females can become pregnant even before they are fully weaned! Voles can be contrasted with a relatively K-selected group of rodents, the squirrels. Squirrels do not generally breed in the year of their birth, nor do they occur in successional habitats. They are mainly to be found in two environments of relatively consistent productivity—forests and open areas such as prairies and semideserts. By each of the criteria in Table 8-1 squirrels are indeed more K-selected than voles: They breed later, develop more slowly, are more stable in numbers, have fewer offspring, and so forth.

TABLE 8-1
Some of the reproductive and life-historical differences
between r- and K-strategies

r-*strategist*	K-*strategist*
Many offspring	Fewer offspring
Low parental investment in each offspring	High parental investment in each offspring
High infant mortality (mitigated during population explosions)	Lower infant mortality
Short life	Long life
Rapid development	Slow development
Early reproduction	Delayed reproduction
Small body size	Large body size
Variability in numbers, so that population seldom approaches *K*	Relatively stable population size, at or near *K*
Recolonization of vacated areas and hence periodic local superabundance of resources	Consistent occupation of suitable habitat, so that resources more consistently exploited
Intraspecific competition often lax	Intraspecific competition generally keen
Mortality often catastrophic, relatively nonselective, and independent of population density	Mortality steadier, more selective, and dependent upon population density
High productivity (maximization of *r*)	High efficiency (maximization of *K*)

Source: From the *American Naturalist,* 104: 592-597 by E.R. Pianka by permission of The University of Chicago Press. Copyright © 1970 by The University of Chicago.

Always remember that description of a species as *r*- or *K*-selected is a relative statement, though any comparison may be implicit rather than explicit. We may draw such comparisons finely or broadly. Tree squirrels are more extreme *K*-strategists than ground squirrels, but all rodents are *r*-strategists in comparison with elephants. Martin Cody describes several cases in which bird populations vary along the *r*-*K* continuum within a single species; the *r*-strategy is generally more pronounced as we go from the tropics to temperate zones [108]. At the opposite extreme of comparative breadth Eric Pianka characterizes vertebrates in general as *K*-strategists in contrast to the more *r*-selected insects [497].

The idea of an *r-K* continuum is richly detailed, and many comparative studies comply with the overall scheme. A recent review article, for example, concludes that "there is a continuum among raptors [birds of prey] from small short-lived species which have relatively large eggs, large clutches, short breeding cycles, and early maturity, to large, long-lived species, which have relatively small eggs, single-egg clutches, protracted breeding cycles, and deferred maturity" [460]. This continuum is conspicuous both among raptors as a whole and in finer comparisons within genera. Comparative studies of many other animal groups fit the picture too.

Some experimental studies, in which researchers have studied the life history traits of populations subjected to different selection pressures, also fit the scheme. Perhaps the best such study does not involve animals at all, but dandelions! Plants can be particularly suitable subjects for studies of variations in reproductive effort, since their gentle habits save us from the immense difficulties of measuring behavioral effort and of deciding which behavior constitutes reproductive effort and which does not. For plants, we can use a simple measure: The proportion of available resources that is channeled into reproductive structures, such as flowers and fruits.

The scourge of the dandelion is the lawnmower. Mowing imposes mortality that is catastrophic and relatively nonselective. Frequent mowing therefore selects for a rapid life cycle with early, intense reproduction, or in other words, for *r*-strategy. Madhav Gadgil and Otto Solbrig [207] compared dandelion populations from frequently mowed versus relatively undisturbed sites. They found that the plants indeed exhibited different levels of reproductive effort, even if transferred to the laboratory and reared under identical conditions. In particular, one biotype was a clear *K*-strategist, allocating resources to leaf biomass at the expense of seed, and thereby gaining a competitive advantage at high density by its capacity to shade and choke out the relative *r*-strategists. As one might expect, the dandelion *K*-strategist was found only in relatively undisturbed sites.

The *r-K* concept thus derives support from both comparative surveys and experimental studies. But it is too much to hope that such a unidimensional explanation of life history variation could be completely general, and we shall now turn to some of the criticism that *r-K* theory has engendered.

THE ATTACK UPON *r-K* THEORY

The concept of *r* and *K* strategies is a handy way to categorize and contrast life histories, but it has been widely criticized in recent literature, for two main sorts of reasons. In the first place, the theory ascribes the two sets of alternative life history characteristics listed in Table 8-1 to particular selective factors, namely stable versus fluctuating environments. Several critics of r-K theory have remarked that the degree of environmental stability has virtually never been assessed separately from the life history traits themselves. The correlations implied by the list in Table 8-1 may therefore be genuine and yet have nothing to do with the selective circumstances to which r-K theory attributes them. Indeed, there are alternative life history models that do not begin from environmental stability and yet tend to predict similar correlations. Thus, the existence of life history "syndromes" like Table 8-1's r-strategy and K-strategy is not necessarily evidence for r-selection and K-selection. The second type of criticism has focused on the limited generality of the syndromes themselves, offering some quite different analyses of cases where the life history correlations of Table 8-1 do not hold up. We shall consider these two classes of criticism in turn.

The main alternative to the r-K formulation has been called "bet-hedging theory" [580, 581]. The basic idea is that high or unpredictable mortality of juveniles will often favor a conservative parental strategy of iteroparity with a low expenditure of reproductive effort in any particular episode of breeding. At the same time, rather intensive nurture of individual young may be necessary for them to have any chance of surviving the hazardous juvenile period. If, on the other hand, adult mortality is high relative to juvenile mortality, selection will favor greater reproductive effort and, in the most extreme cases, semelparity. Thus, bet-hedging theory predicts either K-strategy or r-strategy according to the relative rates of mortality of adults and juveniles, with no reference at all to variations in r or K, nor to population fluctuations, nor to environmental stability.

Stephen Stearns, the most influential critic of r-K theory, points out that r (the intrinsic rate of natural increase of a population) is a population parameter that is highly sensitive to the values of such life historical parameters as fecundity and age of maturation. On the other hand, "K is not a population parameter, but a composite of a population, its resources, and their interaction" [581, p. 155]. This leads him to conclude that r-K theory has a "fatal flaw": Selection for maximization of r may indeed lead to the traits called r-strategy in Table 8-1, but the opposite traits are then called K-strategy "in the absence of either evidence or deductive logic" [581, p. 155] that they have evolved under K-selection, that is to say, under conditions where mortality increases with population density and there is intense competition for scarce resources. Stearns therefore calls for an alternative approach in life history theory, one that would derive falsifiable predictions from measurable (and experimentally manipulable) rates of adult and juvenile mortality. Bet-hedging theory is just a first step toward this goal.

Stearns's critique of *r-K* theory seems to have won many converts [see, e.g., 652], but his program for life history theorists can be challenged. If the problem is to explain, and eventually to predict, variations in life histories, it would seem that the explanatory variables must be extrinsic to the animals themselves. That was the logic of comparative socioecology (Chapter 7): Species differences in social behavior are attributed to ecological differences. Juvenile and adult mortality rates, however, are *themselves* life history traits. They are part of what is *to be explained* by an adequate life history theory, and therefore cannot be the basic elements of the explanation. This is a thorny theoretical problem. Even apparently extrinsic ecological variables are in part properties of the organism rather than simply properties of its environment; the spatial dispersion of food that is encountered by an animal, for example, is only partially extrinsic, depending as it does upon the animal's dietary preferences, perceptual abilities, search strategies, and so forth. Similarly, age-specific mortality rates may be selective forces upon other life history traits such as age at maturity, but such rates are also results of the species' evolved schedule of reproductive effort expenditure. We conclude that no current theory can claim to predict a set of life history traits from an extrinsic set of prior causes. Theory may instead develop in the direction of "systems" models, with multiple feedback loops representing the mutual influences of the various traits [e.g., 279].

EMPIRICAL EXCEPTIONS

For convenience, we shall continue to refer to the alternative constellations of life history traits in Table 8-1 as *r*- and *K-strategy*, bearing in mind Stearns's caveat that they may or may not have been produced by *r*- and *K-selection*. How these alternative patterns come about is certainly a basic question. But are they even general? We claimed earlier that a great deal of comparative evidence complies with the scheme in Table 8-1, but some does not.

One case in which the usual correlations among life history characteristics do not obtain has been described in a study of periwinkles, little marine snails that live on rocks in the tidal zone [260]. *Littorina rudis* is an ovoviviparous species, which means that the eggs hatch inside the mother's body but the young are not nourished there beyond the original provisioning of the eggs. Emerging as miniature adults, the young *L. rudis* do not disperse on the tide as do related oviparous species but settle near their mothers, a fact that leads to local populations with local adaptations. *L. rudis* occurs both on boulder-strewn shores and in crevices in the tidal zone. Comparison of a boulder versus a crevice population turned up life historical differences that do not seem to fit the usual pattern. The boulder population produced more offspring of smaller size than the crevice population. These are apparently *r*-strategic traits, but another characteristic of the boulder-dwelling winkles is clearly not: They mature later at a larger size.

The explanation for these differences must be sought in ecological details of the two habitats. Boulders roll with the waves, and winkles can be crushed. A result is that damaged shells are much less frequent in the crevice population than on the boulders. Another is that boulder-dwelling winkles are subject to a mortality factor that may operate independently of their density. Rolling boulders also mean that vacant, colonizable surfaces occasionally crop up. These are the *r*-selection circumstances that would seem to explain the production of relatively large numbers of relatively small young. But rolling boulders also impose a quite different selection pressure: Large size and thick shells confer some protection against being crushed. Thus, it is not surprising that the boulder-dwelling winkles delay reproduction in order to grow thicker shells and a larger total size. The atypical relationship between maturation and parental allocation of resources in these periwinkles seems attributable to ecological peculiarities.

Not all departures from the usual strategic types can be so neatly explained away. We shall consider one more example that is considerably more perplexing. This is the relationship between adult size and parental brooding behavior in marine invertebrates. We might expect to find, in comparisons within families or genera, that the larger species would exhibit more intensive or prolonged parental nurture (see Table 8-1). Here just the opposite is true. The smaller species are the ones that brood their young, and this is not just a local anomaly but a major trend that has been documented within each of a large number of taxonomic groups of marine invertebrates ranging from oysters to octopi. A recent review of this problem [585] lists a number of hypotheses that have been offered to explain these trends. One hypothesis is that egg-production capacity increases faster with body size than does brooding capacity, making brooding impossible for larger species. Another hypothesis is that small adults may be better able to disperse *as adults,* hence derive less advantage from dispersal as larvae, hence stay and are brooded. And there are several more possible explanations. Unfortunately, no single hypothesis seems to fit all taxonomic groups, and so current opinion is that different explanations apply to different taxa [585]. But we are left wondering why the direction of the correlation is so consistent. A reverse correlation (large size with parental brooding, as we might initially have anticipated to be typical) has apparently never been documented in any group of marine invertebrates.

THE FORMS AND FUNCTIONS OF PARENTAL CARE

We have hitherto discussed parental strategies primarily in terms of the parent's capacity and inclination to invest. But the optimal allocation of parental effort also depends upon the offspring's needs, that is to say, the extent to which parental care contributes to offspring fitness. It is therefore essential that we understand the functions served by parental care if we are to understand how parents budget reproductive effort. Sometimes the functions of parental care are poorly understood, as in the case of parental brooding in marine invertebrates. But numerous functions of parental care have been identified. We cannot review the subject exhaustively here; we shall simply describe a few examples from among the vertebrates in order to afford some appreciation of the diversity that exists.

In oviparous (egg-laying) animals, parental activities must first assure that the environment will be such that the eggs can develop. A first concern is with moisture and the chemical milieu. Oviparous animals in general, and vertebrates in particular, originated in water. Reminders of that origin persist. Amphibians must return to water to lay their eggs, and reptilian eggs must generally be buried in a moist place. The birds have made greater evolutionary progress; eggshells and membranes enable avian eggs to develop in dry air. This is quite a feat of design. Simply coating the egg with an impermeable covering would be no solution, since all eggs need to breathe. There is much metabolic work in the developmental processes of eggs, and they consume oxygen and excrete carbon dioxide just as adult animals do. The provision of optimal temperature for the eggs and their protection from predators are also basic parental concerns.

The upshot of all these various requirements is that there may be considerable parental activity even before oviposition in order to prepare optimal sites for the receipt of the eggs: Fighting fish build floating nests; turtles dig oviposition pits; crocodiles accumulate piles of vegetation to heap upon eggs; most birds construct elaborate nests.

Parental care may or may not persist after oviposition and after hatching. The maintenance of the egg's environment may or may not require continued attention. Many fish lay their eggs in little nest depressions on the bottom; one or both parents then attend the eggs and aerate them frequently by directing a stream of water over them with fin movements. Without this care the eggs would suffocate and die. Certain South American fish lay their eggs out of the water altogether, leaping up to spawn on overhanging leaves, to which they glue their eggs [344]. The parent then regularly splashes water up at the eggs to prevent their drying out. In any such case where eggs are attended, the parent can also provide them with some protection from predation. In many fish the eggs are brooded and aerated in the mouth of one or both parents, and in several of these species the young continue after hatching to enter a parent's mouth in response to parental signals of danger. Certain species of frogs exhibit surprisingly elaborate parental care [657], sometimes brooding eggs in special pouches, carrying eggs and tadpoles to water; and, in at least one case, feeding tadpoles [665]. There is even a species in which the young are brooded through the tadpole stage inside their mother's stomach, and are then "born" orally [626]!

Birds' eggs must be incubated in order to develop [165]. This is almost always achieved by the parents' sitting on the eggs and thereby warming and insulating them with their own bodies and feathers. At least one bird, however, the mallee fowl, incubates its eggs as crocodiles do: A compost pile of rotting vegetation is heaped atop the eggs, and they are incubated by the heat of decomposition. Even this organic powerplant requires some parental attention if it is to run smoothly. The mallee fowl checks the temperature by plunging its beak into the pile; then it turns, adds, or removes material accordingly [204].

Many birds—including songbirds, seabirds, and birds of prey—have to be fed by the parents after hatching. Only gradually do they develop the capacity to feed themselves. But in others, such as ducks and gamebirds, the precocious young feed themselves, though they may still have some food presented to them by the mother, as is the case among domestic chickens. Also, they are still likely to be led to food and to be generally escorted by the parents, as are young ducks and geese. Here, as in the care of eggs, parental care serves multiple functions, and protection from predators may well be more basic than guidance to food.

Maternal care in mammals is of course extremely intensive, involving both placental nurture before birth and suckling after it. Historically, milk came first; it is in fact a defining characteristic of the class Mammalia. In the few surviving egg-laying mammals, namely the platypus and echidnas of Australia, milk is secreted from mammary pores; discrete teats are a more advanced specialization. We have already briefly discussed the physiology of lactation (p. 175). In most familiar mammals the young are gestated inside the mother's body through an ingenious system of placenta and umbilicus, which connects the embryo to the mother's physiological systems by keeping the two circulatory systems separate but in sufficiently close contact to exchange dissolved materials. This elaborate mechanism brings oxygen and nutrients to an embryo that is in no position either to breathe or to eat, and carries away its wastes, all of which permits prolonged internal development in a safe and stable environment. This advanced placental system is absent in the marsupials, but this group, which includes kangaroos, koalas, wombats, and opossums, can hardly be said to be less specialized: Their young are born at a relatively early embryonic stage and then suckled while sheltered inside a maternal pouch. We shall consider some of the very interesting physiological mechanisms of kangaroo reproduction on p. 220.

Parental care serves multiple functions, and these functions tend to proliferate in evolution. We can illustrate how and why functions proliferate with an example of a derived secondary function of maternal care—

the stimulation of elimination. In several small mammals the young are confined for some time after birth to a maternal nest in a burrow or den. The mothers do a great deal of licking and grooming of their nursing pups, particularly their hind ends, cleaning up the helpless pups' urine and feces in the process. Experimental attempts to rear some young mammals, such as kittens and rat pups, without their mothers, often failed before it was understood that the infants were unable to defecate and urinate until stimulated in a manner similar to their mother's licking. It is easy to imagine how this dependence of elimination upon an external stimulus could have evolved. Given that maternal care and confinement to a burrow nest already existed, any antiseptic behavior by the mother, such as cleaning the pups, would be of value in disease prevention and would enjoy a selective advantage. This advantage would then surely be augmented by the pups' eliminating only when licked and hence never fouling the nest. Thus could the observed phenomenon of obligatory stimulation of elimination arise.

No one would argue from this example that the stimulation of elimination is a function for which mammalian parental care evolved. But there is no question that it is now a function of parental care in some mammals, inasmuch as eliminatory failure is a cause of death when a rat pup is deprived of its mother. The function is secondary in two senses: First, it evolved after other selection pressures had already brought about the evolution of maternal care; second, pups deprived of their mother at an age when they could not yet eliminate independently would die anyway for other reasons. Maternal stimulation of elimination does not remain obligatory past the age at which pups could otherwise survive without their mother. Only then, when pups have a chance on their own, could there be any selection pressure against the necessity for maternal stimulation of elimination if the mother were lost.

Such secondary functions are not the ultimate causes of parental care. The true ultimate causes of a phenomenon are the selection pressures that were crucial to its original evolution. These may be called primary functions. It is not always easy or even possible to tell primary and secondary functions apart. One clue is the generality of the function. Suppose that parental care is important for some function such as the development of food recognition by the young of one species, but in related species isolate-reared young can achieve the same end without parental care. We would then strongly suspect the recent secondary evolution of that particular function of parental care. We can propose with some confidence that the primary functions of parental care are protection of the young from predators and from other environmental exigencies such as temperature shock, and that other functions are secondary.

Once intensive parental care has evolved—as, for example, in the first mammals—the parent-young interaction becomes a dependable part of the total situation within which natural selection operates. All future evolution —as, for example, the radiation of the mammalian classes into every manner of habitat and life-style, including even a return to the sea—must then occur within a selective context that always includes this early parental care. It is hardly surprising, therefore, that virtually every characteristic of every mammalian species is inextricably bound up in a web of complex causality with the fact of maternal care. This being so, a list of the functions of parental care is limited only by our capacity to enumerate them. Internal gestation provides a haven from predation risk, from cold, from ultraviolet radiation. Milk not only nourishes the infant; it serves immunological functions as well. The parental environment is an essential part of the species-typical milieu for the development of all infantile behavior in any animal with intensive parental care. Disruption of that environment can disrupt virtually any aspect of development. In a sense, then, the entire process of development is a function of parental care. But this proposition is too general to direct research. A more accessible question is that of the costs incurred by parents.

ENERGETIC ASPECTS OF PARENTAL CARE

Parental expenditure of energy can sometimes be assessed by direct methods, as we saw in the house martin study employing the doubly labeled water technique (p. 197). It is also of interest to experimentally manipulate the energy and other nutrients available to breeding animals and to observe how they budget limited resources. If pregnant mice are switched to a poor diet, for example, they are able to resorb their embryos and await a more propitious breeding opportunity [513]. Rats often become pregnant again in a *postpartum estrus* period immediately after birth, and hence nurse one litter while gestating another; if food availability is then restricted, the mother can concentrate resources in the first litter while delaying the implantation (see p. 220) and development of the second [693].

Animals seem to be "programmed" to cope with the physiological expenditures that reproduction demands. It is very common, for example, to fatten up or otherwise lay in stores in anticipation of the breeding season. The reproducing animal may then expend those stores according to a rather fixed schedule. An interesting example comes from the study of weight loss in incubating birds.

We have already described the seventeen-day fast and 20 percent weight loss that are normally endured by an incubating grey-faced petrel (p. 186). There are in fact many species that perform similar feats. A male Adélie penguin fasts for over a month while hiking across ice to the breeding sight, mating, and taking the first turn at incubation; in that time the male loses almost a third of his body weight, almost all the loss being fat that was laid on before breeding [319]. In gamebirds and waterfowl, it is common for the female to perform the entire incubation herself, and she remains there almost constantly, for if she moves, the ground nest is highly vulnerable to predation. Female eider ducks have been reported to lose as much as 45 percent of their body weight during egg-laying and incubation, and it is perhaps not surprising that as many adult females die during the two-month breeding season as in the remaining ten months of the year [341].

A precipitous weight loss is usually enough to motivate an animal to search for food, but incubating penguins, petrels, and eiders generally sit tight (although the most stressed may indeed abandon the nest). How is the regulation of energy balance modified during incubation? David Sherry of the University of Toronto has studied this question in the Burmese red junglefowl, the ancestor of the domestic chicken [562]. Even in a laboratory cage, a junglefowl hen sits on her eggs almost continuously for a twenty-day incubation period, reducing her food intake to about a fifth that of nonincubating birds. If food is placed so near that she need not leave the nest to eat, she still doesn't eat much and still loses weight steadily. If totally deprived for part of the incubation period, however, the hen eats voraciously when food is returned, so hunger must not simply be switched off. In fact, the temporarily food-deprived hen then eats just enough to increase her body weight to match that of control hens at the same stage of incubation who have had food freely available throughout (Figure 8–9). It appears that the weight loss during incubation reflects a programmed decline in the bird's target body weight or level of body fat. A similar sort of change in target weight characterizes stags defending a harem, hibernating squirrels, and molting penguins [445]. In all these cases, feeding behavior would be detrimental to fitness, either by reducing reproductive success (risking predation of the hen's eggs or loss of the stag's harem), or by costing even more energy (expended by hibernators in warming up to feed and by molting penguins in heat lost to cold water). The physiological solution in each case is to modify the animal's target energy level so as to avoid detrimental conflict between feeding and other activities.

Most studies of parental resource partitioning concentrate upon energy, simply because body weight and caloric content of food are readily measured. It is probably often the case, however, that other specific nutrients limit reproductive capacity. In the guppy, for example, a viviparous (live-bearing) fish, increasing the proportion of protein in the diet induces the female to produce ova that are both more numerous and larger [126].

Effects of specific nutrients can be more subtle still. In one study [148], lactating rats were placed on a calcium-deficient diet, and both the

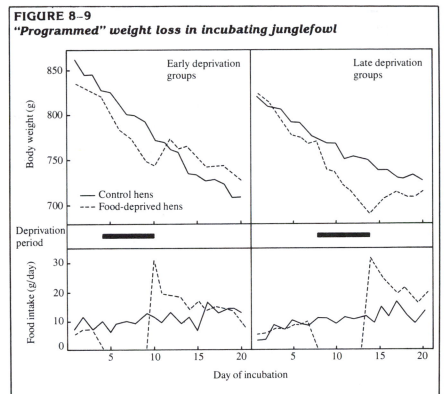

FIGURE 8-9
"Programmed" weight loss in incubating junglefowl

During the twenty-day incubation period, "control" females with ad lib food lost about 15 percent of body weight. "Deprived" females had all food removed for six days (bars), early (left), or late (right) in incubation. When food was restored, deprived females ate only about enough to return their body weight to the control weight-loss trajectory. (After Sherry, 1981, Figure 1 [561].)

pups and the mothers appeared to fare well. Only by the method of x-ray spectrophotometry, a technique that can be used to assess certain mineral contents in tissues, could it be seen how the mothers had coped. During lactation, they had sacrificed half the calcium in their own skeletons in order to produce milk for their pups! Returned to an adequate diet, the mothers took three weeks to recover to 78 percent of initial skeletal calcium. By all overt signs they were fully recovered, but in fact they were still severely depleted. As we remarked in the last chapter, the extent to which such covert maternal depletion of scarce essential resources might occur in natural populations, and might thereby affect the quantity or quality of subsequent progeny, is almost entirely unknown. Our ignorance of these physiological costs of reproduction constitutes a serious constraint upon our abilities to describe and explain maternal strategies.

SEASONALITY

Fat deposition, migration, territorial and sexual behavior—all these are components of strategies whose object is successful reproduction. Their timing should therefore be decided according to the demands and dangers inherent in the various stages of the reproductive cycle. This is clearest in highly seasonal habitats. Birds breeding in the Arctic must get their young fledged in time for the fall migration. Carnivores want the maximum energy demands of their litters to coincide with the best hunting season.

A number of environmental factors act upon the reproductive mechanisms of animals to ensure that breeding takes place when conditions are propitious [299, 540]. Different species exploit different cues, according to which is the best predictor of future availability of crucial resources. Daylength is an effective stimulus for a great many organisms, although its particular effects vary. Increasing daylength activates gonadal function in many animals that breed in the spring and summer. In rhesus monkeys [635], red deer [387], and sheep [196], species that mate in the fall in order to give birth in the spring, decreasing daylength appears to have the same effect. The widespread use of daylength as a temporal cue depends on the fact that it is a precise function of the time of year and hence is as reliable a predictor of the seasons as is available.

Terrestrial organisms rely less on temperature than they do on light, and there is a simple reason why this should be so. Today's temperature is a very poor predictor of anything about tomorrow, let alone next week or month. Any animal that comes into breeding condition in response to the early spring attainment of some temperature threshold, for example, may then find herself with youngsters to feed before much food is available. Or she may wait too long, if the particular weather pattern of that spring brings warm temperatures later than usual. Nevertheless, there is at least one species in which a temperature effect has been demonstrated. In *Anolis carolinensis,* a common American lizard, daylength and temperature have interactive effects upon reproductive condition: Testes grow faster in the spring if the temperature is consistently warm than if it fluctuates, and they regress relatively rapidly in late summer if temperatures are not stable [381].

Water temperature, unlike air temperature, is a good seasonal cue. Bodies of water warm up slowly and steadily in the spring in contrast to the greater fluctuations of air temperature; the water temperature that has been reached by a certain date is probably even better than daylength as a predictor of the future availability of nutrients, whose growth depends partly on temperature. Accordingly, aquatic animals rely upon temperature as a proximate cue controlling reproductive physiology [e.g., 469] to a much greater extent than terrestrial animals, although daylength may also be important [e.g., 383, 491].

Other cues can be used too, particularly where the reproductive strategy is more opportunistic than seasonal. In many deserts, for example, rainfall is crucial if there is to be a food supply adequate to permit breeding, and yet the timing of a rainy season is not reliable. In such cases the animals may cue upon the rainfall itself. Australian desert birds have been observed to begin courtship within minutes after the end of a long drought [298]. Likewise, frogs begin to sing in semiarid central California almost from the outset of a shower. In rapid-breeding desert animals, such as kangaroo rats and other rodents, it is probably more common for the appearance of new vegetation after the rains to be the immediate causal stimulus for ovulatory activity. There is then a good supply of green food for pregnancy and lactation and a new seed crop by the time the young are weaned [520].

Animals outside of deserts sometimes cue upon moisture and vegetation too. In the lizard *Anolis sagrei,* increase in humidity stimulates ovarian development; this cue is generally a consequence of rainfall and a predictor of abundant insects for food [120]. In goldfish, long days stimulate ovarian development, but the floating vegetation in which the fish spawn is itself the primary environmental cue triggering ovulation [491]. In the vole *Microtus montanus*, substances ingested in fresh green sprouts act as chemical triggers of the vole's reproductive development, allowing the animal to anticipate an imminent abundance of herbaceous food [44].

OPTIMAL TIMING: THE FRASER DARLING EFFECT

A population of animals can often be observed to produce their young at the same time. This is largely because the individual animals are using the same environmental cues in the same ways to predict the time of optimal resource availability. But there is quite another selective pressure that can favor synchrony per se, namely predation. A helpless young animal, newborn or hatched, may be an especially likely candidate for predation when there are few young in the vicinity. When there are many, they may derive some safety from numbers. The local predators are swamped by a prey species that breeds in great density and so can take relatively few young during their most vulnerable stages. Synchronization of breeding by social stimulation is called the Fraser Darling effect after the naturalist who first suggested that its adaptive value lay in the overwhelming of predators [136].

Wildebeest cows bear their calves in the midst of huge herds. Hyenas and other predators follow the herds and prey upon them heavily, but there are only a brief few days when very vulnerable calves are available. Thereafter, the predators must find other food. An individual calf's chances of surviving that dangerous period are fairly good [183].

A study of black-headed gulls by Oxford zoologist I. J. Patterson [486] provides a particularly fine demonstration of the predator swamping effect. These small gulls nest on the ground in tightly packed colonies that contain up to several thousand pairs of birds. Eggs and chicks are consumed in great numbers by foxes and other predators, so that most of the gulls fail to fledge any young at all. The only birds with any prospects of success are those that breed in synchrony with many others. Most pairs lay within a single week; all do so within a month. Those that lay outside the peak time for the colony soon lose their chicks (Figure 8–10). Direct observations con-

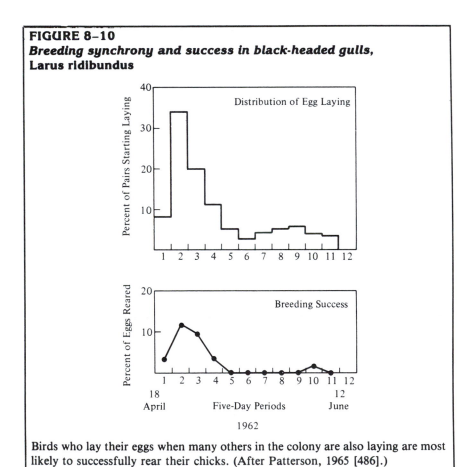

FIGURE 8–10
Breeding synchrony and success in black-headed gulls,
Larus ridibundus

Birds who lay their eggs when many others in the colony are also laying are most likely to successfully rear their chicks. (After Patterson, 1965 [486].)

firmed the predators' role in this loss. In such circumstances there is a powerful selection pressure upon each breeding pair to lay their eggs and hatch their young in synchrony with as many of their neighbors as possible.

One of the proximate mechanisms by which colonial birds achieve such synchrony is group stimulation of the ovarian development of the colony's females. In many such birds, including gulls, weaver finches, pigeons, and pelicans, egg-laying occurs with a high degree of synchrony in a colony or one area thereof, whereas dates of oviposition vary among colonies or areas within them. This suggests that the birds are synchronized by social stimuli rather than by merely responding similarly to the same environmental cues. In ring doves the facilitation of the reproduction of a mated pair by stimuli from neighbors has been demonstrated experimentally [398]. It has further been suggested that the extinction of the American passenger pigeon may have been due in part to the population's falling so low, as a result of hunting pressure, that adequate group stimulation of reproduction could no longer be maintained.

Social stimuli can produce synchronization of reproductive processes in mammals too, although the adaptive significance of such effects is by no means clear. Group-maintained female mice exhibit a disruption of their estrous cycle (p. 216) in comparison with solitary females (the Lee-Boot effect). If the odor of a male is introduced, the females come into the state of sexual receptivity simultaneously (the Whitten effect) [68, 667]. This synchronization is not known to occur in wild mice, but there are similar phenomena in natural populations of other mammals. It is not known if the synchronous delivery of wildebeest calves is partially dependent upon social stimuli. Similarly, in wild lemur troops [320] all females come into heat at once, and births occur within a remarkably short interval; this seems to be as true between neighboring troops as within troops, so the effect may be explicable as a simultaneous response to a common environmental cue rather than a social effect.

Surprisingly, there is a similar synchronization in our own species, and its proximate causation depends on social factors. It is commonly observed that the menstrual cycles of women living together come to coincide over time. One careful study of this phenomenon was conducted by Martha McClintock at an American women's college in 1969–1970 [425; see also 224]. Over the course of the academic year the menstrual cycles of roommates and close friends converged until they were very often coincident. Evidently the cycles came into step with one another as a result of frequent personal contacts between the women, not because of any extrinsic cue in common. Nor did the convergence depend upon explicit knowledge of the friend's or roommate's cycle. McClintock suspects that odor cues are relevant, but both the causal mechanism of the phenomenon and its functional significance, if any, remain to be discovered.

OPTIMAL TIMING: NONPREGNANT CYCLES IN
FEMALE MAMMALS

If a female mammal misses an occasion to become pregnant, then she may undergo a series of cyclic events in order to return to a fertile state. The cycle may take anywhere from a few days to several weeks, according to the species.

In the majority of mammals the nonpregnant ovarian cycle is manifested in the estrous cycle [490]. The word "estrus" refers to a relatively delimited period of female sexual receptivity, a stage defined by that receptive behavior rather than by hormonal or other physiological factors. Estrous behavior is generally restricted to a brief time interval: A female rat, for example, will reject the male on all but about one half of one day out of her four- to five-day cycle, and a cow is in heat for only about half a day out of every three weeks [389]. With few exceptions, the brief period of estrus corresponds more or less precisely with the timing of ovulation and hence with the fertile period. The female is interested in mating at that time when a mating could prove fertile; she is uninterested and moreover is usually unattractive to males, when a mating would prove fruitless for want of ripe ova.

Nonpregnant cycles have been much studied in the laboratory [187, 389, 493] where they are of great interest and value for investigating the causal connections of successive events in the reproductive physiology of mammalian females. However, successive cycling of this sort is probably a rarity in nature. In most mammalian species a very high proportion of females become pregnant when conditions permit. Where food is inadequate for reproduction, cycling is apt to be disrupted. Animals do not lightly allow reproductive opportunities to pass them by, and nonpregnant cycles are an unproductive luxury. A missed reproductive opportunity means lost time, and this can entail real fitness penalties. The whole elaborate synchronization of female reproductive physiology is directed toward successful impregnation at ovulation [656].

In each mammalian ovarian cycle [493], one or more follicles grow from a small inert group of cells into a large mass secreting estrogen. Shortly before the ovum is released from the follicle (ovulation) its meiosis resumes. Ovulation itself is triggered by a surge of the pituitary hormone LH, which results either from the attainment of a threshold level of blood estrogen ("spontaneous ovulation"; see below) or from extrinsic stimulation. After ovulation, the follicle may be transformed into a structure called the corpus luteum, which secretes progesterone, a hormone important for implantation of the conceptus in the uterus and for the maintenance of pregnancy. Sexual receptivity is generally confined to a part of the cycle near the time of ovulation.

TABLE 8-2
Types of non-pregnant cycles in female mammals

	First type	*Second type*	*Third type*
Ovulation	spontaneous	spontaneous	induced
Corpus luteum formation	spontaneous	induced	induced
Characteristics	relatively long repeated cycles (several days or weeks)	relatively short repeated cycles (less than a week)	prolonged estrus until mating occurs
Examples	dogs, goats, people	rats, mice, hamsters	cats, camels, rabbits

Mammalian nonpregnant cycles can be usefully categorized into three basic types according to the factors that control ovulation and corpus luteum formation (Table 8-2). In the first type, both ovulation and the formation of a functional corpus luteum occur "spontaneously" in the absence of copulation. The uterus must then inform the ovary if no fertile conceptus has been implanted so that the ovary can mature another egg or eggs. Such a cycle is characteristic of dogs, cows, guinea pigs, goats, monkeys, and people [111, 389].

In the second type of cycle, ovulation again occurs spontaneously, but a functional corpus luteum does not form unless mating occurs. This type of cycle, evidently far less widespread than the first, is characteristic of rats, mice, and hamsters.

In the third type of cycle, ovulation and luteal function both occur as responses to copulation. Animals with this sort of cycle are commonly called induced ovulators, by comparison with the spontaneous ovulators of the preceding cycle types. In induced ovulators—an odd assortment from several mammalian orders including rabbits, ferrets, cats, opossums, lemmings, voles, and camels—the egg is shed as a result of stimulation from the act of mating [e.g., 112]. In the absence of such stimulation, the estrous state may persist for days or even weeks until mating occurs and ovulation is induced.

These three different cycle types demand explanation at the level of reproductive strategies [111]. The dimension of *r*- versus *K*-strategy seems to be relevant to the differences between the first two types of nonpregnant cycle. Ovulation is spontaneous in both, but they differ in whether a functional corpus luteum develops spontaneously or must be induced by mating. In the former case the cycle tends to be longer, and the female is slower to return to estrus if she misses the occasion to mate. This is because the luteal phase is initiated even in the absence of pregnancy, and it must be terminated before the next ovulation can occur. This termination awaits a signal from the uterus that no conceptus has implanted. Thus, the ovary has to wait longer than the species' normal implantation latency before it is known for certain that pregnancy has not occurred. On the other hand, where a functional corpus luteum does not form in the absence of mating, a return to estrus can occur quite soon. The lengths of the estrous cycle in rats, mice, and hamsters are all four to five days, whereas the guinea pig cycle with spontaneous luteal function takes sixteen days. Rats, mice, and hamsters are all relatively *r*-selected small prey animals. Their short cycle appears to be an *r*-strategy: If they somehow miss out on mating in one estrous period, they get back into estrus rapidly. (Neither cycle type affords protection against sterile mating. In both, the likely result is a luteal phase or "pseudopregnancy.")

The most *r*-selected of all mammalian species are the voles and lemmings. Their huge population fluctuations are unrivaled [342]. They are the only rodents known to have a cycle of the third type—induced ovulation. And it would appear that this can be a still more extreme *r*-strategy than the short cycles of other small rodents. By remaining in estrus until mated, a vole or lemming that misses a mating opportunity today is still receptive tomorrow. Lemming and vole populations regularly increase explosively after a crash. At minimal density, as the population growth begins anew, the situation of pure *r*-selection commences: It is a race among the survivors to see which will contribute most to the ensuing explosion. And it is at that minimum, with the animals unusually scarce, that a female might fail to contact a male on her estrous day. She would then gain in the race by being continuously receptive rather than wasting even the few days required for a rat-type cycle. In the explosive phase of these animals' population growth, even days are crucial in determining an individual's eventual genetic contribution to the peak population.

Thus, among the rodents in which all three cycle types occur, the *r-K* dimension nicely matches the observed distribution of cycle types. It remains to be explained, however, why any species should have a slow cycle. Why not all be like lemmings? We cannot answer this with certainty, but we can suggest that there must be some costs associated with frequent or continuous estrus. Estrus may be an energy-expensive physiological state; it may entail a heightened predation risk too, since estrous females commonly increase their roamings. (Cyclic rather than continuous receptivity in unmated females is not peculiar to mammals [384, 492].)

Induced ovulation is a tactic of other mammals than just voles and lemmings. The same explanation of extreme *r*-selection may be applicable to rabbits and hares. (Indeed, the population cycles of hares are surpassed in amplitude only by those of voles and lemmings.) But we need a different explanation to account for the many carnivores that also exhibit this type of cycle.

Carnivores live at lower densities than herbivorous animals; if most predators were not less numerous and more spread out than their prey, they could not find enough to eat. A solitary female carnivore may not encounter even one potential mate on any given day. In solitary carnivores, such as mink, ferrets, and several other members of the family Mustelidae, the female is an induced ovulator that remains in heat for several days until mated. We suggest that the strategy of induced ovulation is therefore an adaptation to solitariness in a low-density species. Several facts support this idea. Only one mustelid carnivore lives with the sexes in permanent association, the group-living European badger [348]. She also turns out to be the only spontaneous ovulator known in the family. Other carnivore families also exhibit the expected cycle types. Solitary cats are induced ovulators. Dogs, descendants of the very sociable wolves, are spontaneous ovulators, and so are pair-forming foxes. Lions do not match our hypothesis, however, for they are sociable cats and yet exhibit induced ovulation like their solitary relatives.

Camels are unusual among ungulates in being induced ovulators. The explanation for this anomaly may be the same as that just proposed for solitary carnivores: Camels wander over vast distances in small groups that do not always contain a male. Induced ovulation may again be an adaptation to low density and irregular contact between the sexes.

OPTIMAL TIMING: DELAYED IMPLANTATION

A sophisticated timing device in the reproductive physiology of many female mammals is the suspension of offspring development at a preimplantation stage [684]. This capacity to delay implantation permits the female to adjust the interval between mating and birth. Delayed implantation, which has presumably evolved independently several times, has two main uses: First, it allows females of some species some flexibility in the temporal spacing of successive births; second, it allows females of certain annually breeding species to stretch out their complete reproductive cycle so that it is in harmony with a seasonal pattern of activities.

The first strategy—that of birth-interval adjustment—is exemplified by the many mammals that ovulate immediately after giving birth, then become pregnant during the postpartum estrus that accompanies this ovulation [e.g., 113]. The female may delay implantation, thereby halting the second pregnancy at an early stage to resume it at her leisure. The time schedule from implantation to birth is invariant, but the adjustable latency of implantation allows the mother to distribute her own reproductive effort optimally. The larger the litter she is nursing, for example, the later she implants. If food supplies are good and the present lactation is not exacting too great an energetic price, then she implants relatively early. If not, she waits [693].

A variant on this timing strategy is practiced by kangaroos. The first young, or joey, is born at a very early stage of its development and fastens itself to a teat in the mother's pouch. As long as it remains there nursing, the next conceptus remains in a state of suspended animation (diapause), alive but developing no further than the 70- to 100-cell blastocyst stage. The suckling joey prevents the implantation of the blastocyst by means of a neuroendocrine reflex from teat to pituitary to ovary. When it begins to relinquish the nipple, the blastocyst implants and resumes development. The older joey is not fully weaned until long after the second is born, however, and the mother nurses both, each from its own teat. It is a remarkable fact that the mother feeds milk of differing compositions to the two joeys, each mixture appropriate to the joey's stage of development. If a joey is lost early in the nursling stage, the reserve blastocyst can be implanted immediately. Thus, the mother kangaroo loses considerably less time than she would if she had to start a new reproductive cycle all the way back at ovulation and conception. In the swamp wallaby (*Wallabia bicolor*), a close relative of the kangaroos, the female actually comes into heat, mates, and conceives the second joey some two to three days before the birth of the first! Her two ovaries each have their own uterus, and successive ovulations alternate strictly between them. Thus a new embryo can be launched before the previous joey is born [556].

The second strategy—that of stretching the reproductive cycle to fit a seasonal schedule—is exemplified by seals. Females haul out on beaches only once a year; in that single visit to land they deliver this year's pup and conceive next year's. The active gestation period is only four to six months, but implantation is delayed for six to eight months. Most species that breed annually and employ this mechanism of delay time their birth season by implanting in response to a daylength cue.

The adaptive significance of delayed implantation in some other species is more difficult to discern. It is common but not universal in the carnivore genus *Mustela*, for example: Stoats, fishers, and mink use the device in order to mate in summer or fall and litter the following spring, whereas ferrets and weasels do not [684]. It may be suspected that the species that delay implantation are relative *K*-strategists by virtue of taking larger prey, for they are indeed the larger species of *Mustela*; they may therefore be practicing a less productive and more intensive parental strategy, but this is only a guess.

How organisms allocate reproductive effort over time has been the concern of this chapter, and we have seen great variability in the characteristic life histories of different animal species. Explanation of this diversity is an active area of theory and research. But there is a strategic aspect of reproductive effort allocation that we have hitherto ignored: How should parents partition that effort between daughters and sons? That is the subject of Chapter 9.

SUMMARY

A life history strategy is a species-characteristic adaptive schedule for the expenditure of reproductive effort over the lifetime. Semelparous organisms die after a single debilitating reproductive effort; iteroparous organisms save themselves to garner further resources and breed again. *r*-strategists mature early and produce many young; *K*-strategists mature later and produce relatively few, more intensively nurtured offspring.

There are trade-offs between survival and reproduction, between growth and reproduction, and between parental nurture and fecundity. The resolution of these trade-offs varies between species and between the sexes within a species. Contrasting life history strategies can sometimes be understood as adaptations to contrasting ecological circumstances.

Animals use various devices to schedule the expenditure of reproductive effort adaptively, including response to seasonal cues predictive of food resources, response to social cues that synchronize breeding, and physiological mechanisms that permit flexible control of the timing of reproductive events like estrus and implantation.

SUGGESTED READINGS

Stearns, S. C. 1976. Life-history tactics: a review of the ideas. *Quarterly Review of Biology,* 51: 3–47.

Williams, G. C. 1966. *Adaptation and natural selection.* Princeton: Princeton University Press.

9

Sex allocation

Females generally invest more in individual offspring than do males, as we saw in Chapter 5, so that males are often veritable parasites of female parental effort. Why males exist at all is a serious question, which we discussed in Chapter 4, where it was seen that the adaptive significance of sexual reproduction is still a subject of lively debate. But let us take sex and the female/male dichotomy as givens. The question remains why parents produce roughly equal numbers of female and male offspring. In animals as diverse as fruit flies and elephant seals, it is the number of breeding females, not males, that will determine the size of the next generation. Many females can be bred by a single male, whose only essential function is fertilization. So why not produce mainly females and cut down on superfluous males?

FISHER'S SEX RATIO THEORY

There is a rather simple answer, first perceived by the British biostatistician Sir Ronald Fisher [191]. Imagine a population in which females greatly outnumber males. Female X is about to become a mother. If she could choose her child's sex, what should it be? A male, for the following reason. Every future offspring in the population will have both a mother and a father. Females greatly outnumber males, yet each sex leaves the same number of total offspring, so the average male produces more offspring than the average female. This being the case, X will do better to have a son than a daughter, for sons will, on average, leave more offspring than daughters. If males are rare, a son should produce more grandchildren than a daughter. What this means is that there will be a selective advantage to any genotype that favors the production of males, and such selection will raise the proportion of males in the future progenies.

The argument is symmetrical. The sexes in the last paragraph can be reversed without changing the logic: If females are rare, there will be a selective advantage for female-producing genotypes. We must conclude that the rarity of either sex will automatically set up selection pressures favoring the production of the rare sex, thus driving the **sex ratio**—the number of males produced divided by the number of females—back toward unity.

This conclusion is in no way altered by the fact that one male can serve many females. Suppose that there are ten females and ten males and that each female produces two offspring. One male sires all twenty, while the nine other males die without issue. The *average* production per male is two offspring, just as in the females. If the events are typical, the average expected payoffs are two grandchildren per son and two grandchildren per daughter. Although a daughter is more likely to breed than a son, a few sons will have greater reproductive success than any daughter, and the average payoff for sons and daughters is the same. It is only when the sexes are produced in equal numbers that there is this equilibrium, in marked contrast to the selection pressures provoked by unequal numbers, and hence unequal average reproductive success, of the two sexes. We conclude that parents should produce sons and daughters with approximately equal likelihood, and this expectation is abundantly confirmed.

It may be protested that this is an unnecessarily devious explanation for a simple phenomenon. Chromosomal sex determination (Figure 9-1) and the 50:50 coin toss of meiosis guarantee equal numbers of females and

FIGURE 9–1
Chromosomal sex determination

MALE HETEROGAMETIC

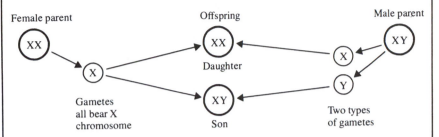

Female parent | Offspring | Male parent

Gametes all bear X chromosome

Daughter

Son

Two types of gametes

In an XY system (characteristic, for example, of almost all mammals), males are the *heterogametic* sex, whose two types of gametes determine the sex of the off-spring. Females are the *homogametic* sex. The X and Y chromosomes are usually quite distinct in size and structure, and they do not exchange material by cross-over during meiosis. Typically, the Y chromosome is much smaller and contains few genetic loci.

When the female is heterogametic and the male homogametic (as, for example, in all known birds), then the sex chromosomes are referred to as Z and W.

FEMALE HETEROGAMETIC

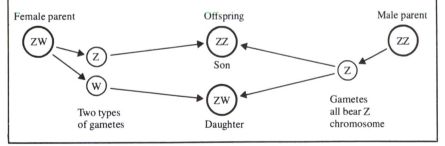

Female parent | Offspring | Male parent

Son

Two types of gametes

Daughter

Gametes all bear Z chromosome

males. However, not all organisms have sex chromosomes, as we shall see shortly, and even in those that do, sex ratios often depart from unity in ways that appear to be adaptive.

Fisher's formulation of the sex ratio problem was actually more general than we have yet indicated. He began by noting that parents invest "a certain amount of biological capital" [191, p. 158] in each offspring, and furthermore that the total reproductive value of all the male young in the entire population must be equal to the total reproductive value of all the female young, "because each sex must supply half the ancestry of all future generations" [191, p. 159]. From this reasoning, Fisher concluded that parental strategies ought to evolve toward *equal investment* in offspring of the two sexes. This does not always mean equal numbers, since daughters and sons may be differentially costly to produce.

Suppose, for example, that two males can be produced with the same effort that it takes to produce one daughter. Then parents would do better to raise sons rather than daughters, *unless* a daughter's own reproductive prospects are as good as the reproductive prospects of both sons put together; in such a case, one female would cost as much as two males but would also pay off, in grandchildren, as much as two males. Costs and payoffs will reach this equilibrium when males are exactly twice as numerous as females. At that sex ratio, each male will on average leave only half as many offspring as each female, but we have already specified that he is only half as expensive to produce. In general, the unbeatable parental strategy is to invest equally in daughters and sons, and a sex ratio of unity is the special case of this general strategy when daughters and sons are equally costly. So it is not the *sex ratio* itself that should equal unity according to Fisher's theory, but the **ratio of parental investment in the two sexes**.

As Fisher noted, the sexes are not produced in equal numbers in our own species, and the departures may be explicable by the equal investment argument. Throughout the long period when people invest in their children —including pregnancy, nursing, and subsequent nurture—boys are likelier to die than girls. When they die, the period of parental investment is of course curtailed. If parents are to invest equally in sons and daughters, they should *start more boys,* since boys will, on average, use up less parental investment than girls. This is just what happens: In early pregnancy human males outnumber females six to five. By birth this disparity is already greatly reduced. (There is reason to suspect that females actually outnumber males right at conception, and that this difference is reversed almost immediately, owing to differential mortality of female and male zygotes [295]. This does not affect the argument, however, since parental investment of time and energy up to this stage is relatively minor.)

PARENTAL FAVORITISM ACCORDING TO OFFSPRING SEX

It does appear that something more is going on in the determination of sex than just a Mendelian toss of a coin. Parents may have no control over the sex of zygotes at fertilization, but they can certainly do things to modify the sex ratio later. They might choose to nurture one sex and abandon the other, for example, although this sort of strategy seems a poor one unless offspring sex can be detected at a stage when the parental investment already expended constitutes a trivial proportion of the total required to bring the offspring to maturity. In a fish with little post-zygotic investment, for example, it would make little sense to selectively eliminate young of one sex. But in a mammal with prolonged gestation and postnatal care, there are circumstances in which such discriminative behavior might be worthwhile.

Fisher's argument for equal parental expenditure upon daughters and sons describes an equilibrium circumstance for an entire sexual population. An addendum to the theory was introduced by Trivers and Willard [622], who noted that there are circumstances in which *individual* parents might profit by biasing the sex of their own offspring. Recall that males of polygynous species exhibit a greater variance in fitness than do females, with both a higher ceiling on reproductive success and a higher probability of total failure (pp. 151–152). It follows that a given increment in the viability or competitive skill of an individual male may lead to greater fitness gains than a comparable increment in the viability or competitive skill of an individual female. From the parent's point of view, extra investment in a son of good quality may yield greater returns than comparable investment in a daughter of good quality. A mother might therefore prefer to produce sons when she has the resources necessary to give them a better-than-average competitive ability, and daughters when she does not.

At first sight, the implications seem clearest for those polygynous species in which litters are small and parental care is prolonged. Indeed, Trivers and Willard particularly had in mind mammals who raise a single offspring annually. Among deer, for example, they suggested that a doe in good condition might produce a male fawn while one in poor condition produces a female. In fact, there is very little evidence for such effects [454]. However, mammals that produce a single offspring annually are probably just the sort of organisms that are constrained *not* to be able to modify offspring sex ratios adaptively: The chromosomal sex determination mechanism prescribes fetal sex so that a female could probably adjust the sex ratio of her total progeny only by aborting or neglecting certain offspring and thereby passing up entire seasons. (In deer, however, offspring sex is strongly related to the day of the female cycle on which conception occurs, with early matings producing mostly daughters and late matings mostly sons [641]. Whatever the mechanism for this effect, it would appear to offer does a means to control offspring sex, which makes the lack of evidence for adaptive modification of offspring sex in deer the more disappointing.)

It is in organisms with nonchromosomal sex determination mechanisms, as we shall see shortly, that there is clear evidence for elaborate strategies of adaptive control of offspring sex. Even in species with XY or ZW systems (Figure 9-1), however, there is still the possibility of sex-biased culling of broods. Thus, it is in species that produce litters or clutches larger than a singleton that we might better seek evidence of parental adjustment of offspring sex ratio. A recent study of the Florida packrat provides just such evidence [426]. In laboratory cages, well-fed control females invest equal amounts of energy in sons and daughters, rear them in equal numbers, and wean them at the same weight, although the males grow larger in adulthood and are apparently dependent upon large size for success in aggressive competition over females. Mothers who were food-deprived during lactation, unlike the control mothers, channeled 68 percent of transferred energy into daughters and only 32 percent into sons, with the result that many male pups died and those that survived grew more slowly and were weaned at lighter weights than their sisters. This discrimination against male pups was observed to result from direct rejection: Mothers suffering food stress removed sons from their nipples while permitting daughters to suckle.

Even with litters of one, it might sometimes pay a mother to jettison an offspring of one sex in order to get on with a subsequent pregnancy sooner, especially if the offspring dependency is even longer than a season, as it is in some primates. It must be very unusual, however, for one sex to be so much more valuable than the other as to favor such a strategy; for one thing, the replacement pregnancy still has only a 50 percent probability of being of the preferred sex. (We are again supposing that the mother cannot detect the nonpreferred sex early and abort at low cost in time and energy. If she *can* so disciminate, then the story changes, but there is as yet no solid evidence for such abilities.) Sex-discriminative treatment of offspring, including sex-selective infanticide, is well established and widespread in human parents, however, and this is a topic about which we shall have more to say in Chapter 12.

HAPLODIPLOIDY AND THE CONTROL OF OFFSPRING SEX

Chromosomal sex determination limits parental options for adaptive control of progeny sex ratio, but other modes of sex determination can be considerably more amenable to parental ploys. We have already discussed the best-studied such mode, haplodiploidy (Figure 3-6, p. 49). Under this genetic system, which is characteristic of the immense insect order Hymenoptera and several lesser groups, males are haploid, females diploid, and offspring sex is determined either by fertilizing the egg (female) or neglecting to do so (male).

The distinction is not simply between mating and failing to mate (although virgin females may indeed lay male-producing eggs). A queen bee or other hymenopteran female stores sperm from past matings—sometimes many months past—in a storage organ called a spermatheca. When she is about to lay an egg, she may or may not fertilize it with a sperm from the spermatheca, and although some early authors thought fertilization rather a haphazard affair, it is now clear that females of at least some hymenopteran species have quite precise control over the fertilization of individual ova, as we shall see. There are circumstances in which that control proves valuable. Recall that Trivers and Willard suggested that mothers might profit by specifying offspring sex according to maternal condition. This proves to be a specific case of a much more general principle: Control over offspring sex will be of value wherever an environmental cue is indicative of circumstances differentially favorable for daughters and sons.

In many bees and wasps, including both social species with sterile workers and nonsocial species, the females are larger than the males. Mating takes place during a brief season, and males pursue fertilizations by scramble competition more often than by combat. Females carry out the elaborate patterns of nesting and brood care without male help. Large size apparently contributes to female fitness in several ways, especially by size-dependent fecundity and longevity; in some species large size provides an advantage in subduing large prey to provision offspring, or an advantage in female-female competition for nest sites or for primary rank in multiple-foundress nests. It is therefore probably quite general that female fitness is a more extreme function of size than is male fitness (though direct evidence on this point is rather thin [99]). The upshot of these various factors is that females are usually larger, and often several *times* larger, than conspecific males.

The size of an adult bee or wasp is dependent upon the food it receives in preadult stages. If the species is a parasite that lays its egg in a host animal or fruit, then the size of the host may dictate the size to which the parasite can grow. Often food is provided by maternal provisioning of a burrow, crevice, or artificial brood cell in which one or more eggs are laid. Thus a digger wasp paralyzes caterpillers by chemical injection and buries several in a burrow with an egg that will hatch to feed upon them; a solitary bee provisions a brood cell with a loaf of pollen moistened with nectar. The provisions required by daughters and sons often differ considerably, and where food is cached with a single egg, adult females can allocate resources effectively by provisioning two different types of cells and fertilizing only those eggs laid in the more heavily provisioned cells (e.g., Figure 9–2). Many species of bees and wasps construct series of provisioned cells within elongate beetle borings or other holes in wood or earth. The larger cells are usually toward the back of the nest, and there the eggs are fertilized to

FIGURE 9-2
Differential provisioning and control of offspring sex in bees

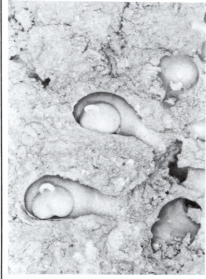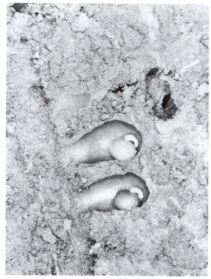

In the primitively eusocial sweat bee, *Evylaeus malachurus,* brood cells are provisioned with pollen balls that are eaten by the larvae. The large pollen balls on the left are for production of the large reproductive females; the queen therefore fertilizes all eggs laid in such cells. The small pollen balls on the right are for production of the little males, and the queen has not fertilized those eggs. Cells for worker production are provisioned with a small pollen ball and a fertilized egg, but the combination of a large pollen ball and an unfertilized egg does not occur. (Photographs by G. Knerer.)

develop into females, the males in front developing faster and emerging sooner. Besides laying fertilized and unfertilized eggs in consecutive series, females of these hole-nesting species have been shown to adjust progeny sex ratios facultatively. If holes of varying calibers are drilled in blocks of wood and offered as nest sites, adult females produce increasing proportions of female offspring the larger the diameter of the hole [e.g., 99, 346, 393].

In many parasitic species, such as the fig wasps discussed in Chapter 5 (p. 106), females do not provision a nest but simply oviposit in a host that provides all the nutrients for development. *Lariophagus distinguendus* is a tiny wasp that lays a single egg in a weevil larva, which is itself devouring from within a single grain of wheat. If individual wasps are offered a series of hosts of a single size, then their offspring sex ratio is a function of that size: Eggs laid in weevil larvae less than about 0.6 mm in length are all unfertilized and develop as males, and the proportion that are male drops to about 20 percent for weevil larvae longer than 1.1 mm. The maternal strategy is, even more reasonably, relative: When 1.4 mm hosts were offered successively, the wasps produced 15 percent male progeny, but 1.4 mm hosts offered in alternation with 1.8 mm hosts received 30 percent males, whereas 1.4 mm hosts offered in alternation with 1.0 mm hosts received only 2 percent males [99]. Evidently the female keeps track of the size distribution of available hosts and concentrates males in the smaller ones. At least a few other species of parasitic wasps exhibit similar behavior.

LOCAL MATE COMPETITION

An important extension of Fisher's sex ratio theory was proposed in 1967 by W. D. Hamilton [251], who noted that the prediction of equal investment in daughters and sons rests upon a hidden assumption that is actually rather improbable. This is that competition for mates is randomly distributed throughout the population. How this assumption is imbedded in Fisher's argument can best be seen by considering an extreme case where it does not hold.

In some species of mites, young reach sexual maturity inside their mothers' bodies; and in certain of these species, males complete their life cycle, fertilize their sisters, and die before they are born! The young females then parasitize their own mother, and emerge as adults, already fertilized, later to be devoured by their daughters in their turn. How is such a mite to maximize her fitness? Clearly, she should send forth the maximal number of inseminated daughters. Sons should be produced only in sufficient numbers to guarantee that all daughters are fertilized. Extra sons are superfluous, since they will only compete for the same limited pool of fertilization opportunities that are *already* going to sons.

This, then, is local mate competition. The contribution of sons to parental fitness does not increase linearly as the number of sons is increased, because sons compete with one another for local mates, and one's gain is another's loss. Under these circumstances, parents may be expected to bias the sex of offspring toward daughters. And that is exactly what the ghoulish matriphagous mites do: Broods typically consist of dozens of females and only a few males. The mites achieve this sex ratio control by the same device as do the Hymenoptera, namely haplodiploid sex determination. Hamilton was able to list sixteen families of arthropods in which a female-biased sex ratio was achieved by haplodiploidy and was associated with regular predispersal brother-sister matings and hence local mate competition between brothers.

Where there is some mortality within broods, it probably pays the female to produce a few extra sons as insurance against death of a designated stud male and resultant failure of daughters to be inseminated. Where whole broods are likely to survive or fail together, however, the mother may eliminate wasteful local competition altogether by producing only a single son. Thus, in the wasp *Goniozus gallicola*, a parasite of moth larvae, the number of females in a brood ranges from zero to fourteen with a median of six, but the overwhelming majority of broods include *exactly* one male [227; see also 643].

If several females parasitize the same host (superparasitism), the problem of optimal sex ratio becomes a bit trickier. An ovipositing female with the host to herself might ideally produce a single son and many daughters, but what should a female do on an already parasitized host? With all those emerging females to be fertilized, the second female will want her own sons in the scramble, and two sons *are* better than one now, since they are competing not just with each other but with the first female's son as well. But then the first female might anticipate these competitors by producing more than one son too. Still there is local mate competition between brothers, and it still must therefore be adaptive to produce female-biased progenies, but female-biased to a lesser degree than in the single-parasite case. To identify the precise optimal strategies here we should have to know the values of several parameters, including the incidences of singly, doubly, triply, and so on parasitized hosts. Without such detailed knowledge, we can still make an interesting prediction: There should be an increased production of males in response to signs of multiple parasitism.

Increased male production on superparasitized hosts is well known in many hymenopteran parasitoids, and has been extensively investigated in *Nasonia vitripennis*, a wasp that parasitizes fly pupae. Several environmental cues, which are indicative of prior parasitization of the host, inspire ovipositing females to fertilize a smaller proportion of their eggs and thus to increase the proportion of males in the progeny. These cues include encountering another wasp walking on the fly pupa [699], a high density of female wasps in the environment [646], and finding on the host a telltale hole drilled by a previous wasp's ovipositor [699].

It is possible to make a more precise prediction. The second female on a host should not merely increase the proportion of males in her progeny but should do so to varying degrees according to the number of eggs laid [658]. If she were to lay just one, for example, it should be a male who could compete with the first female's son(s) for her many daughters. But as the

FIGURE 9–3
Sex ratio adaptation to local mate competition in a parasitic wasp

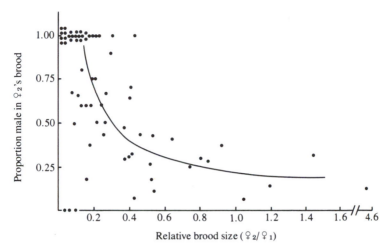

The first female *Nasonia vitripennis* (♀₁) to parasitize a blowfly pupa produces a female-biased brood: an average of 24.9 daughters and 3.7 sons. This bias is adaptive because of local mate competition among the males.

If a second female (♀₂) parasitizes the same pupa, she produces a much higher proportion of sons to compete with ♀₁'s sons. Furthermore, ♀₂ adaptively modifies her progeny sex ratio: When her brood is small relative to ♀₁'s, ♀₂'s progeny is highly male-biased, and the proportion that is male declines as ♀₂'s brood size increases relative to ♀₁'s.

The curve shows the optimal sex ratio for ♀₂ if she is to maximize her fitness. The points show actual sex ratios. (After Werren, 1980, Figure 1b [658]. Copyright 1980 by the American Association for the Advancement of Science.)

second female lays more eggs, the marginal value of each additional son declines, since they will compete with one another as well as with the first female's sons. John Werren [658] has shown that the second *Nasonia* female to parasitize a blowfly pupa produces a decreasing proportion of sons as the size of her own progeny increases relative to that of the first female (Figure 9–3).

The control of offspring sex is so effectively employed by so many haplodiploid species that several authors have concluded that such control must be the adaptive function for which haplodiploid genetic systems have evolved.

Local mate competition should in principle affect the optimal parental strategy in diploid species too: *Wherever* brothers compete more than sisters, we might expect female-biased investment. The reverse is also to be expected: If, for example, females disperse less than males so that sisters compete for local resources such as food, then parents might bias effort toward production of sons [102]. It has been suggested that this is the explanation for the male-biased sex ratio in galagos, small nocturnal primates of West African forests [102]. More generally still, parental strategies for allocating effort between daughters and sons should be sensitive to any sex-differential tendency for siblings to influence one another's reproductive success, either competitively or cooperatively [599]. However, local mate competition in the Hymenoptera is the only special case of this general principle that has received extensive empirical confirmation.

WORKER-QUEEN CONFLICT OVER THE SEX RATIO

So far we have been primarily concerned with sex ratio control in solitary Hymenoptera. The presence of sterile workers, as in the ants and honeybees, leads to a more complex situation: The queen who lays the eggs and the workers who tend them may disagree about the sex ratio to be produced [621].

Queens are Fisherian strategists, who may expect to derive equal fitness from sons and from reproductive daughters if they invest equally in them. (Nonreproductive worker daughters are in effect asexual and are not counted in the sex ratio of offspring; for simplicity, we ignore those cases in which workers lay some of the male-producing eggs.) To the workers, on the other hand, the female reproductives, their sisters, are more valuable as inclusive fitness vehicles than are the males, since a sister carries three quarters of a worker's genes and a brother only one quarter (see Figure 3–6, p. 49). If workers control the sex ratio, then the theoretical equilibrium situation is one in which three times as much effort is invested in female reproductives as in males [621]. Who has the power to control sex ratio is an open question, and the answer probably varies between species. Queens appear to control the primary sex ratio by choosing whether to fertilize eggs, but it is the workers who then care for the brood and who often control which female larvae are to be workers and which potential queens. And if brood cells appropriate for female versus male larvae are of different sizes, then workers may manipulate the distribution of cell sizes to control the primary sex ratio too.

Ants indeed seem to invest more energy in female production than in males, and the investment ratios are sometimes quite close to the theoretical optimum value for workers, that is 3:1 [621]. It would appear, then, that workers are at least sometimes victorious in this conflict with their queen. At other times the queen seems to win; in social wasps of the genus *Polistes*, for example, the investment ratio is very close to unity [435, 464]. Sometimes the ratios evidently vary systematically within species; for two species of ants, for example, one recent paper reported female-biased investment ratios in preferred habitats where colonies occurred at high density and equal investment in females and males in marginal habitats where nests were sparse [67]. Whether this means that power switches from queen to worker as density and habitat quality change is not yet clear, but this sort of ecologically based variation in sex ratios holds out promise for finer analyses of worker-queen conflict.

Even where investment ratios are female-biased, however, worker control of the sex ratio is not necessarily the explanation. Recall that *local mate competition* among brothers favors female-biased maternal investment even in nonsocial species (pp. 231–234). In social species, the same phenomenon may be happening, to a lesser extent than in sib-mating parasitoids, to be sure, but still enough to have selected for female-biased broods [14]. In the event of local mate competition, both the workers' and the queens' preferred ratios should be shifted toward females. There will still be conflict, then, but it becomes much more difficult to make precise quantitative predictions of the ratios to be expected.

If workers lay some of the male-producing eggs, as they do in some social species, or if the queen has been inseminated by more than one male so that not all sisters are full sisters, then calculation of the workers' and queen's "preferred" sex ratio becomes more complicated still. The effects of these complications constitute an area of active research. Since Robert Trivers and Hope Hare first proposed worker-queen conflict theory in 1976, research and further theorizing have burgeoned, and the last word on sex ratio in social Hymenoptera will not soon be written.

ENVIRONMENTAL DETERMINATION OF SEX

Bonellia viridis is a burrow-dwelling marine worm, about 8 cm in length, which extrudes an enormous proboscis ten times the length of its own body in order to feed. That, at least, is the foraging method of the female *Bonellia* who thereby manages to remain permanently in her burrow. The tiny male is even less adventurous: He lives as a parasite inside the female's uterus! As larvae, *Bonellia* are free-floating and sexually bipotential. If a larva contacts a female's proboscis, then it settles, enters her, and develops into a spermatogenic dwarf male in about three weeks. If no female is encountered, the same larva will develop into a female, requiring about two years to become sexually mature [377]. Here we have an animal in which sex is determined not by the individual's genotype but by extrinsic stimuli.

Under what circumstances is it adaptive for sex to be environmentally determined? Eric Charnov and James Bull [97] have reviewed a number of cases, at first glance very different one from another but with an abstract commonality: Environmental sex determination appears to be the strategy of choice where an environment is spatially heterogeneous ("patchy") in a way that affects female and male fitness differently and where both offspring and parent have little ability to predict whether the patch type that the offspring will enter will be favorable for females or for males. In the case of *Bonellia,* the "patchy" environmental variable is dichotomous: a conspecific female (a male-favorable patch) or none (a female-favorable patch). In several parasitic animals, the relevant environmental variable is

---

FIGURE 9–4
Sex determination by incubation temperature in reptiles

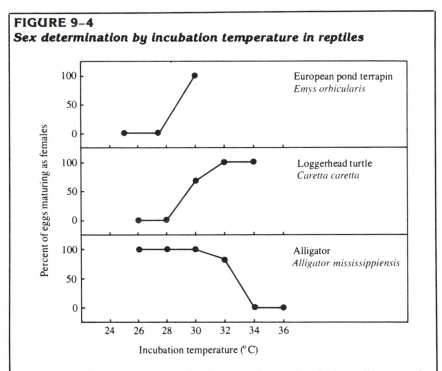

A turtle or alligator egg may develop into a male or a female depending upon the temperature at which it is incubated. Natural nest temperatures vary such that some broods will be all female, some all male, and some mixed. (After Mrosovsky & Yntema, 1980, Figure 1 [446]; Ferguson & Joanen, 1982, Table 1 [188], reprinted by permission from Nature, vol. 296, No. 5860, pp. 850–853. Copyright © 1982 Macmillan Journals Limited.)

the spatial density of conspecifics: Sex is determined by the degree of crowding experienced in early larval stages within the host, the larvae adopting whichever sex matures sooner at lower energetic demand when they are crowded, and the slower, costlier sex when uncrowded. In orchids, to take an example from the plant kingdom, the relevant "patchy" variable is sunlight. Orchids can grow larger in a sunny spot than in the shade, and profit more from large size as females; quite sensibly, they therefore develop as females in bright sunlight and as males in shade.

There is a peculiar mode of environmental sex determination in certain reptiles (and at least one fish [114]), whereby the temperature at which eggs are incubated in the nest is the determining factor (Figure 9-4). Most turtles, alligators, and some lizards exhibit this system, while other reptiles have sex chromosomes instead [75]. A clutch of turtle eggs can thus emerge as all males, all females, or a mixture of the two, according to weather, shade at the nest site, and depth of the nest pit (Figure 9-5). Whether there might be adaptive value to such incubation-temperature dependency of sex

---

FIGURE 9-5

A female green sea turtle (*Chelonia mydas*) digging her nest on a beach in the Sabah Turtle Islands, Malaysia. The sex of the hatchlings will depend upon the temperature at which the eggs are incubated. Whether turtles adaptively modify offspring sex ratios by adjusting nest-site selection is unknown. (Photograph by N. Mrosovsky.)

is not known. The situation does not seem to parallel that of *Bonellia* or an orchid, in which the environmental cue controlling sex is directly related to the fitness prospects of females versus males; turtles and alligators do not reproduce until long after they disperse from the nest site, so why use *incubation* temperature? Perhaps early rates of growth and development, which are influenced by incubation temperature, have different optima for the two sexes [188]. Nor is it yet clear whether nesting female reptiles might somehow exploit environmental sex determination to bias their sex ratios adaptively [444]. What *is* clear is that there are crucial implications for conservationists trying to save endangered sea turtles; some artificial incubation programs, instituted in order to protect eggs and increase populations, may instead be producing all-male broods as a result of unnaturally low incubation temperatures [446].

HERMAPHRODITISM: THE BEST OF BOTH WORLDS?

For an orchid or a *Bonellia* worm, being a female is a better bet in some circumstances, and being a male is better in others. Mightn't an individual organism profit from the ability to reproduce both as a female *and* as a male? In Greek mythology, Hermaphroditus, the son of Hermes and Aphrodite, became united in a single body with the nymph Salmacis, and it is to this bisexual creature (though just to the male half!) that we owe our name for an animal combining the characteristics of both sexes.

A true **simultaneous hermaphrodite** possesses the functional sexual organs of both sexes and can be at once mother and father. This is true of numerous invertebrates, including earthworms and many snails, as well as many plants. Such an hermaphrodite may sometimes indulge in self-fertilization, but there are often rather elaborate mechanisms to prevent this. Avoidance of self-fertilization is not surprising when we consider that it would carry with it many of the costs of both sexual and asexual reproduction without enjoying all the benefits of either (cf. Chapter 3).

The taxonomic distribution of animal hermaphroditism has been reviewed by Michael Ghiselin [210]. Among vertebrates, hermaphroditism is normal only among fish, which are more often *sequentially* than *simultaneously hermaphroditic*, usually a female first and later a male (**protogyny**). Among invertebrates, sequential hermaphroditism more commonly occurs in the opposite sequence (**protandry**).

Many coral reef fish are protogynous hermaphrodites, and the process of sex reversal is often under social control. *Labroides dimidiata,* for example, is a cleaner fish, which lives in groups consisting of one male and several females on the Australian Great Barrier Reef [529]. The male actively displays to his females and thereby suppresses their tendency to change sex. Should the male die or disappear, the largest and most dominant female assumes the male role immediately, and over the next several days she changes sex physiologically [see also 554, 555]. In other protogynous fish, such as the bluehead wrasse *Thalassoma bifasciatum,* the mating system is again highly polygynous, but spawning occurs, not in stable harem associations, but either on male display territories (leks) or in multimale groups [654]. Here, as in the sunfish in North American lakes (Chapter 5, pp. 102–103), an alternative male strategy of sneak fertilization becomes an option. Accordingly, there are two alternative life histories: Some individuals are males for life, maturing as little sneakers; others reproduce first as females and only later, at large size, as males [654].

Why be a sequential hermaphrodite? Ghiselin [212] has argued that if the advantage of large size is greater for sex A than for sex B, then the animal might maximize lifetime fitness by reproducing as sex B first and switching to sex A when large enough. In the case of protogynous harem-guarding coral reef fish like *Labroides,* the largest individual is much the fittest because it monopolizes male function and spawns with all the females. With egg-production capacity imposing the upper bound on female fitness, it will rarely pay a female to monopolize several males; that, we have already argued, is why polyandry is rarer than polygyny (Chapter 5, pp. 79–82), and it seems also to be the reason why despotic protandry should be rarer than despotic protogyny.

There *is,* however, a substantial fitness advantage in large size for females: The bigger they are, the more eggs they can produce. It follows that we might expect protandrous life histories where large size is of no great advantage to males. In the protandrous shrimp, *Pandalus jordani,* female fecundity increases with age and size. As in the coral reef fish, the sex reversal process is sensitive to social stimuli, but control is presumably not achieved by aggressive dominance in the shrimp. If there are many large, older females about when the fall breeding season approaches, then small first-time breeders are males. But the same young shrimp who will reproduce as a male if large females are plentiful is prepared to change sex early and breed only as a female if older, larger animals are few [95, 98].

Protandrous life cycles are rare in vertebrates, but not unheard of. The little coral reef fish *Amphiprion* lives in monogamous pairs within the shelter of a sea anemone's tentacles. A pair is likely to be joined by several small juveniles, not their offspring, who are tolerated as long as they remain sexually immature. If the female is removed, then the male metamorphoses into a female and the largest, most dominant juvenile becomes the new male [202].

Although some invertebrates and plants can change sex in both directions [see 201], sequential hermaphroditism is a one-way street in all the fish that exhibit it. At least among vertebrates, sex reversal appears to be a complex, time-consuming process: It is common for a month or more to pass while the gonads of sequentially hermaphroditic fish convert into functional producers of the alternative gamete type. Among terrestrial vertebrates, sex reversal as a normal event in the life cycle is unknown, nor can we induce it artificially. Human sex change operations have become almost commonplace [440], but there is no imminent prospect of transforming anyone into a reproductively functional member of the opposite sex.

Why more organisms are not *simultaneous* hermaphrodites is an intriguing question. It would seem that it might often be useful to be prepared to reproduce as female *or* male on short notice. Furthermore, there must often be diminishing returns from increased investment in either male or female function alone, especially in sessile (immobile) organisms: If I broadcast a number of male gametes (sperm or pollen grains), for example, and thereby achieve a certain success in fertilizing whatever ova are within reach, I am unlikely to double my success by broadcasting twice as many gametes. This is because there will be local mate competition between my own gametes [95, 100, 424]. So why not invest some effort in female function at some point where the returns on male effort are plummeting? Well, in fact, simultaneous hermaphroditism is actually very common in sessile organisms, both plants and animals [210, 212]. It is in highly mobile animals that **gonochorism** (separate individual females and males) predominates, perhaps because there are positively accelerated fitness returns from investment in direct male-male competition and combat in highly polygynous breeding systems (see Chapter 5, pp. 92–93). This may have led to such divergent specializations of females and males that a bisexual mutant must fail. There *are* simultaneously hermaphroditic vertebrates, again among the fishes [see e.g., 190], but to the best of our knowledge, no adequate socioecological rationale for the presence of this reproductive system in some fishes and not in others is yet forthcoming.

The female-male phenomenon has inspired wistful thinkers, from the ancient Greeks to the present, to dream of an androgyne who could bridge the gap and fully experience both sexual identities. Alas, we mammals are thoroughgoing gonochorists, and communication between women and men will surely remain imperfect.

SUMMARY

Reproductive effort must be budgeted between the production of sons and daughters. According to Fisher's theory, parents should invest equally in the two sexes across the population as a whole. For individual parents, however, it may sometimes be advantageous to invest preferentially in one sex. Haplodiploidy in the Hymenoptera allows mothers to control offspring sex adaptively in relation to environmental variables. In eusocial species, haplodiploidy leads to worker-queen conflict over the sex ratio to be produced by the colony.

If siblings of one sex compete for local resources, equal investment is no longer optimal. When brothers compete for a limited pool of fertilizations (local mate competition), broods tend to be strongly female-biased.

Some organisms lack sex chromosomes and are able to develop as females or males according to environmental cues. Many organisms are hermaphrodites, able to reproduce as both female and male; the allocation of reproductive effort between female and male function is then an adaptive problem analogous to that of offspring sex ratio.

SUGGESTED READINGS

Charnov, E. L. 1982. *The theory of sex allocation.* Princeton: Princeton University Press.
Hamilton, W. D. 1967. Extraordinary sex ratios. *Science,* 156: 477–488.

10

Sexual development and differentiation

"A hen is only an egg's way of making another egg." So said Samuel Butler, utopian novelist and philosopher of biology, a hundred years ago [81, p. 134]. We tend usually to think of the adult as the reproducer, but the succession of generations is a cyclical process without beginning or end. Gametes are produced by adults for the purpose of achieving sexual union and reproductive posterity. Reverse the words *gametes* and *adults* in the foregoing sentence and it still makes sense.

But neither our usual perspective nor Butler's seems just right in the context of modern biological knowledge. Sexual organisms "reproduce," yes, but neither gametes nor adults produce copies of themselves: Each individual in a sexual species is unique, and the taxonomic "type specimen" is an anachronism. The entities with sufficient copying fidelity and transgenerational stability to be called "replicators" are *genes* [142]. An organism can therefore be thought of as the genes' way of making gene copies. A successful gene is one that contributes to its own replication, and that usually means contributing to the reproductive success of the organisms it helps to build and operate. The history of evolution on earth is a history of the successes and failures of genes, some of which have been fruitful and multiplied while others have perished.

To this point we have been concerned with the evolutionary functional concepts of sociobiology and, to a lesser extent, with the behavioral control mechansims ("causation" *sensu* Tinbergen [613]; see p. 17) underlying sociobiological strategies. In this chapter, we shall shift attention to ontogeny and enquire how it is that immensely complex behaving organisms develop from single cells and how developmental processes differ between the sexes. We shall consider in some detail the processes by which two morphological types, the female and the male, diverge from a common fetal plan, and we shall consider a number of syndromes that involve disruption of normal sexual differentiation.

Sociobiologists have not generally paid much attention to developmental processes. Ontogeny is often lumped together with immediate causation as "proximate causation" and scorned as merely "descriptive" in contrast to the more "explanatory" analysis of "ultimate causation." However, attention to developmental processes seems to us essential. A life history "strategy" is a developmental chronology, which can never be fully understood without attention to the causal processes in ontogeny. Moreover, any attempt to model "strategic" behavior, such as mate choice or sex ratio allocation, inevitably raises the question of the organism's capacity to acquire, process, and retain experientially given information; thus, sociobiologists cannot ignore analyses of the learning process. Evolution may be portrayed as a phylogenetic series of adult phenotypes, but "phenotype" as it is usually understood is an impoverished conception of the organism, a static cross-section at one stage of the life history; evolution itself must be intelligible as a modification of developmental process over generations [400]. These are all reasons why some attention to ontogeny is appropriate. But above all, we hope to defuse the misguided controversy of nature versus nurture, and thus to facilitate a multilevel understanding of the sexes, with special attention to our own species.

DIFFERENTIATION AND EPIGENESIS

When a sperm has won its way into an ovum, several processes begin. The two cells become one by the progressive fusion of their membranes and the gradual envelopment of the sperm within the egg's cytoplasmic mass. Up to this point the ovum has been dormant in the midst of meiosis for a long time, but now development resumes and the arrested second meiotic division is completed. The two parental nuclei remain separate for awhile, twelve hours or so in the human case, while male and female pronuclei grow. These finally come into contact, their nuclear envelopes disintegrate, and the parental chromosomes unite in a single cell, the zygote [22].

Soon the zygote begins mitotic divisions, and the free blastocyst stage is under way, ending (in the case of mammals, upon which we shall focus) with implantation in the uterine wall. But even before implantation the blastocyst's few dozen cells manifest a certain structural organization. They are not just a blob of cells but a hollow sphere, and by the time of implantation the constituent cells have differentiated into two distinct types: trophoblast cells on the outside of the sphere and an inner cell mass at one "pole" of the blastocyst [430].

Further description of the detailed stages of early structural differentiation is beyond the scope of this book. Such descriptions can be found in embryological texts. But without belaboring the details of embryogenesis, it is important to our themes that we make certain general points about the nature of developmental processes.

Embryology is a richly descriptive science. Its foundation is a detailed account of the observable sequence of events that takes place as an organism proceeds from the unicellular zygote stage to adulthood. During that development differentiation occurs, which is to say that there arise distinct types of cells that differ from one another in structure and in function [e.g., 236, 645]. A liver cell differs from a nerve cell. They don't look alike, and they don't act alike: The metabolic processes and products of one cell type are not the same as those of another cell type. In each type of cell a different set of genes is active, each gene busy directing the synthesis of its particular protein.

A basic question about biological organization is how it is that cells of a single organism differ functionally. One possible explanation would be that cells engaged in different metabolic processes differ genetically from one another. Thus, certain genes might be altogether absent from some cell types, and other genes might be multiply represented in those cells in which they are particularly active. In fact, such nuclear genetic differences between the cells of an individual organism have been documented in certain cases, but it is much more common for the various cell types of an individual to have identical complete complements of nuclear genes [236]. Their differential function then depends on cytoplasmic and local environmental differences, such as differences in the availability of particular enzymes which facilitate the synthesis of particular gene products.

A mature, complex animal such as a person is composed of hundreds of distinct cell types. There are, furthermore, hundreds of distinct multicellular structures—such as particular bones, brain structures, and organs—each composed of several different cell types. A full set of structures is species-typical (though perhaps sex-specific), and so are the proportions of the cell types constituting these structures. All this complex structure arises in an orderly, species-typical fashion from a single-celled zygote. We call this development "differentiation" because what we observe is an emergence of structural organization whereby cell types and structures differ from one another, rather than just a growing mass of homogeneous protoplasm. Already in the preimplantation blastocyst there is a recognizable structure: The trophoblast, which will give rise to part of the placenta, contains cells different in form from those of the inner cell mass, which will give rise to embryo, amnion, allantois, and yolk sac.

Once differentiation has begun and structure is apparent, certain parts of the embryo can be identified as precursors of later structures: The direct descendants of some particular set of cells will be some particular organ (Figure 10-1). Primordial germ cells, for example, are the ancestors of gametes. They cannot be induced to become blood or bone cells instead. (We also refer to differentiation between organisms as well as within a single organism, as for example when we refer to the contrasting developmental chronologies of male and female as "sexual differentiation.")

Early in development there may be a good deal of **equipotentiality** among cells. If those cells that are normally the precursors of some particular organ are removed, that organ may develop all the same. Other cells, not originally destined to give rise to the organ in question at all, take over the job of producing it. A normal mouse can be grown from half a mouse blastocyst, for example. But that half blastocyst must contain part of the inner cell mass. The outer cells (the trophoblast) alone will not produce an embryo. We thus see that equipotentiality often wanes or disappears as differentiation progresses. A mammalian intestinal cell, for example, contains a full complement of genes, but no known manipulation can induce that intestinal cell to produce a nerve or liver or lymph cell. All it can produce is more intestinal cells. The cell has differentiated irreversibly.

We introduce these facts—the multiplicity of cell types, the genetic identity of functionally dissimilar cell types, equipotentiality and its disappearance—in order to persuade the reader to a certain view of development. Development is a process of continual alteration of preexisting structure within a particular environment. Structure at time t is always a determinant of structure at time $t + 1$. The environment in which that structure exists is also always a determinant of the developmental change in the structure. For one thing, cells differentiate in a milieu of other cells, and these surrounding cells are part of the environment and modify the functions of the cells they surround, a fact abundantly demonstrated in experimental embryology.

FIGURE 10-1
Differentiation of nervous system cells

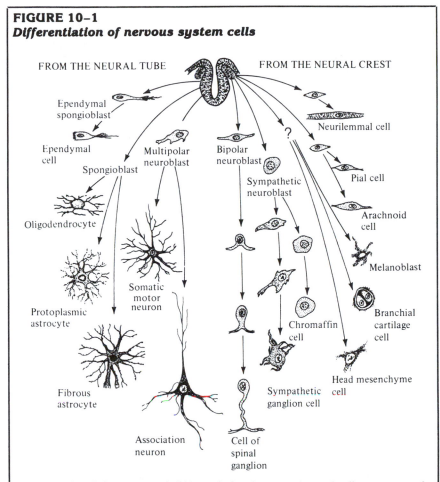

FROM THE NEURAL TUBE   FROM THE NEURAL CREST

Ependymal spongioblast

Neurilemmal cell

Ependymal cell

Multipolar neuroblast

Bipolar neuroblast

Spongioblast

Pial cell

Sympathetic neuroblast

Oligodendrocyte

Arachnoid cell

Melanoblast

Protoplasmic astrocyte

Somatic motor neuron

Chromaffin cell

Branchial cartilage cell

Head mesenchyme cell

Fibrous astrocyte

Sympathetic ganglion cell

Association neuron

Cell of spinal ganglion

An example of the course of differentiation in one subset of cell types, namely mammalian nervous system cells. As embryological development progresses, the number of distinct cell types in the nervous system proliferates. Similar differentiation occurs in nonneural cells too. (From Young, 1975 [703].)

The growth and proliferation of the trophoblast, for example, depends upon the presence of the inner cell mass. In addition to this cellular environment, the environment external to the organism is also relevant to every detail of differentiation, for it is from that external environment that the chemical and thermal necessities of life must be dependably and steadily delivered.

It is a common abstraction to speak of environmentally induced developmental events as opposed to genetic effects, but this dichotomy, so attractive at first glance, always fails us. *Nothing* occurs in the absence of environmental influence. And since gene action is intimately involved in the metabolic processes of every cell throughout life, it too is relevant to *every* developmental process in the organism, to language acquisition no less than to the manifestation of eye color. This being so, the nature-nurture issue is not satisfactorily resolved even by insistence that all developmental processes depend upon some combination of genes and environment if the impression is left that there is a variable balance so that some developmental events are mostly genetically determined and others are mostly environmentally determined. On the contrary, developmental processes are entirely dependent upon both genes and environment. This is true in that the processes would be altered or would fail completely if either genetic structure or environment were altered beyond particular rather narrow limits.

The perspective upon development that we are here espousing is an **epigenetic** one: organismic structure changes over time as an interactive function of the preexisting structure and the environment in which that structure finds itself. An alternative perspective is **preformationism**, in which each structural element in an organism is considered to have its separate precursors, so that even in the zygote there are elements corresponding to every adult structure. Epigenesis versus preformationism is an ancient embryological dispute that has long since been resolved in favor of epigenesis. There is a grain of truth to preformationism in the complex genetic organization in the zygote and in the fact that certain genes, such as those determining eye color pigments, are highly localized in their effects. However, there is no component in the zygote, no genetic or other material, that is the precursor of the eye or of any other particular organ. As we have learned more about the mutual influences of cellular components, of cells, and of multicellular structures during ontogeny, we have seen the vindication of the epigenetic view: Development is an interactive process of structural elaboration.

Behavioral scientists have often tried to distinguish between behavior that is "innate" and behavior that is "acquired." The distinction is untenable, for the reasons we have discussed above. A more subtle approach is to concede that behavior cannot be dichotomously classified in this way, but to argue that we might be able to abstract out two distinct types of developmental process: "experience" versus "maturation," the latter being somehow more intrinsic than the former. Since experience necessarily operates upon a preexisting structure and maturation can occur only in the context of reliable environmental inputs, this dichotomy seems no more practical than "innate"-"acquired." This is not to say that rigorous dichotomies of developmental process are impossible. A scientist might, for example, define *learning* with sufficient precision that changes in behavior could then be reliably classified as "learned" and "unlearned." For example, learning might be defined according to some experimental paradigm such as the experimenter's modifying behavior by making rewards contingent upon certain acts, or alternatively according to a criterion of structural changes in nerve cells as a result of neural activities. Such operational definitions will not necessarily all identify the same phenomena as examples of learning, nor will they necessarily exclude the development of behavior classically considered "innate," which is behavior that is manifested in appropriate circumstances without specific practice. "Learning" as defined in psychology or neurology is a developmental process, but it is not the opposite of maturation.

The nature-nurture debate has been pronounced dead many times, but it won't stay buried. The persistence of this vexatious issue is partly a result of a bad habit of conflating questions of ontogenetic causation with questions about the sources of population variability [cf. 142, ch. 2]. One may note, for example, that the observed variability in behavior A correlates rather nicely with experiential variations between individuals, whereas the observed variability in behavior B is strongly correlated with genetic differences. It is a short step to suggest that A is more "environmentally determined" than B or even more "modifiable" and hence less "innate." It is the variation *between* individuals, however, that can properly be called more or less "environmentally determined," not the behavior *of* individuals (cf. discussion of "heritability," pp. 32–35). Similarly, it is the *difference between* individuals that is more or less "genetically determined" by allelic differences. Knowing the extent to which behavior is environmentally versus genetically "determined" in this populational sense tells us nothing at all about its ontogeny. The term "genetically determined" *is utterly meaningless when applied to individuals.*

It must be apparent that we consider the nature-nurture dichotomy an unproductive distraction. We now intend to discuss sexual differentiation without recourse to it.

SEXUAL DIFFERENTIATION

The morphological differences between conspecific mammalian females and males are legion, ranging from overt sex-specific structures such as mammary glands and beards, through skeletal proportions and the sizes of internal organs [28], to neuroanatomical measurements. Virtually all these differences between the sexes are induced by the differential action of female and male gonads.

FIGURE 10–2
Sexual differentiation of human internal reproductive organs

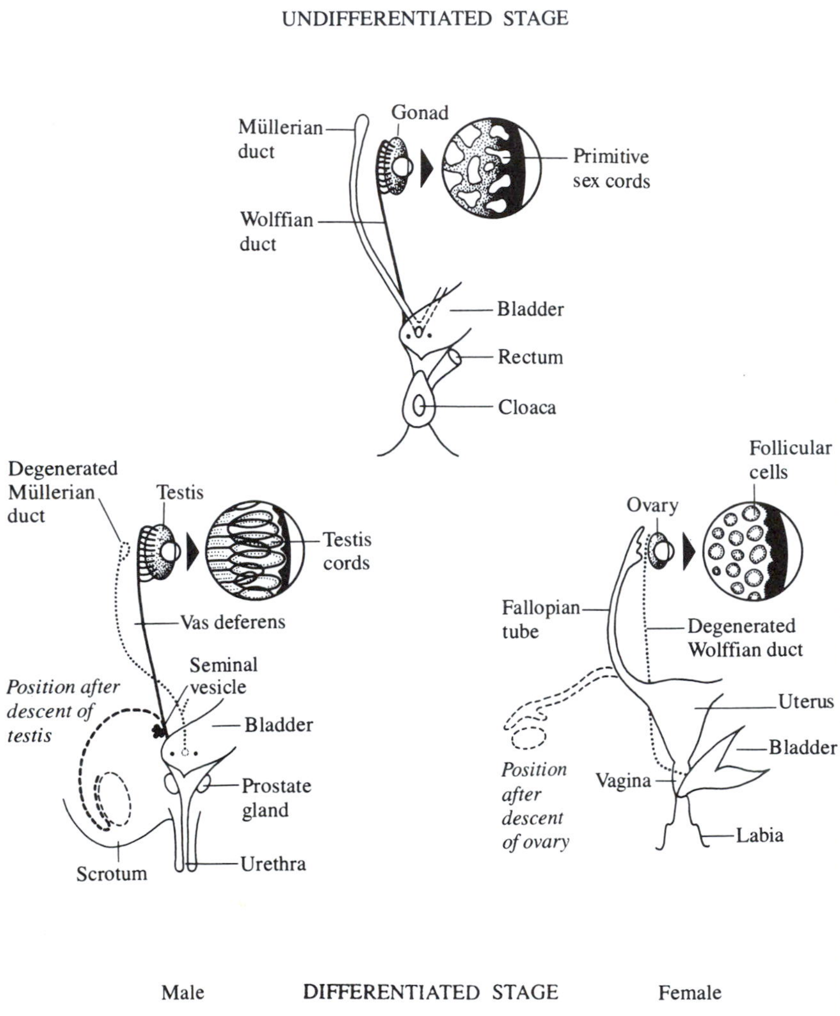

UNDIFFERENTIATED STAGE

Müllerian duct

Gonad

Primitive sex cords

Wolffian duct

Bladder

Rectum

Cloaca

Degenerated Müllerian duct

Testis

Testis cords

Vas deferens

Seminal vesicle

Position after descent of testis

Bladder

Prostate gland

Scrotum

Urethra

Follicular cells

Ovary

Fallopian tube

Degenerated Wolffian duct

Uterus

Bladder

Position after descent of ovary

Vagina

Labia

Male   DIFFERENTIATED STAGE   Female

The top figure shows sexually indifferent internal organs at seven weeks postconception. The bottom figures show differentiated internal organs. Enlarged details illustrate the structure of the indifferent gonad at seven weeks, the testis at four months, and the ovary in the fifth month. (After Langman, 1975 [363]; Young, 1975 [703].)

The classic experimental investigations of the role of the gonads in sexual differentiation were conducted by French developmental biologist Alfred Jost [322]. Jost found that the removal of gonads from undifferentiated embryonic rabbits led to the development of female phenotypes regardless of the embryo's genetic sex. Conversely, the transplantation of an embryonic testis or the injection of testosterone induced the male phenotype in embryos of both sexes. Thus, Jost showed that the testis induced masculinization, whereas feminization did not require a functional gonad. Although a great deal of detail has been filled in since, some of which we shall briefly discuss, Jost's basic explanation of sexual differentiation stands firm.

Sexual differentiation in development depends primarily upon the same sex-typical gonadal steroid hormones that are stimulative factors for sex-typical behavior in adulthood. It is for this reason that endocrinologists speak of the **organizing** functions of gonadal hormones that induce permanent sexual differentiation of various organs during embryonic and later development, as distinct from the **activating** functions of gonadal hormones in the mature organism.

There are a number of internal structures that differ between human females and males besides ovaries versus testes. The female's internal anatomy includes the Fallopian tubes, uterus, and cervix; male internal anatomy includes the epididymis, vas deferens, seminal vesicles, and prostate gland (Figure 10–2). These organs derive embryologically from two early tubular structures—the Wolffian and Müllerian ducts. At first the Wolffian ducts function as excretory passageways. The Müllerian ducts develop alongside them by the sixth week. Embryos of both sexes have both sets of ducts, but in the female the Müllerian ducts develop into the internal female organs just mentioned and the Wolffian ducts degenerate, whereas in the male the Wolffian ducts develop into internal male organs and the Müllerian ducts degenerate. This internal sexual differentiation becomes apparent in the human embryo by the end of the second month.

The source of testosterone, both embryonic and adult, is the interstitial cells of Leydig, which lie between the seminiferous tubules in the testis. In humans these cells are first detectable in the eight- or nine-week-old fetus, and they are already capable of testosterone manufacture [569].

The Wolffian ducts develop into male internal organs under the organizing influence of testosterone. In its absence, as in the embryonic female or an experimentally castrated male fetus, the Wolffian ducts degenerate. In the embryonic female, and in fetuses of either sex that lack functional gonads, the Müllerian ducts develop into female organs. The degeneration of the Müllerian ducts in the intact male does not occur spontaneously. This effect too is induced by the testis, although not by steroid

FIGURE 10–3
Human prenatal sexual differentiation

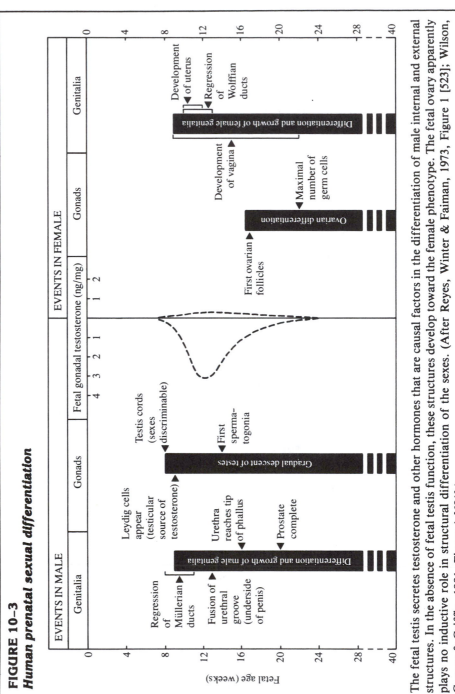

The fetal testis secretes testosterone and other hormones that are causal factors in the differentiation of male internal and external structures. In the absence of fetal testis function, these structures develop toward the female phenotype. The fetal ovary apparently plays no inductive role in structural differentiation of the sexes. (After Reyes, Winter & Faiman, 1973, Figure 1 [523]; Wilson, George & Griffin, 1981, Figure 1 [681].)

hormones; the anti-Müllerian hormone is a glycoprotein produced in the seminiferous tubules [321]. Another testicular androgen closely related to testosterone, namely dihydrotestosterone (DHT), also plays a role in prenatal differentiation, masculinizing the urogenital sinus and external genitalia. The time course of prenatal sexual differentiation in *Homo sapiens* is outlined in Figure 10-3.

Although it is the testis that has an organizing influence in the embryology of the several mammalian species that have been investigated, this is not the story universally. In birds, sexual differentiation is under the influence of early ovarian function rather than early testicular function. Female structures develop in the presence of early estrogen, and male structures develop in its absence. The situation is the same as in mammals, but mirror-imaged. It is perhaps relevant that birds also differ from mammals in the chromosomal determination of sex: In birds, females are the heterogametic (ZW) sex and males the homogametic (ZZ; see Figure 9-1, p. 225). In other vertebrate classes there is considerable variability in chromosomal sex determination, and present evidence suggests that it is always the heterogametic sex whose gonad is functional in sexual differentiation [4].

Differentiation of the external genitalia is also under the influence of the gonadal hormones—hence androgens in people and other mammals. Stages in this differentiation process are illustrated in Figure 10-4.

If a genetic male is deprived of early androgen, he will develop external genitalia of female appearance. The same result occurs in the androgen-insensitivity syndrome, a clinical condition in which androgen is present but the tissues that normally respond to it lack sensitivity to the hormone. Internally, both the Wolffian and the Müllerian duct systems degenerate. Externally, the appearance is female, and so the condition may not be detected until the time of puberty arrives and menses fails to occur [441].

A genetic female who is exposed to early androgen will develop external genitalia of male appearance. This occurs in the syndrome called congenital adrenal hyperplasia (CAH, also known as adrenogenital syndrome), a clinical condition in which a malfunction of the adrenal glands leads them to produce, not the normal adrenal hormone cortisol, but a substance that is structurally and functionally similar to testosterone. The extent of masculinization of the external genitalia is variable, and the baby may be identified as either male or female [441].

The sex differences that develop at about the time of puberty are also under the control of gonadal hormones (Figure 10-5). In men, testosterone is necessary for the development of such attributes as a masculine distribution of body hair and musculature as well as the change of voice. In women, estrogen is necessary for the development of breasts and of other aspects of

FIGURE 10–4
Sexual differentiation of human external genitalia

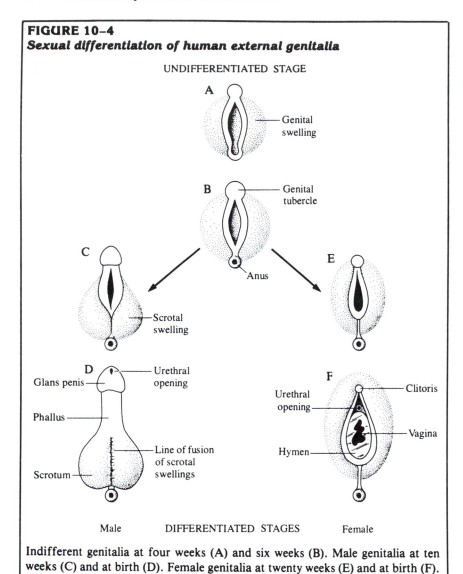

UNDIFFERENTIATED STAGE

DIFFERENTIATED STAGES

Male · Female

Indifferent genitalia at four weeks (A) and six weeks (B). Male genitalia at ten weeks (C) and at birth (D). Female genitalia at twenty weeks (E) and at birth (F). (After Langman, 1975 [363].)

typical female form. Here we have an important difference between the control mechanisms of early sexual differentiation and later sex-typical development. In the pubertal changes estrogen plays an active role: Feminization does not result from the mere absence of testosterone.

FIGURE 10–5
Puberty

Conspicuous pubertal event	Female	Male
Maximal rate of skeletal growth	12 years	14 years
Maximal rate of muscle growth	12	15-16
Adult pubic hair pattern	13-14	14-15
First menstruation	13	—
Breast development	11-15	—
Adult voice	—	15-16
Facial hair (upper cheeks & below lip)	—	16

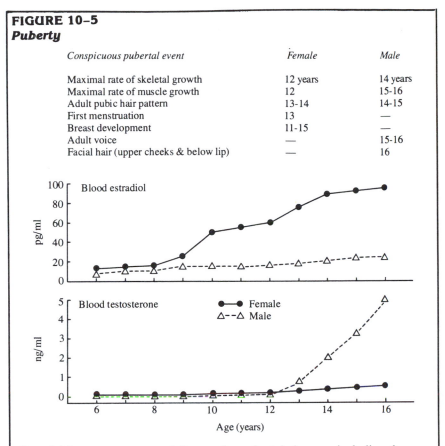

Gonadal hormones are causal factors for pubertal changes, including the conspicuous events listed above. Increased production of estradiol (an estrogen) in the ovary precedes the resumption of testosterone production by the testis. The precise timing of pubertal events varies between populations; these data are representative of postwar European people. (After Tanner, 1981 [595]; Billewicz et al., 1981, Tables 1 and 2 [52]; Sizonenko, 1978, Figure 4 [573].)

THE PARTIAL SUBSTITUTABILITY OF STEROID HORMONES

We have hitherto alluded somewhat blithely to the organizing (and activating) effects of testosterone and other steroid hormones, and it is indeed the case that particular effects are normally achieved each by their own hormones. It is also true, however, that steroid hormones have to some extent

FIGURE 10-6
Chemical kinship of steroid hormones

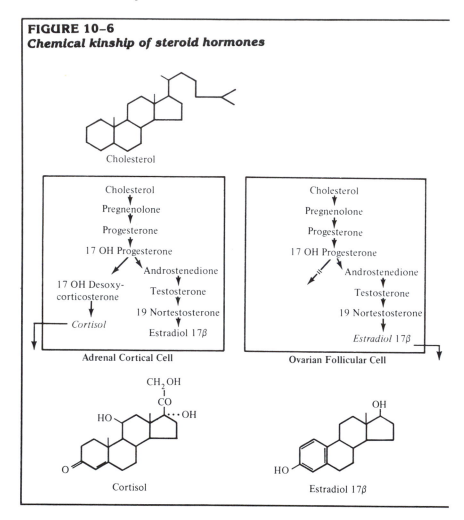

the potential to duplicate the effects of one another [495]. In an adult already differentiated as male or female, the activation of behavior appropriate to that animal's sex can be achieved by injections of either androgen or estrogen. This commonality of effects applies not only to the activating influences of these gonadal hormones but to their organizing influences as well. As we have seen, it is normally androgens and other testicular secretions that masculinize embryonic or neonatal mammals, whereas the absence of any embryonic or neonatal gonadal function leads to feminization. If a young mammal is exposed by injection to quantities of

The secretory cells of different endocrine glands use similar biochemical pathways to produce a closely related set of steroid hormones. (From Tepperman, 1973 [602]; copyright © 1980 by Year Book Medical Publishers, Inc.; Bentley, 1976 [43].)

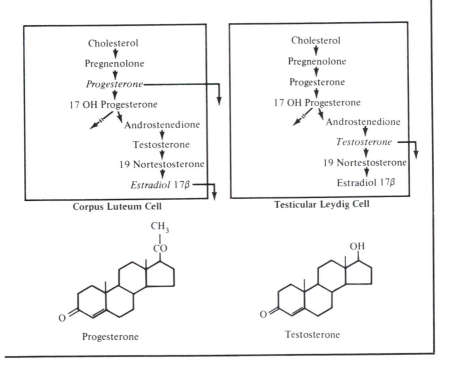

estrogen greater than ever occur naturally, some *masculinization* remains likely. So although it is normally the androgens that are involved in both organizing and activating masculine behavior and the estrogens that are involved in activating feminine behavior, the sex-typical behavior resides not in the sex typicality of the hormones themselves so much as in the early differentiation dependent upon the presence or absence of early gonadal function. The fact that the steroid hormones are to some degree functionally interchangeable presumably depends upon their close chemical similarity (Figure 10–6).

PRIMARY DIFFERENTIATION OF SEX

If gonadal function is essential to sexual differentiation, the question remains how the gonads are themselves originally differentiated ("primary differentiation"). In human embryos, the sexes are histologically indiscriminable until well into the second month. At six weeks, the sexually indifferent gonad contains structures called primitive sex cords, and it is by the continued growth of these cords in the male, and their degeneration in the female, that the sexes finally become discriminable by about seven weeks (Figure 10-3). The ovary remains histologically indistinguishable from the indifferent gonad until about week 12, but the biosynthetic pathways for estrogen production are in place much sooner [681].

The substance that is believed to be responsible for primary gonadal differentiation was discovered serendipitously, in the course of immunological research with mice. Highly inbred strains of mice are genetically uniform, but even within such strains, females reject male skin grafts. The reverse is not true, implicating an antigen genetically coded on the Y chromosome as responsible for the incompatibility. This substance is called **H-Y antigen** [218, 264, 644, 692].

The H-Y antigen has been found to be universal in mammalian males, and it evidently characterizes the heterogametic sex of other vertebrates as well. Its role in gonadal differentiation is indicated by studies of individuals with inconsistencies between phenotypic and chromosomal sex: In mammals, wherever a testis has been found, there too is H-Y—in androgen-insensitive XY "females" (p. 253), for example, as well as in XX "males" in whom a translocated bit of Y chromosome has evidently become attached to another (autosomal) chromosome. H-Y antigen is therefore strongly suspected of being the primary gonadal differentiator, but it has not yet proved possible to isolate or synthesize this protein. When that is achieved, the definitive experiment of administering H-Y and observing its effects will become possible.

SEXUAL DIFFERENTIATION OF BRAIN AND BEHAVIOR

In most animal species, females and males exhibit numerous behavioral differences. Some are absolute. In many songbird species, for example, only males sing and only females build nests. In many hoofed mammals only males bump heads and only females form adult friendships. Far more common than behaviors exclusive to one or the other sex, however, are quantitative sex differences. Almost anything we measure is likely to show some differences—depth of burrows, time patterns of eating and elimination, migration dates, speed of walking in a laboratory environment, duration of sleep, frequency of scentmarking, of vocalizing, or of grooming. All these behavioral measures show dependable average differences between the sexes within species.

What is the substrate underlying sexually dimorphic behavior? Is the brain itself a sexually differentiated organ? As we shall see, there is a good deal of evidence of structural differences between the brains of females and males, but it is important to realize that behavioral sex differences are not in themselves adequate evidence that this is the case. An undifferentiated brain might instead produce sex-typical behavior as a response to sex-typical hormones. In fact, a few behavioral sex differences seem to work in just this way: We have already mentioned the induction of song in normally silent female white-crowned sparrows by an activating injection of testosterone in adulthood. A great deal of sexually differentiated behavior, however, has now been shown to derive from brain dimorphism.

Perhaps the most obvious of sexually dimorphic behaviors is copulation. In mammals, males generally mount females from behind. In a few species, notably humans and pygmy chimpanzees, the partners usually face one another during copulation. Mammalian males generally exhibit one or more pelvic thrusts during each mount, and the thrusts usually serve to intromit the penis into the vagina; there is considerable variation among species in the temporal patterning and precise form of this behavior but considerable stereotypy among the males of a single species [149]. The copulatory behavior of many female mammals consists mainly of assuming and holding a posture ("lordosis") that permits the male to mount and intromit [33]. Normal adult males rarely assume female copulatory postures, whereas normal females commonly exhibit at least part of the male pattern, mounting animals of either sex from the rear, although they are much less likely to perform pelvic thrusting [32].

The performance of copulatory behavior is under a stimulatory influence from the gonadal hormones. Castrated males and ovariectomized females forsake copulation (though complete cessation may take quite awhile). This dependency of copulation upon gonadal hormones does not imply, however, that merely injecting testosterone into a castrated animal will necessarily induce male copulatory behavior. Besides the hormone's activating function in adulthood, it has an organizing function in early development too.

If testosterone is not present in rats within the first four days after birth, they will not manifest complete male copulatory behavior in adulthood [35, 666]. Within those first few postnatal days some relevant process of development is under way, presumably in the infant rat's nervous system. If a male rat is castrated when one day old, he will not copulate as an adult, and no amount of hormones at maturity will make him do so. Hold off castrating him until he is a week old, however, and he *will* copulate, providing he is given an activating dose of testosterone in adulthood. Likewise, a genetic female will exhibit male copulatory behavior in adulthood if she is given both an organizing shot of testosterone in infancy and an activating shot in adulthood. Neither injection will do the job by itself.

It therefore appears that the neural organization underlying male copulatory behavior develops under the influence of early testosterone [259]. This is another organizing effect of early testicular function, similar to its developmental influences upon the internal and external male sexual organs. The effect on copulatory behavior occurs a little later than the overt morphological effects. In the rat the behavioral effect can be induced immediately after birth, but the rat is not necessarily typical in this regard. In guinea pigs and rhesus monkeys, for example, the critical period for sexual differentiation of copulatory behavior capacity is entirely prenatal.

A genetic female rat will react to early testosterone like a male. She will both acquire the capacity for male behavior—which she exhibits when given adult testosterone—and lose the capacity for female lordosis—which she does not exhibit even though given adult estrogen and progesterone. She lacks a penis, so she cannot intromit; nor can she receive all the stimulatory feedback that a male enjoys. It is not surprising, then, that her male performance may be somewhat deficient, but she thrusts like a male and may even exhibit a malelike orgasmic response [648]. Once again, it is early *testicular* function that is relevant to the sexual differentiation; removal of the ovaries does not affect these developmental processes. The female that is ovariectomized early still develops the normal female capacity for lordosis behavior, given estrogen and progesterone in adulthood.

We can summarize these effects as follows: In mammalian copulatory behavior, as in the development of external genitalia and internal reproductive structures, early testicular function leads to masculinization, whereas the lack of any early gonadal function leads to feminization.

Sexually differentiated responsiveness suggests a sexually differentiated brain. Sex differences in brain function (as opposed to gross anatomical differences; see below) have been known for some time. Electrophysiological stimulation of cells in the amygdala, for example, produces different patterns of activity in "downstream" brain structures in female versus male rats; neonatal castration of males produces a female-like response, and early testosterone treatment of females makes them respond like males [169]. The implication is that there are sex-typical "wiring" patterns whose differentiation is under neonatal hormonal control.

In recent years, there has been a spate of published descriptions of sex differences in brain anatomy—in the thickness of a layer of cells here, in a detailed dendritic branching pattern there [e.g., 219, 359, 410, 427, 463]. Many of the differences are striking. The medial preoptic area of the hypothalamus, for example, is four or five *times* as large in male rats as in females [219], a sex difference that develops rapidly between birth and about ten days of age under androgen influence [219, 311]. This brain structure is involved in the control of LH secretion and hence ovulation in the female and in the control of sexual behavior in the male. This and other dimorphic brain structures [e.g., 418] are areas that were specifically examined for sex differences because they were already known to contain receptor sites for steroid hormones and to be involved in the control of sexually differentiated behavior.

In the canary, brain nuclei that control singing behavior are almost three times larger in males than in females, and furthermore wax and wane seasonally, growing large in the spring and regressing in the autumn when the birds fall silent [465, 466]. Administration of testosterone to adult females induces song and growth of the song control nuclei, both of which normally occur only in males [145]. Even more remarkably, the size of the song control region in the brain is correlated with the number of distinct songs in the repertoires of individual males!

Some sexually differentiated behavior patterns, then, depend upon both organizing and activating effects of steroid hormones (e.g., copulatory behavior in rats). Other sexually differentiated behaviors evidently do not depend on early differentiation but can be activated in adults of either sex by the appropriate hormone (e.g., singing in the testosterone-injected white-crowned sparrow). The third logical possibility is that a sex difference depends *only* upon organizational effects of gonadal hormones: Prenatal or neonatal hormones organize sexually dimorphic behavior that is manifested even in the later absence of sex-typical hormones. Examples are the urination postures of dogs and the rough-and-tumble play of male rhesus monkeys [223].

SEX DIFFERENCES IN HUMAN BEHAVIOR

Before we say more about the development of sex-typical behavior in humans, we would do well to consider just how the sexes actually differ. There have been hundreds of reports germane to this question, too many to review here. Others have reviewed this subject for us, and we shall now consider two very different summations of the subject.

The first is a review by a cultural anthropologist, G. P. Murdock, who

TABLE 10-1

Division of labor between the sexes in a sample of 224
human societies. M: Exclusively male; Both: Some
participation by both sexes; F: Exclusively female.

	Number of societies		
	M	Both	F
Metal working...	78	0	0
Weapon making	121	1	0
Pursuit of sea mammals..	34	1	0
Hunting...	166	13	0
Manufacture of musical instruments	45	2	1
Boatbuilding...	91	8	1
Mining and quarrying	35	2	1
Work in wood and bark..	113	15	1
Work in stone	68	5	2
Trapping or catching of small animals	128	18	2
Work in bone, horn and shell....................................	67	7	3
Lumbering...	104	8	6
Fishing..	98	56	4
Manufacture of ceremonial objects............................	37	14	1
Herding...	38	12	5
Housebuilding	86	60	14
Clearing of land for agriculture	73	44	13
Netmaking...	44	12	11
Trade..	51	56	7
Dairy operations..	17	8	13
Manufacture of ornaments..	24	49	18
Agriculture—soil preparation and planting	31	76	37
Manufacture of leather products................................	29	15	32

surveyed the available descriptions of human societies in order to see how certain tasks were apportioned between the sexes [451]. The results of his survey are presented in Table 10-1.

One of the few great generalizations of cultural anthropology is this: Gender is always highly relevant to the division of labor in any human society. Sex-typing is always strong (though not necessarily absolute) for at least some of the important economic activities and ritualistic activities in every society. However, the nature of the division of labor by sex is by no

	Number of societies		
	M	Both	F
Body mutilation, e.g., tattooing	16	80	20
Erection and dismantling of shelter	14	13	22
Hide preparation	31	10	49
Tending of fowls and small animals	21	13	39
Agriculture—crop tending and harvesting	10	89	44
Gathering of shellfish	9	19	25
Manufacture of nontextile fabrics	14	11	32
Fire making and tending	18	53	62
Burden bearing	12	59	57
Preparation of drinks and narcotics	20	22	57
Manufacture of thread and cordage	23	23	73
Basketmaking	25	19	82
Matmaking	16	12	61
Weaving	19	10	67
Gathering of fruits, berries, and nuts	12	31	63
Fuel gathering	22	30	89
Pottery making	13	16	77
Preservation of meat and fish	8	26	74
Manufacture and repair of clothing	12	20	95
Gathering of herbs, roots, and seeds	8	19	74
Cooking	5	38	158
Water carrying	7	12	119
Grain grinding	2	22	114

Source: Modified from G.P. Murdock, 1965 [451] and reprinted from *Social Forces*, Vol. 15, May, 1937. Copyright © by the University of North Carolina Press.

means consistent. For the great majority of activities listed in Table 10-1 the same activity is exclusively the province of women in some societies, of men in others, and is not rigidly sex-typed in still others. Only a few activities seem to be sex-typed in a consistent fashion in all cultures. Besides those in Murdock's table, the most consistently sex-typed activities include the care of infants by women and the waging of war by men.

There are two main problems with such a tabulation. The first is what anthropologists call "Galton's problem"—the question of whether cultures can properly be considered independent of one another and then tallied in this way [456]. Certain technological practices, for example, have surely spread from culture to culture, and their sex-typing may have been passed along with them. If we find that boat building is exclusively the province of men in ten Amazonian cultures, for example, this fact may represent a transmission of boat-building technology between men in different cultures. The ten might warrant no greater emphasis than a single culture elsewhere. Such technological transmission, as well as the descent of multiple cultures from common ancestral cultures, calls into question the independence (in the statistical sense) of different cultures. Murdock dealt with Galton's problem by sampling a large number of cultures from around the world, but this can never be a complete solution.

The second problem is that most of the data upon which Murdock based his table were collected by male anthropologists from male informants. The accounts of women's work are therefore likely to be incomplete and perhaps biased. A number of different anthropologists conducted the original research, and they set about it with different interests, so that some facts are necessarily missing in this survey of 224 cultures. It may well be that sea mammals are pursued in only 35 of the 224, but we are told who bears burdens in only 128 and even who cooks in only 201 of the 224. It seems plausible that women's work was more likely to go unreported than men's—for example, trade may not be such a male preserve as it appears to be, and other activities may be even more female-dominated than they appear. But despite weaknesses of the ethnographic record, Murdock's summary permits the important conclusion that sex-typing of activities, while universal, is highly variable in its specific manifestations.

A different approach to behavioral sex differences was taken by Eleanor Maccoby and Carol Jacklin in their 1974 book, *The Psychology of Sex Differences* [408]. These authors sifted through a large body of psychological studies in order to identify the most reliable male-female contrasts. They categorized the data according to abstract attributes, such as "quantitative ability" and "task persistence and involvement," and according to more strictly behavioral criteria, such as "crying" or "donating to charities." When a sex difference showed up consistently in a large proportion of the studies listed under one of these headings, Maccoby and Jacklin accepted it as a "real" sex difference.

This approach to the literature led Maccoby and Jacklin to identify just four basic sex differences as real:

- Girls have greater verbal ability than boys.
- Boys have greater visual-spatial ability than girls.

- Boys have greater mathematical ability than girls.
- Males are more aggressive.

A number of other alleged sex differences were listed under the heading "Open Questions: Too Little Evidence, or Findings Ambiguous." Still others were considered to be unfounded myths and stereotypes.

Maccoby and Jacklin's is an outstanding review of a confusing mass of data, and it deserves the serious attention of everyone interested in sex differences, but their psychological approach to the issue has some serious flaws. One important problem concerns the traits by which the data were organized. Consider, for example, the popular notion that girls are more "social" than boys. Maccoby and Jacklin considered this proposition unfounded and spent pages debunking it, concluding that "Any differences that exist in the 'sociability' of the two sexes are more of kind than of degree. Boys are highly oriented toward a peer group and congregate in larger groups; girls associate in pairs or small groups of age-mates, and may be somewhat more oriented toward adults, although the evidence for this is weak" [408, p. 349]. This is a fair and interesting summary statement, but it does not justify tossing "sociability" into the ashcan of mythical sex differences. There are some real behavioral differences here, most notably the tendency of little boys to form big groups, a phenomenon that has been observed in a number of cultures and settings. If other measures of "sociability" do not show the same effect, then they are not measuring the same thing. It is the concept that will not bear scrutiny, not the data.

The same tendency to throw out the baby with the bathwater of an overly broad concept arises elsewhere. For example, Maccoby and Jacklin include "dominance" among their "open questions," yet they concede that "Dominance appears to be more of an issue within boys' groups than girls' groups. Boys make more dominance attempts (both successful and unsuccessful) toward one another than do girls. They also more often attempt to dominate adults" [408, p. 353]. The alleged ambiguity of the evidence resides in the fact that males do not necessarily dominate females. This is another matter, and subsuming these several phenomena under the single label "dominance" is unenlightening.

What we are criticizing here is a widespread vice of psychologists in particular: the abstraction of all-encompassing "traits" from a set of very different behaviors, which are then "explained" in terms of those same traits. In suggesting that girls are more "oral" than boys, for example, one author tells us that newborn girls open their mouths when their hands approach their faces more than do newborn boys and that adolescent girls who are referred for psychiatric help more often have alimentary problems than do boys [200]. Now, for all we know there may actually be some link between these two phenomena, but it is far from obvious what that link might be! Calling both "orality manifestations" is not helpful. Abstract explanatory concepts like "orality"—and "dominance" for that matter—are usefully postulated only when there is some justification for believing that the observed phenomena are really related, whether causally, developmentally, or functionally. If, for example, individual differences in some objectively measured infantile oral activity were shown to be predictive of the later psychiatric conditions, then the orality concept would gain some validity. But even a valid trait is only a partial predictor of behavior, so that a real sex difference in some dominance behaviors, for example, need not imply any such difference in others.

A further problem with Maccoby and Jacklin's review, and indeed with much of the sex difference literature, concerns the meaning of their quest for "real" sex differences. This "reality" is a woolly concept. It excludes not just statistical artifacts, bad experiments, and unreplicable results, but also those differences that are age-specific, situation-specific, or culture-specific. It is a quest for immutable human nature, and it implicitly resurrects the false dichotomy of nature versus nurture. Consider the insistence that a sex difference should be replicable across cultures before it is to be believed. This universalist criterion ignores most of the actual sextyping of human behavior. Murdock's tabulation (Table 10-1) clearly shows that the same activity can be strongly sex-typed in opposite ways in different cultures. If a sex difference observed in one society reverses itself in another, this indeed suggests that the nurture of a specific socialization process is involved in its development, but the phenomenon is no less "real" (or natural) for being culture-specific and reversible. Conversely, if a sex difference were found to be truly universal, we might surmise that there is some fundamental constraint upon the division of labor in the nature of the sexes, and yet socialization is still likely to be relevant to the nurture of the observed sex-typical behavior. Thinking in terms of nature versus nurture is nowhere more mischievous than in the study of sex differences.

Ultimately, by whatever developmental processes, women and men behave differently. In the United States, 85 percent of murderers and 93 percent of drunken drivers are men [634]. In the economic sector called "Transportation and public utilities" just 1 percent of the labor force is female [629]. These statistics represent enormous differences in the actual behavior of men and women. Such differences may or may not be culturally arbitrary, and they may or may not disappear in the near future, but they are facts. Sex differences like these are no less "real" than differences in "aggression" or "verbal ability."

HUMAN SEXUAL DIFFERENTIATION: SOME CLINICAL RESULTS

It is not at all easy to tease apart the contributions that different developmental factors make to the sexual differentiation of brain and behavior. This problem is especially acute when we consider our own species, for we cannot then appeal to the results of controlled experimentation. Nevertheless, nature and medical malpractice have each provided us with some unintentional experiments on humans, and we can learn a good deal from them.

A case of twins

In a unique case described by John Money of Johns Hopkins University [441], the penis of a normal male child, seven months of age, had to be removed because of a mishap when circumcision was being performed. The boy's parents, who were from a rural American background and had only grade-school education, made the decision to reassign the child's gender and to raise him as a girl. And so, at seventeen months of age, the child's name, clothing, and hairstyle were changed. Four months later, the child was castrated and a first step in genital plastic surgery was undertaken. What makes this case especially interesting is that the child in question had an identical twin brother. So two genetically identical children, each of whom had presumably been exposed to a normal male prenatal environment including fetal androgens, were raised together in the same household, one as a boy and one as a girl!

The mother set about the task of feminizing her daughter with determination. Eighteen months after the gender reassignment, the mother wrote that she had made a special effort to keep her daughter almost exclusively in dresses. "I even made all her nightwear into granny gowns and she wears bracelets and hair-ribbons." In another year the little girl expressed clear preference for dresses over slacks and took pride in her long hair. Interviewed when the children were four and one half years old, the mother reported, "She likes for me to wipe her face—she doesn't like to be dirty and yet my son is quite different. I can't wash his face or anything. . . . She seems to be daintier. Maybe because I encourage it." Elsewhere in this same recorded interview the mother said, "One thing that really amazes me is that she is so feminine. I have never seen a little girl so neat and tidy as she can be when she wants to be . . . she is very proud of herself. When she puts on a new dress or I set her hair she just loves to have her hair set, she could sit under the dryer all day long to have her hair set. She just loves it'' [441, pp. 119–120].

For Christmas the girl wanted and received dolls, a dollhouse, and a doll carriage. The boy wanted and received a garage, cars, gas pumps, and tools. But not all was smooth sailing for the determined mother. The little girl was described as having several tomboyish traits, including explosive physical energy, a stubborn streak, and a tendency to dominate both her brother and her playmates. The mother made no bones about her attempts to modify such behavior: "Of course, I've tried to teach her not to be rough. . .she doesn't seem to be as rough as him. . . of course, I discourage that. . . I never did manage, but I'm going to try to manage them to—my daughter—to be more quiet and ladylike" [441, p. 122].

This famous case has sometimes been cited as proof that socializing experiences are of overwhelming primacy in the childhood development of sex roles. Such a conclusion is excessive. It must first be recalled that the child was surgically deprived of androgenic influences from an early postnatal age [cf. 591] and therefore cannot be considered an exemplary little boy differing from his brother *only* in having been socialized as a girl. Moreover, the latest information on this case in fact suggests that the sex reassignment may not have been entirely successful: In her young teens, the girl is reported to be unhappy, and is teased by her unwitting peers for her masculinity; despite the mother's zealous efforts at traditional feminine socialization, the daughter aspires to be a mechanic [153].

Congenital adrenal hyperplasia

Congenital adrenal hyperplasia (CAH) is a clinical condition of endocrine malfunction in which the adrenals produce an androgen instead of cortisol (cf. Figure 10-6). A genetic female with this recessive defect is likely to exhibit some degree of genital masculinization at birth. Enlargement of the clitoris is common, and labial fusion may also occur. If the condition is detected at birth, adrenal androgen production can be controlled with lifelong cortisone treatment, and the partially virilized external genitalia can be surgically feminized. Puberty, secondary sexual characteristics, and reproductive function can then be normal, although menarche is apt to be late [441].

CAH girls, detected early and unequivocally identified and raised as girls, are of special interest to the student of sex differences. Such girls differ from normals in their prenatal exposure to androgens. The cortisone therapy eliminates this difference soon after birth. What, then, are the effects of such prenatal androgenization on behavior?

This question has been carefully studied by Anke Ehrhardt and Susan Baker in Buffalo, New York. They compared seventeen CAH girls (all of whom were detected early and raised as girls) with their unaffected sisters and mothers in interviews conducted at ages between four and twenty years old [174]. By a variety of measures the CAH girls were tomboys. They engaged in more rough play and preferred boys as playmates far more than did their sisters and far more than their mothers recalled having done. They claimed fewer daydreams of wedding and marriage, pregnancy and motherhood. Fourteen of the seventeen professed "little or no" interest in dolls compared to just one out of eleven unaffected sisters. The CAH girls expressed less interest in clothes, in jewelry, in makeup, in their hair, in babies.

It is worth stressing that these girls, while decidedly tomboyish, were not in conflict about their sexual identity. Their behavior was within the range of acceptable female behavior. The parents were generally unconcerned about their daughters' interests and behavior and showed no signs of treating their CAH daughters as special. As teenagers the CAH girls became romantically interested in boys, and there was no reason to suspect that CAH girls were likely to become lesbians. The same tomboyish syndrome was found in an earlier study of CAH girls in Baltimore by Ehrhardt and Money [441]. The consistent results "suggest strongly that it is the fetal exposure to androgens that contributes to the typical profile of behavior exhibited by AGS [CAH] females" [174, p. 48].

Progestin-induced hermaphroditism

Even stronger support for this conclusion comes from the study of another clinical condition—progestin-induced hermaphroditism [441]. During the 1950s pregnant women were sometimes given synthetic progesterone in order to prevent threatened miscarriages. The hormonal manipulation had more effects than just the desired one. We have already remarked the partial functional substitutability of steroid hormones (Figure 10–6), so it might be anticipated that the early progesterone would have an androgenic effect. And that is just what occurred. The fetal daughters of women given this prenatal treatment were partially virilized in a manner very like the less severe cases of CAH. In interviews conducted by Money and his colleagues, nine of ten such girls called themselves tomboys, a label their parents confirmed. Like the CAH girls, they enjoyed competing against boys in energetic sports and were minimally concerned with feminine frills, dolls, and babies [441, 442; also 521]. These cases provide still stronger support for the notion that fetal androgens can somehow predispose a genetic female to a tomboyish behavior pattern. In the CAH syndrome that conclusion is more questionable: The presence of early androgens might be correlated with other effects of the recessive gene, and the effects of the condition are not separable from the effects of the postnatal cortisone therapy. Cases of progesterone-induced hermaphroditism are more strictly analogous to animal experiments on early organizing hormone effects. It is hard to doubt that the fetal hormones are indeed the critical factor in inducing tomboyishness. As in the case of the CAH syndrome, this virilizing effect did not extend to sexual orientation; twelve of twelve such girls developed heterosexually [442].

Not all girls suffering from CAH are detected early, and those with symptoms mild enough to escape medical attention may be reared as girls or boys according to the degree of masculinization of the external genitalia. Evidently, such children usually embrace the gender identity of rearing. If at some time after birth the condition is detected, it has been the fashion, both in CAH and in other types of genital abnormalities, for attending physicians to recommend "gender reassignment." The prognosis for such childhood sex change cases is not always good: A great deal of psychological distress and lifelong resentment of the reassignment sometimes results, especially if the decision is made after about two years of age. It seems that sex-role differentiation is already sufficiently under way by this very early age that it cannot be easily reversed [441]. The immutability of the older child's sexual identity cannot be put down to direct effects of the constitutional abnormality, for children with the same clinical conditions *can* be successfully reassigned if the decision is made very early.

It is certainly the case that small children identify themselves as members of their own sex, and this component of their identity is very important to them. A two-year-old may not be able to answer many questions, but the question "Are you a girl or a boy?" is one that the child can handle [603]. In fact, children categorize themselves by sex before they have much of a notion of what that categorization means. Two-year-olds sometimes have no more idea of the difference between girls and boys than that their clothes and hairstyles differ, yet are insistent about their own sex. There can hardly be a surer way to provoke an outburst of childish indignation than to teasingly suggest the wrong gender identity. The idea that gender identity is virtually immutable by age two or three has therefore been widely accepted. Recent studies of yet another clinical syndrome, however, call this proposition into question.

5α-Reductase deficiency

The enzyme 5α-reductase facilitates the synthesis of DHT from testosterone. A genetic deficiency of this enzyme thus impedes those aspects of sexual differentiation that normally proceed under DHT influence, especially the prenatal masculinization of external genitalia. Boys with 5α -reductase deficiency are likely to be identified as girls and raised accordingly until puberty, at which time the elevation of testosterone levels that is normal in pubertal boys produces change of voice, penile and testicular enlargement, erections, and ejaculations from a urethral orifice at the base of the penis, and masculine patterns of musculature [300, 494].

The 5α -reductase deficiency syndrome is very rare, but 38 cases have been discovered in a rural district of the Dominican Republic [301]. This rash of related cases began at about the turn of the century. The first several cases in each village, 19 in all, were reared unambiguously as girls, but recent cases have been recognized early and treated as special. There are adequate social histories for 18 children raised as girls. Seventeen successfully changed to a male gender identity (although one of these men, heterosexual in orientation, continues to dress as a woman), whereas the eighteenth person retained a female identity, lives alone, and denies any attraction to women. "Although these subjects behave unequivocally as males, they experience certain insecurities because of the appearance of their genitalia. They view themselves as incomplete persons, and this attitude saddens them. They fear ridicule by members of the opposite sex and initially feel anxious about forming sexual relations" [301, p. 1234]. Nonetheless, 15 of these men have entered into common-law marriages, the normal marriage type for the society. The striking thing about the Dominican studies is the apparent ease with which these men adapted to what amounts to a spontaneous sex change in adolescence. This is a society with distinct childhood sex roles: Prepubertal girls help their mothers in the household, while boys accompany the men in agricultural activities. Nevertheless, after more than a decade of socialization to feminine sex roles, most men were able to assume fully male identities in society. Isolated cases of 5α -reductase deficiency in other countries generally indicate a similar capacity for sex-role reversal at a much later age than had been believed possible on the basis of the CAH studies [302, 545].

SEX-ROLE DEVELOPMENT IN CHILDREN

What does all this tell us about the processes by which people develop feminine or masculine self-images and behavior? The 5α -reductase deficiency studies certainly cast doubt on the popular view that sex-specific socialization sends a small child down one of two alternative roads. But these studies do not, as has sometimes been suggested, conclusively demonstrate an early testosterone-induced masculinization of the brain that is unmasked at puberty. Instead, they might reflect the bisexual brain's capacity to change gender identity if external pressures (one's appearance and its social consequences) demand. The CAH and progesterone-induced hermaphroditism syndromes, on the other hand, do seem to indicate some organizing hormonal effects upon sex-typical behavior and attitudes.

The development of neuropsychological measures that discriminate the sexes holds out some promise for future assessment of the extent to which these clinical syndromes might be accompanied by sexual differentiation of the brain at various ages. Recent research has focused upon sex differences in laterality, the extent to which the two hemispheres of the brain are specialized for different tasks [428]. It is well known, for example, that language functions are primarily controlled by the left hemisphere in most people. There are several indications that this sort of laterality is more extreme in men than in women. Men are more severely disabled by unilateral brain injuries; women evidently have a more bilateral representation of speech and other functions and hence are better able to cope with, and recover from, brain damage. For the investigation of laterality in intact people, neuropsychologists make use of the fact that sense organs and limbs are connected to the opposite hemisphere; by presenting distinct sensory information to each ear, for example, and measuring latencies to respond with each arm, an experimenter can infer what sort of information is processed in each hemisphere. The evidence of a variety of sex differences is mounting, though there remains much controversy on the subject [303, 428]. If reliable sexually differentiated measures can be developed, they should provide a valuable tool for the study of the neural ''sexual identity'' associated with the various clinical syndromes.

So what of experiential influences? The simple fact that sex roles vary from one human society to another demonstrates that the acquisition of behavior appropriate to one's sex is in part the same process as the assimilation of one's own culture. And it is clear that children learn about their natal culture from a number of sources—from parents, peers, and relatives, from mass media and institutions such as schools and churches.

While we say that children learn about their natal culture, that does not imply that they have any conscious appreciation of the culture's norms or the ability to articulate them. We infer the learning from the performance: The child who behaves in accordance with the cultural definitions of appropriate behavior is an encultured child. It is not enough to say that the learning is imitative or observational, since the imitation is selective. Children raised by transsexual or homosexual parents, to take a dramatic example, grow up to be ordinary heterosexuals [226].

Most of what a child does, at least from the age of a year or so, is to some extent culture-specific. All children play, but the games vary. What does the child want to wear, to eat, to watch on television? Neatness, aggression, and reactions to strangers are behavioral aspects on which children differ, and the differences are in part predictable from the particular culture and the child's sex. Much of this must surely be picked up from parents, but a good deal of childish sex-typical behavior is part of a children's subculture [472] of which adults may be only dimly aware.

Although it may be clear that parents and peers influence a child's behavioral development, it is not entirely clear what is the best way to conceptualize this influence. In American psychological science, an elaborate "learning theory," derived primarily from studies of animals in restricted environments, has dominated thought about development for decades. According to this theory, behavior changes as a result of extrinsic rewards contingent upon behavioral performances. (The rewards are taken as "givens" and are in practice often surmised from the behavioral change, a catch that makes it possible to apply this "reinforcement" theory to any and everything.)

No doubt children learn sex roles, and psychologists have been inclined to trace behavioral sex differences to differential parental administration of rewards and punishments. But this approach is surely too simple. The evidence of differential treatment is often surprisingly thin [172, 408], and where it occurs, the parents may be *reacting* to sex differences as much as creating them. A pattern that appears to be cross-culturally robust, for example, is for three- to six-year-old girls to be more willing to mind younger children than boys; mothers don't necessarily assign more child care duties to girls, but girls are likelier to comply when asked [172]. Similarly, boys are punished for aggressive behavior more often than girls because they aggress more [408]. This sex difference seems to develop despite parental behavior rather than because of it. Before dismissing discriminatory parental influence, however, we should note the probable effectiveness of subtle differences in parental communication of approval and disapproval; children easily recognize amused indulgence lurking behind a parental scolding.

The sociobiological analysis of the parent-offspring relationship [8, 620] offers a radically different perspective that has yet to be extensively applied in developmental research: Rather than viewing the parent as merely an enculturating agent and the child as putty, we may consider the two parties as self-interested actors with some commonality and some conflict of interests (Chapter 3, pp. 55-58). Differential treatment of sons and daughters may serve parental interests, as we shall see (Chapter 12, pp. 336-337).

STRATEGIC ASPECTS OF SEXUAL DEVELOPMENT

The history of developmental processes within the lifetime of an individual organism explains observed behavior at one level (cf. Chapter 1). A major intention of this book is to relate different explanatory levels to the functional or strategic level. Developmental processes have clear functional significance insofar as they are steps toward an end. Embryological description inevitably consists in part of purposive-sounding statements: This structure is a precursor of that organ, this event an antecedent of that process. Embryological concepts may encapsulate the scientist's awareness that the developing organism is going somewhere: A noteworthy example is "canalization," which denotes the fact that diverse developmental pathways may lead to the same endpoint [645].

We have seen that a great deal of mammalian sexual differentiation takes place under the influence of a single simple mechanism—the early production of steroid hormones by the male only. Given the existence of such a mechanism that controls the differentiation of reproductive organs, it is not surprising that other sex differences should come under its control too, in the course of evolution. The reproductive strategies characteristic of the two sexes vary among species in many ways; in highly polygynous species, for example, males exhibit more severe aggression than in monogamous ones. The proximate control mechanisms for such sex-typical behavior, such as androgen-sensitive neural structures subserving aggressive behavior, develop under the influence of the same basic sex-differentiating early steroid mechanism that is responsible for the differentiation of sexual organs. This makes sense, since the linkage between the individual's sex and behavior is of selective importance. Optimal male strategy differs from optimal female strategy, and the developmental processes that are favored by natural selection should assure that individuals behave in ways appropriate to their own sex.

It also makes good functional sense that sexual differentiation of brain and behavior should take place later in development than sexual differentiation of gonads and other sexual organs. How and whether the sexes differ neurally and behaviorally is highly variable between species. We have seen, in the discussion of socioecology (Chapter 7), that taxonomy is often a poorer predictor of mating systems and other aspects of sociosexual behavior than is ecology, so that there may be all manner of different reproductive strategies among closely related species. What this means is that sex-typical behavioral strategies are readily altered during a species' evolution. It is a maxim of comparative embryology that the evolution of species differences is primarily due to the accumulation of mutations that cause only small changes acting late in development [645]. It is only such small, late changes that stand a reasonable chance of being beneficial. Organisms are complex adaptive machines, and any novel mutation that alters their normal development is unlikely to improve their adaptive functioning. A small random change in a complex machine will probably harm it; a large random change certainly will do so. Early random modification of a developmental program will have widespread consequences that are sure to be destructive. (The embryological conservatism of natural selection is the truth behind the ancient claim that ontogeny recapitulates phylogeny. Early embryos of various species do indeed look remarkably alike, for their evolutionary divergence has occurred as the result of changes in later developmental processes.) In the present context, it is to be expected that the most evolutionarily malleable aspects of sexual reproduction, namely the sex-typical behavioral strategies subserving reproduction, would differentiate relatively late, and indeed they do.

The social development of immature animals, and in particular the developmental differences between the sexes, also warrant consideration from a functional perspective. We have seen in earlier chapters how the mature behavior of the two sexes often differs—how females are commonly more discriminating in their choice of mates than males, for example, and how males are commonly more aggressive. These differences can be given functional explanations, since the optimal reproductive strategies of the two sexes differ, dictating a whole complex of dimorphisms, behavioral and otherwise (cf. Chapter 5). But why should the sexes differ behaviorally before they are reproductively mature?

The answer must be that immature behavior, like adult behavior, is under the selective influence of eventual reproductive success. This success depends upon the entire developmental history of the individual because the mature phenotype depends upon that history. Hence, developmental variations are subject to the effects of natural selection. From this perspective it seems probable, for example, that the relatively rough play of young male monkeys as compared with females is not just a functionless by-product of sexual differentiation but is instead useful experience for the males, whose adult reproductive success will depend upon their competitive and aggressive skills, which are partially dependent upon this youthful play experience [592]. Likewise, the interest that young females show in infants is apparently functional in the development of maternal skills. In fact, parental experience is known to be relevant to reproductive success in several vertebrates (e.g., fulmars, Figure 8-4, p. 193), and not just the "higher" mammals.

The sexes are also apt to differ in the extent to which their sexual behavior and sexual preferences are modifiable as a result of individual experience. In guppies, for example, males have to learn to direct their courtship to females of their own species, whereas females require no such experience [382]. This sort of sex difference appears to be widespread. There is also evidence that some mammalian males must learn efficient copulatory behavior through experience, whereas female performance tends to be more nearly perfect on the first occasion. These sex differences may be partially attributable to a greater selection pressure upon females than upon males for an error-free sex life, since a mistake can cost a female considerable time and energy (cf. pp. 113–115).

A final strategic consideration in sexual development concerns mechanisms by which the genes are protected. Ova are remarkable in that they begin their final meiosis long before they complete it. The primary oocyte may remain dormant in an early stage of meiosis for many years, while somatic cells are repeatedly dividing and replacing themselves. It is plausible that this special behavior of the germ cell has been selected specifically to minimize replication errors. Mutational changes occur during the process of cell division, and the larger the number of successive cell divisions, the more likely it is that there will be a mutation at any particular locus. Accordingly, mutation rates are ideally measured per division rather than per generation. The early embryonic appearance of primordial germ cells, which thereafter remain isolated from the complex process of somatic differentiation, seems a device peculiarly suited to the task of minimizing the number of cell divisions between generations and hence maximizing the faithful transmission of parental genes.

A functional perspective upon developmental processes underlies all our ideas about the relevance of early events to later ones. It allows us to formulate hypotheses about the significance of phenomena as diverse as suspended meiosis and juvenile play. Without a functional perspective, the study of development would be a barren descriptive exercise. With such a perspective, that study illuminates and is illuminated by the whole of biology.

SUMMARY

Organismic development is an elaborative (epigenetic) process, rarely reversible, in which tissues differentiate in form and function. At every stage this involves complex interactions among genes and the intracellular, extracellular, and extraorganismic environments, so that the attempt to separate genetic and environmental processes in development is misguided and futile.

The internal and external sex organs and certain sex-typed aspects of brain and behavior all differentiate in male or female direction as a result of the presence (leading to masculinization in mammals) or absence (leading to feminization) of early gonadal function. Early castration can make a genetic male phenotypically female, and early androgenization can make a genetic female phenotypically male.

Sex differences in human behavior can be hard to pinpoint. The data of cultural anthropology indicate that sex roles are important everywhere but that few activities are sex-typed consistently across all cultures. A psychological approach suggests a few consistent sex differences in abstract abilities (e.g., "verbal skill") and inclinations (e.g., "aggressiveness"), but such abstraction of traits can obscure behavioral differences. Children acquire their notions about sex-appropriate behavior from parents, peers, and others. They attach great importance to their own gender identities, which may be irreversible as early as two years of age.

Developmental processes have evident directionality and hence lend themselves to functional interpretation. Developmental differentiation functions to bring about phenotypic sex differences in maturation age, aggressiveness, selectivity, and so forth—sex differences with adaptive functions that were discussed in earlier chapters. The developmental buffering of primordial germ cells may function to protect parental genes from excessive risk of replication errors.

SUGGESTED READINGS

Goy, R. W. & McEwen, B. S. 1980. *Sexual differentiation of the brain.* Cambridge, Mass.: MIT Press.

Science, 1981. March 20 issue (#4488) of vol. 211. (A special issue on sexual development and differentiation.)

11

Human reproductive competition

The relationship between women and men—the complementarities and the conflicts—here is an inexhaustible theme for the arts and the sciences. Yet this immense subject is only a part of a larger issue: the significance of sex and sociality in the biological world. Any view of women and men that fails to situate their relationship within this larger comparative framework must be incomplete.

People, like other organisms, are the products of natural selection, a process of the competitive ascendancy of features contributing to fitness. It follows that human attributes should be describable, at some level of abstraction, as the tactics of fitness-promoting strategies. We will approach what is usually called human reproductive biology—the physiological facts of puberty, pregnancy, nursing, and so forth—from the reproductive strategic perspective in the next chapter, but first we intend to apply this perspective to some social and psychological phenomena, with special attention to sex differences.

Human cultures are so variable that one might despair of arriving at a valid general description of our species' behavioral attributes. The problem is to find that level of abstraction at which phenomena in diverse cultures can be described in common terms. Details differ, but *all* human societies exhibit music, competitive games, kinship terminology, and a host of other "cross-cultural universals" [450]. Among the cross-culturally general features of human society are certain aspects of male-female transactions, and in this chapter we shall analyze these general features in terms of the fitness interests of the actors.

BATEMAN'S PRINCIPLE REVISITED

The body of evolutionary biological thinking that is most relevant to this task is Bateman's principle and its corollaries, which we discussed in Chapter 5, and will briefly reiterate: In most animal species, the female's greater investment in each offspring means that her maximal reproductive potential is lower than the male's. Males therefore compete among themselves for fertilization opportunities. Investing little in each offspring, males are selected to sow their seed wherever opportunity arises. Investing considerably in each offspring, females are selected to exhibit greater selectivity in their choice of mates. One feature on which females may exercise selectivity is the male's willingness or ability to make an effective parental contribution. But wherever males do in fact invest parentally, they are under selective pressure to protect themselves against cuckoldry, and therefore males have a greater concern than females over the fidelity of their mates.

One expected consequence of this selective scenario is that males should have evolved to be more promiscuously inclined than females. Attitude surveys are unanimously confirmatory. One survey of middle-aged, middle-class American couples, for example, found that 20 percent of the men and 10 percent of the women claimed to have engaged in extramarital sexual intercourse. The opportunities for one sex are of course constrained by the willingness of the other: 48 percent of the men and just 5 percent of the women said that they would *like* to engage in extramarital sex in the future [318]. The results for a group of young, single, working-class Germans with steadies are strikingly similar: 46 percent of men and 6 percent of women said that they would take advantage of a casual opportunity for sexual intercourse with someone attractive [568]. Interviewees of both sexes were inclined to agree that women are "naturally more faithful than men" and that men's sexual desires exceed women's. A study of Israeli adolescents raised in kibbutzim is of special interest in this context, for these children live in coeducational dormitories, have ready access to contraception and abortion, and are exposed to a vigorously advanced ideology of sexual egalitarianism; nevertheless, over 40 percent of boys and less than 10 percent of girls consider casual intercourse "legitimate" [18]. This sort of sex difference is certainly not peculiar to Western society. Although quantitative survey data are rare, the ethnographic record (the corpus of anthropological descriptions of human societies) so abounds with examples of men's promiscuous yearnings and efforts that Alfred Kinsey, the pioneer of American sex research, felt justified in asserting: "Among all peoples, everywhere in the world, it is understood that the male is more likely than the female to desire sexual relations with a variety of partners" [328].

It is not just that men are promiscuously inclined—they seem everywhere to be concerned with controlling women's reproductive capacity [135, 156, 157, 158, 478]. Given male competition for sexual access to females, men will profit (in terms of fitness) from monopolizing particular women and assuring their paternity of those women's children, while remaining alert to extramural, low-cost sexual opportunity. The resultant tendency to long-term pair-bonding could be expected to selectively eliminate the sex difference in fitness variance and concomitant polygynous competition only if men were to content themselves with monogamous bonds. Typically, they do not.

MARRIAGE

In virtually all societies there is a relationship between wife and husband that has the following characteristics: some degree of mutual obligation, some persistence in time, some formalized societal sanction, sexual access (usually, but by no means invariably, supposed to be exclusive), and some form of legitimization of the status of any offspring. The universality of marital alliance has been questioned on the grounds that the privileges and duties of a husband in one society may be divided among several men in another. Certainly it is difficult to devise a simple definition of marriage that can be applied to all cases [528]. Yet a formal alliance between individual women and men, with the characteristics listed above, *is* species-typical. It may seem that we are belaboring the obvious here, but it is worth recalling that in the great majority of mammalian species, alliance is either absent or a much more temporary affair. It is particularly rare for a mammalian alliance to persist through gestation and lactation, yet such persistence is usual in human societies.

Marriage is a panhuman institution, then, and it is furthermore for almost everyone. One review of the ethnographic literature concludes that "in almost all primitive societies there is or was a pattern of early and almost universal marriage" [266, p. 24]. This statement applies to women more than to men. Bachelors are likelier to be found than spinsters. But a career of voluntary celibacy is an unusual and almost certainly a recently invented phenomenon.

Polygynous marriage

Most marriages are monogamous—one woman, one man. However, a substantial majority of human societies legally permit a man more than one wife (Figure 11-1). Polygynous unions remain rarer than monogamous ones, however, even within those societies that permit them. This is hardly surprising given male-male competition for wives and the pressures that are

FIGURE 11-1
Incidence of polygynous, monogamous, and polyandrous marriage practices in a sample of 849 human societies

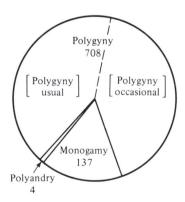

(Data from Murdock, 1967 [452]. Adapted from *Ethnographic Atlas* by George Peter Murdock by permission of the University of Pittsburgh Press. © 1967 by the University of Pittsburgh Press.)

engendered when any substantial number of men are consigned to involuntary celibacy. Hence it is usually only a minority of wealthy, powerful men who have the means to acquire and maintain multiple wives.

Polygynous marriages can be tempestuous. Co-wives have conflicts, especially when each has children who must compete for the same paternal resources and inheritance. These conflicts can be violent, even homicidal. Cases have been reported in which women killed their co-wives, or their co-wives' children, or their husbands, in resentment of such inequities as the man's having diverted the fruits of one wife's labors to the other wife's children [60, 392]. Although senior wives sometimes welcome junior wives as helpers, they more often resent them as competitors, and indeed the women in polygynous unions do seem to suffer reductions in material welfare and in fitness (see Figure 6-6, p. 127).

Little wonder, then, that women so often resist polygyny. That very

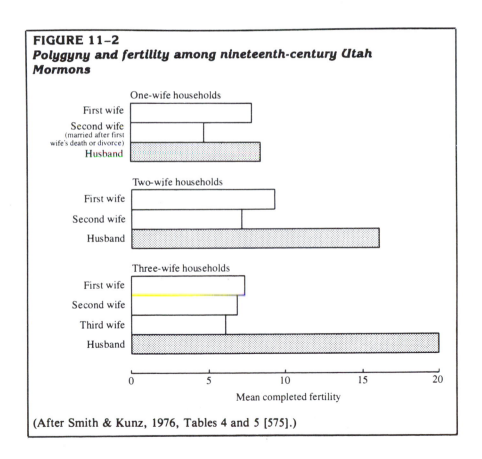

FIGURE 11-2
Polygyny and fertility among nineteenth-century Utah Mormons

(After Smith & Kunz, 1976, Tables 4 and 5 [575].)

resistance must raise the costs for any man who aspires to polygyny. And yet men retain that aspiration. In the Western world, where monogamy is legislated, there is a steady trickle of anecdotal accounts of men of means maintaining multiple mates, usually in separate households, though the prevalence of this phenomenon can only be guessed at. "Isn't one wife enough?" inquires a chronicler of Mormon life [704]. Polygynous Mormons were often obliged to establish separate domiciles for their contentious wives, whose demands included clear title to their own houses. Husbands had to adopt fixed schedules of daily or weekly alternation between their wives' households in order to minimize jealous conflicts. Yet, for all the financial and other costs of maintaining multiple wives, Mormon men, like polygynous men elsewhere, accumulated as many wives as they could afford. Polygyny clearly contributed to male fitness (Figure 11–2), and the highest-ranking churchmen enjoyed exceptional fitness: Nineteenth-century church leaders averaged about 5 wives and 25 children, compared with 2.4 wives and 15 children for polygynous Utah Mormons not in the church hierarchy, and 6.6 children for the monogamous [186].

The establishment of wives in separate households is the recourse of polygynous men in many cultures, but there is an interesting exception. Where co-wives are themselves sisters, then conflict between them is often mitigated so that cooperative cohabitation becomes more feasible. In fact, sororal polygyny may be the marriage pattern preferred by women in societies where women cooperate economically [see 308]. The data in Table 11–1 show that sororal polygyny and the cohabitation of co-wives are highly associated across cultures. Within a single society, such as that of the Mormons, the same phenomenon is apparent, with sisters likelier than unrelated co-wives to dwell together in peace.

TABLE 11–1
Sororal polygyny and cohabitation

The data represent 377 societies in which marriage is "usually polygynous." In 74 societies (20 percent), co-wives are typically sisters ("sororal polygyny"); co-wives are much likelier to dwell together in such societies than in those in which the women are unrelated [452].

	Shared dwelling	*Separate dwelling*
Sororal polygyny	60 societies (81%)	14 societies (19%)
Nonsororal polygyny	96 societies (32%)	207 societies (68%)

Source: Adapted from *Ethnographic Atlas* by George Peter Murdock by permission of the University of Pittsburgh Press. © 1967 by the University of Pittsburgh Press.

Through most of human history, polygyny was probably not so extreme as it became after the invention of agriculture perhaps ten thousand years ago. In preagricultural economies, there was little basis for the sort of disparities in material wealth that would permit a few successful men to monopolize numerous women. Studies of modern foraging ("hunting-and-gathering") peoples provide support for this suggestion. The Kalahari !Kung San, for example, are a people whose way of life is probably as close as any to that of ancestral *Homo* (pp. 324–327); about 5 percent of married men have two wives and hardly any have more [374]. Among Australian aborigines, marital polygyny was more intense than in any other known foraging culture, the older men having several wives; but the degree of effective polygyny in the breeding system was probably not really so extreme, since adulterous affairs between young wives and young unmarried men appear to have been rampant [e.g., 261, 433].

Clearly, the marital structure of a population does not provide us with a direct description of the effective mating structure. If a marriage contract provided a man with a magical guarantee of paternity, the world would be a more peaceable place! No, the degree of marital polygyny in a society is not a perfect, direct measure of the degree of effective polygyny and hence the intensity of male-male competition. But it is an indicator. In all societies, polygynously married men are concerned to "protect" their wives from other men, their reproductive competitors.

With agriculture, settlement in towns, and new heights of stratification of social status and wealth, it became possible for the most powerful men to monopolize harems containing dozens or even hundreds of women, a phenomenon that was invented in emerging state societies the world over. The greatest extremes were reached in the most despotically hierarchical systems, for which there are several reports of royal harems exceeding a thousand women. In the words of one anthropological reviewer, "It may be wondered whether such large numbers of harem inmates could possibly have any reproductive significance. I think it is well not to underestimate their masters' capacities. Van Gulik's . . . description of Chinese Imperial Harem procedures, involving copulation of concubines on a rotating basis at appropriate times in their menstrual cycles, all carefully regulated by female supervisors to prevent deception and error, shows what could be achieved with a well-organized bureaucracy. Given nine-month pregnancies and two- or three-year lactations, it is not inconceivable that a hardworking Emperor might manage to service a thousand women. (Van Gulik records much grumbling on the part of overburdened polygynous males.) An early 20th century observer reported that the Nizam of Hyderabad became the father of four children in the space of eight days with 9 more expected the following week, though how long this rate of achievement persisted is not stated (Cooper . . .)" [156, p. 15].

One consequence of this exaggerated polygyny was the creation of large numbers of disenfranchised celibate men consigned to life as soldiers, brigands, monks, and the like. Their existence must have provided a constant wellspring of disgruntlement and revolutionary potential, to be exploited by the various grass-roots religious and democratic movements that advocated limits on polygyny. The spread of legislated monogamy in recent history may likewise be attributable in part to the relative inability of intensely polygynous systems to command the loyalty of their foot soldiers. It is not farfetched to suggest that much of history may be interpreted as the effects of reproductive competition among men [9].

Polyandrous marriage

What are we to make of those rare cultures in which marriages are regularly polyandrous, a woman having two or more husbands (Figure 11-1)? There is no single answer to this question. The several societies in which polyandry is common are concentrated in two main parts of the world—Nigeria and the Indian subcontinent. Within each of these regions there are important differences among several polyandrous cultures, but as a first superficial descriptive device, we can distinguish between an African and an Asian type of polyandry.

Most of the African examples of polyandry concern the contracting of multiple alliances by a woman's having one husband in each of several clans. In some cases, the woman never cohabits with certain of her nominal husbands, but where she does cohabit with each husband, she does so sequentially. Two or more husbands of the same woman are not members of the same household. The marriage system is polygynous as well as polyandrous: Members of both sexes contract multiple marriages. And whereas a woman does not have two husbands living with her simultaneously, a man may have two wives in his household [449]. In terms of their effective breeding structure, these Nigerian systems are probably more polygynous than polyandrous.

It is on the Indian subcontinent, mainly in the Himalayas, that we find the few recorded cases of simultaneously cohabiting polyandry. The commonest situation is one of fraternal polyandry, brothers taking a common wife. In recent Tibetan society, for example, polyandry appears to have functioned as a wealth-conserving system [see 36]. It was practiced primarily among an élite class of prosperous serfs, the Tre-ba, who had large landholdings and substantial tax obligations, and it was a deliberate device for preventing the partition of family holdings: In each generation of a Tre-ba family there was only one marriage. Sons generally took a common wife and remained in a single household. Sexual jealousy among the brothers was an acknowledged problem, and those involved in such an arrangement were likely to express the opinion that monogamy is a preferable form of marriage, but not if it leads to partitioning land among sons [268]. Polyandry was thus seen as the lesser of two evils. Other Tibetan classes, lacking such property, practiced almost exclusively monogamous marriage.

The Pahari are a north Indian people among whom wives are purchased at a substantial price [46]. Brothers typically get together and buy one wife, then add another later, when they can afford her. The eventual result is group marriage, or **polygynandry**: a single household containing two or more husbands and their two or more wives, with all men considered to be married to all women. The Pahari are the only human society in which polygynandry is the actual majority practice and furthermore corresponds to the expressed preference of the people.

Human polyandrous mating structures are never the mirror-image equivalent of polygynous marriage as it is practiced in most societies. In all polyandrous systems polygynous marriages can occur too, legally and in fact. (Even among the Tre-ba of Tibet a polygynous marriage is contracted when there are daughters but no sons: The daughters take a husband, and the family holding is passed on through them.) The brothers in a polyandrous Himalayan household are not cast in a role similar to that of the co-wives in a polygynous marriage. As in most societies, men acquire wives; women do not acquire husbands. If the marriage fails to produce an heir, it is not the wife but the husbands who are considered to have a grievance, and it is they who take steps to remedy the situation. In the home the men are domineering, and the wife is obliged to defer to each of her husbands [379]. Adultery by a wife may provoke violence or renunciation on the part of her husbands [268, 379]. In all these features male and female roles remain as they are in most polygynous and monogamous societies.

Whether the parties to a polyandrous union are acting in violation of their fitness interests is not altogether clear. In one recent analysis, Tibetan women in polyandrous unions actually produced fewer children than those in monogamous unions, so that neither women nor men seemed to profit reproductively [36]. The authors of this study concluded that the family's economic goals motivate behavior contrary to the predictions of an inclusive-fitness-maximization model. However, the economic and other circumstantial equivalence of the monogamous and polyandrous groups is hard to establish, and for all we know limitations of family size to preserve property and social rank may be an effective *long-range* fitness-maximization strategy [268]. But we should not be too surprised to find that people, in pursuing proximate economic and hedonic goals in evolutionarily unforeseen ecological circumstances, occasionally act against their fitness interests (see pp. 309–310).

Whatever polyandrous unions may be, they are clearly not female-dominated. There is a persistent myth that some societies are or were "matriarchies," but there is no evidence for this belief. Sigmund Freud, Friedrich Engels, Karl Marx, and other thinkers entertained the theory that matriarchy was a stage in the evolution of human society. Modern anthropology has discarded the idea. There are, to be sure, matrilineal societies, in which descent is traced through the women rather than the men. They are less numerous than patrilineal societies, but they are not rare, comprising about 14 percent of human societies [636]. And there are matrilocal societies, in which a married couple takes up residence with or near the bride's kin rather than the groom's. Matrilocality is often associated with matrilineality, and it characterizes about 17 percent of societies [636]. But there are no matriarchies in which men are treated as property by powerful women, and the known polyandrous societies are neither matrilineal nor matrilocal. The kingdom of the Amazons is a fiction.

Lest we overstate the case, we must add that we are not denying that each sex has its spheres of influence in every society nor that women make decisions that affect the lives of men and women alike. Of course they do, every day in every part of the world [e.g., 532]. Any claim that either sex wields more power than the other probably hangs on a rather arbitrary definition of what constitutes power. A man may appear to dominate his wife, may even beat her and order her about, and yet it may well be that she controls his behavior as much as he controls hers. Men usually appear to be running the show, but it can be argued with some justice that they are peripheral hangers-on whose posturing and prancing is largely irrelevant to the essential business of human life and procreation. But even if men are not necessarily the prime movers of society, they certainly make themselves conspicuous! Men are everywhere the more political sex. They wheel and deal, bluster and bluff, compete overtly both for valuable commodities and for mere symbols. Ultimately, these male machinations reflect a struggle for access to female reproductive capacity.

MARRIAGE AS A REPRODUCTIVE UNION

Marriage is of course a rather complex affair. Many anthropologists would protest that it is a union more economic than reproductive, and hence but weakly analogous to the mateships of nonhuman creatures. This argument won't wash. The hornbill who feeds his incubating mate, the pair of beavers maintaining their dams and domicile—we could call these unions, with their division of labor and exchange of benefits, "economic" too. In people, as in other animals, the mundane interactions of mated pairs are seldom of immediately reproductive function, and yet the union can only be understood

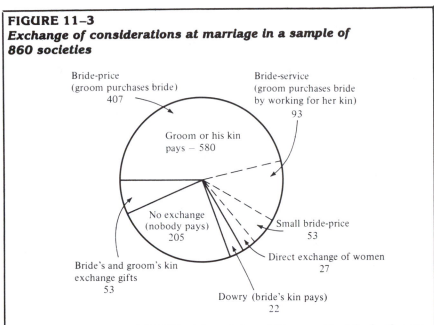

FIGURE 11-3
Exchange of considerations at marriage in a sample of 860 societies

(Data from Murdock, 1967 [452]. Adapted from *Ethnographic Atlas* by George Peter Murdock by permission of the University of Pittsburgh Press. © 1967 by the University of Pittsburgh Press.)

as a fundamentally reproductive alliance. In any species, individual organisms accumulate and allocate resources; their behavioral control mechanisms therefore subserve proximate "economic" functions. But more ultimately, the function of this economic activity is reproductive. To contrast economic and reproductive goals, then, is to confuse proximate and ultimate considerations.

What, after all, is marriage all about? Claude Lévi-Strauss [380] proposes that it is universally a contract between *men,* a formalized exchange of women as commodities. Though the universality of this proposition is debatable, it certainly rings true in a substantial majority of societies (Figure 11-3). Sometimes the exchange is a direct trade of two women; sometimes it involves an exchange of a woman for other valuables; and sometimes no payment is made but an obligation of future reciprocity is incurred. (Many of the 205 "no-exchange" examples in Figure 11-3 are of this latter type.) In our culture and many others a man *gives* his daughter in marriage. Men *purchase* wives the world round, and they commonly demand their money back if the acquisition proves unsatisfactory. (The practice of dowry may spring to mind here as evidence that payment can go in either direction, but dowry and brideprice are not opposites. Brideprice is paid to the bride's kin to purchase her, but dowry is not paid to the groom's kin to purchase him. The dowry generally remains with the newlyweds; it is part of the package. Furthermore, dowry is a much rarer practice than brideprice, being confined mainly to Europeans.)

The degree to which control is exercised by kin groups, rather than the couple themselves, indeed makes human marriage unique among animal mateships, and there can be elaborate political motives in marital negotiations between kin groups. Yet the alliance is no less a reproductive union for being a political alliance as well. What men are buying when they buy wives are breeders. Women are supposed to reproduce. C. S. Ford reviewed ethnographers' accounts of two hundred human societies and remarked that "a particularly striking fact emerging from this survey is the universal concern over the inability of an adult woman to conceive" [194]. If the woman is barren, the man has a grievance. In a great many societies he may divorce her, and in many of these he may demand the return of his bridal payment or even the wife's replacement by a younger sister. In industrial societies, infertility remains a major reason for marriage breakup. Dissolving an unproductive union may of course be as adaptive for the woman as for the man. As in birds who "divorce" after reproductive failure, both parties may try again elsewhere [518].

The exchange of women as commodities has probably become exaggerated since the invention of agriculture. In foraging societies, women often seem to have been more autonomous, perhaps entering a first arranged marriage but later leaving it to make a match of their own choice [70]. Marriage may then be viewed as a reciprocal exchange of benefits between husband and wife, perhaps ultimately as an exchange of paternal investment (given by the man) for paternity of offspring (given by the woman) [308]. But even in strongly patrilineal societies, such as Mediterranean pastoralists, where women are very much treated as male property, the role of mature women as matchmakers and political forces has probably been underreported [70], due in part to a male-biased ethnographic record. Lévi-Strauss's picture of marriage as an exercise in male politics may be somewhat exaggerated, then, but we think it highly significant that men think and talk this way. Men *strive* to control women and to traffic in female reproductive capacity [e.g., 478] with varying degrees of success; this ambition is precisely what one might expect if men have evolved to pursue fitness.

THE DOUBLE STANDARD IN ADULTERY LAW

Men are of course concerned not merely that their wives breed but that their own paternity is assured. We have discussed the evolutionary consequences of paternity uncertainty (Chapter 7, pp. 163–169) and noted the prevalence of mate-guarding and other anticuckoldry tactics in species with substantial paternal investment. *Homo sapiens* is of course such a species: Human fathers commonly make considerable investments in the children attributed to them, and selection must have favored whatever behavioral propensities increase the probability that that investment is not misdirected to nonrelatives.

Some social scientists have suggested that men are not concerned with the biological paternity of children but only with "sociological paternity," that is to say with the right to claim those children and to incorporate them into the kin network of mutual support and obligation. This argument is hard to reconcile with the ethnographic record, which is replete with articulate and powerful conscious expressions of the urge to invest only in one's own offspring. When a Yanomamö man acquires a previously married wife, for example, he may kill her infant or order her to do so. He makes no bones about the reason: The baby is another man's child, not his, and it competes for the time and milk that the woman should be devoting to breeding for him [87]. The practice of adoption may seem anomalous, but the most typical features of adoption are in fact consistent with the view that people are nepotistic strategists: Adoption is primarily the recourse of the infertile or postreproductive, preferentially involves close genealogical relatives, and is frequently accompanied by discriminative favoring of natural children [130, 570]. This is not to deny that the psychological mechanisms of parental love can be redirected to an unrelated child. After all, parental love can sometimes even be redirected to a nonhuman pet! But parental solicitude is far from indiscriminate with respect to relationship (see pp. 42–45).

In a great many societies, suspicion of adulterous conception is grounds for infanticide or disposal of the errant wife. And wherever powerful men have been able to accumulate harems, they have invested heavily in guarantees of their sexual monopolization of the women; there are even reports of entire harems being put to death when the security system has been breached. The very word "cuckoldry" is believed to derive from the name of the cuckoo, the bird that lays its egg in another's nest. It is for the parasitizing of his parental investment that the cuckolded man is ridiculed or pitied.

Wherever people have codified law, they have incorporated limitations upon sexual freedom (beyond the incest prohibitions, to which we shall return). Through the ages and around the world, men have repeatedly invented *adultery laws* with a remarkable consistency of concept that reinforces Lévi-Strauss's view of wives as property: Sexual contact between a married woman and a man other than her husband is an offense, and the victim is the husband. It is important to recognize the typical asymmetry of adultery laws, since modern Western law uniquely (and only recently) treats women and men alike. Numerous ancient and modern legal codes, on every continent, have defined the offense of adultery by the woman's marital status, while ignoring the man's. When a married man has sexual intercourse with an unmarried woman, the man's wife is seldom considered a victim; if any party's *property* is considered to have been violated, that party is the unmarried woman's male kin. Whether the violator is himself married has no bearing on the case. (Moreover, it is much commoner for women than for men to display their marital status with markers such as wedding rings [402].)

Hadjiyannakis [240] has reviewed the history of European adultery laws, which he suggests have moved only slowly and recently toward equality of the sexes. Ancient Egyptians, Syrians, Hebrews, Romans, Spartans, and other Mediterranean peoples defined adultery according to the marital status of the woman and punished both parties severely, often with death. The offending man's marital status was irrelevant. The first known legal provision concerning a husband's infidelity is to be found in a Roman law of 16 B.C., according to which a man lost the right to confiscate his wife's dowry for adultery if he too was unfaithful. Male infidelity was not criminalized until 1810, and then in only a very limited way, when a French law made it a crime for a man to keep a concubine in his conjugal home against his wife's wishes. In 1852, Austria was the first country to institute legal equality between spouses, but even there the criminal code implicitly considered the cuckolded husband to be the offended party by retaining a provision raising the penalties for adultery if the paternity of a subsequent infant was thrown into doubt.

Besides criminalizing adultery, legal traditions commonly acknowledge that when adultery is discovered, a jealous rage on the part of the victimized husband is only to be expected. Jurists may then stipulate that cuckoldry justifies or at least mitigates responsibility for otherwise criminal violence. Even in English law, where homicide by cuckolds has been much less condoned than in continental Europe and elsewhere, observation of the adulterous wife *in flagrante delicto* has been considered a special provocation, reducing homicide to manslaughter of the "lowest degree . . . because there could not be a greater provocation" [57, pp. 191–192]. American juries often decide that even a reduced charge is excessively harsh, and instead vote to acquit homicidal cuckolds altogether on the basis of the "unwritten law," which is, according to Bouvier's *Law Dictionary,* "a popular expression to designate a supposed rule of law that a man who takes the life of a wife's paramour or a daughter's seducer is not guilty of a criminal offense." Such a law is both written and observed in many other countries.

The male inclination to constrain female sexual liberty, by legislation and by threat of violence, is readily interpreted as a consequence of the uncertainty of paternity. However, this interpretation need not imply a conscious link between the two. A bank swallow guards his mate zealously throughout her fertile period, but we need not invoke fear of mistaken paternity as a proximate motive of his actions. Similarly, men may often be "automatically" jealous of their wives. Nevertheless, paternity uncertainty has frequently been invoked as a justification for the double standard in adultery laws. The French revolutionaries, for example, strove to abolish unjust discrimination when they rewrote the laws, but they retained sexually discriminatory adultery legislation, arguing as follows:

> It is not adultery *per se* that the law punishes, but only the possible introduction of alien children into the family and even the uncertainty that adultery creates in this regard. Adultery by the husband has no such consequences. (Our translation of Fenet, 1827, as quoted in [240], p. 502).

To what extent is a double standard concerning adultery cross-culturally general? This question has been badly muddled in the anthropological literature. Several writers have insisted that single standards are widespread, and at least one author has reported a minority incidence of reversed double standards "favoring females" [668, p. 70]. These claims are based on a simple confusion. "Single" or "reversed double" standards have been inferred because adulterous men are reported to be treated every bit as harshly as adulterous women, or more harshly, respectively. Indeed they are, as the original ethnographic sources for the societies in question make plain, but the familiar double standard prevails: The men who are treated so violently are not "adulterers" by virtue of their *own* marital status but by virtue of sexually violating another man's wife. It is the same old story. The offended party is still the wronged husband. The offense is defined strictly in terms of the woman's marital status [135]. Single standards that identically restrict husbands and wives, if indeed they characterize any human society, are assuredly very rare.

Even where the law treats men and women alike, sex differences in behavior and attitude remain. Kinsey [328] found that 51 percent of divorced American men considered extramarital coitus by their wives to be a major factor in their divorces compared with just 27 percent of divorced women citing their husbands' extramarital excursions, even though the men were twice as likely as the women to have committed adultery [see also 604]. So although both sexes have the legal right to divorce for adultery, it appears that men are much likelier to feel that adultery warrants divorce.

SEXUAL JEALOUSY

"Jealousy" may be defined as a state that is aroused by a perceived threat to a valued relationship or position and that motivates behavior aimed at countering the threat. Jealousy is then "sexual" if the valued relationship is sexual. By defining jealousy in this way, we are able to consider behaviors as diverse as violence and vigilance to be manifestations of sexual jealousy.

Sociobiologists might anticipate certain differences between the sexes in the psychological content of sexual jealousy. This is because unfaithful spouses pose distinct threats to women and men. The threat to a man's fitness resides in the risk of alien insemination of his adulterous wife, whereas the threat to a woman's fitness lies not so much in her adulterous husband's sexual contacts as in the risk that he will divert resources away from the wife and family. It follows that male jealousy should have evolved to be more specifically focused upon the sexual act, and female jealousy upon the loss of male attention and resources. Such a sex difference was precisely the result in one study in which American undergraduate couples described their reactions to a jealousy-inducing, role-playing situation; the men were afflicted with fantasies of sexual contact between the girlfriend and a rival, the women with thoughts of losing the boyfriend's time and attention [600]. This finding warrants replication and extension. Also of interest in this context is a common tendency to blame the victim in rape; women who have been raped are often in double jeopardy from accusatory husbands or boyfriends [see 610]. Even where the "infidelity" has been clearly involuntary, men are greatly distressed by the violation of their rights of sexual exclusivity and often appear to devalue the woman accordingly.

Sex differences in the psychological content of jealousy have received rather little attention from researchers. What is much better documented is a behavioral difference: It is men who become violent over infidelity. Male sexual jealousy is far and away the leading motive in spousal homicides in North America, and almost certainly throughout the world [135]. If we include disputes between men over women, then male sexual jealousy may well be the number one motive in *all* homicides. Surveys of battered women indicate that male sexual jealousy is the leading motive in nonfatal wife beating too [135].

Again, we should like to know how cross-culturally consistent are these phenomena. And again, this is a point of some contention. Several anthropologists have implied that there are at least a few societies in which violent male sexual jealousy is absent. In one influential volume [195], for example, it is alleged that in 7 out of a sample of 139 societies, "the customary incest prohibitions appear to be the only major barrier to sexual intercourse outside of mateship. Men and women in these societies are free to engage in sexual liaisons and indeed are expected to do so provided the incest rules are observed" [195, p. 113]. An idyllic sexual liberty, it would seem. Yet when we consult the original ethnographic sources, we find that extreme violence by cuckolded men is reported to be characteristic of every one of the seven societies [135]! The "freedom" to engage in sexual liaisons apparently refers only to an absence of societally prescribed legal penalties. The threat of male violence provides a very real "barrier" to the exercise of that hypothetical freedom. Human societies are highly diverse in their sexual restrictions, but the diversity is not without limits and has often been overstated. In fact, constraint of female sexuality by the threat of male violence appears to be cross-culturally universal.

COERCIVE CONSTRAINT OF WOMEN

Men's proprietary attitude toward women's reproductive capacity is manifest not only in violent sexual jealousy but also in a variety of other confining "claustration" practices. Women of reproductive age are commonly clothed more modestly than men, up to the most extreme forms of veiling; and they are commonly chaperoned, guarded, and even incarcerated in veritable prison fortresses, for the acknowledged purpose of protecting their chastity. Mildred Dickemann, an American anthropologist, has reviewed the prevalence of these practices and made a convincing case for their interpretation as confidence-of-paternity tactics [158].

Claustration practices tend to occur in societies that are characterized by an ethic of "honor and shame." These concepts embrace a number of aspects of moral rectitude and reputation including bravery versus cowardice and high versus low birth, but most authors agree that the most essential component of familial honor is female chastity. A man can gain or lose honor by his deeds and by the deeds of his kin, though lost honor is not easily or quickly regained. A woman, however, can only lose honor for herself and her kin by unchaste or immodest behavior. For her, lost honor can never be regained. A woman thus has numerous male relatives whose honor is bound to her chastity. Their threats and vigilance over her virtue may be as constraining as any jealous husband could be. Not infrequently, men salvage some of their lost honor by killing an unchaste wife, sister, or daughter (and the male seducer). Shrinking from such vengeance may even add to their dishonor [345, 541, 542].

That men should endeavor to control the sexuality of female relatives, as well as that of wives, follows from the treatment of female reproductive capacity as a valued commodity. Chaste sisters and daughters make marketable wives. Dickemann points out that honorable families are in effect guaranteeing much more than virgin goods. More than merely guarded, the woman "must be socialized to value feminine modesty, to submit to the goals and demands of her future husband and his kin, to maintain the honor of his family, and finally to socialize her own daughters into the ideology of female purity. This is what the public reputation of the bride's family and the formal symbolic displays are intended to attest . . . for men, honor is the capacity and willingness to defend and to enforce the purity of their female kin, which allows them to achieve reproductively successful matings for their daughters and sisters; for women, it is the possession of that which is defended, chastity and fidelity, sexual morality itself. The core of a family's honor is its ability to produce such women'' [158, p. 427].

Claustration practices tend to be most extreme in the upper strata of complex hierarchical societies. Dickemann cites the example of northern India, where the more aristocratic the family, the higher and smaller are the windows in the women's quarters, to which the women are increasingly confined until the greatest ladies are prisoners who never see the sun. One of us (MD) encountered an almost identical system in a Saharan oasis town. Several reasons for this status-graded intensity of claustration can be suggested. One is the exaggerated polygynous competition that exists in highly stratified societies, so that men of means have a real concern to sequester women. As Dickemann aptly remarks, "It is important to recall that the societies we are examining were characterized not only by arbitrary sexual rights of lords and rulers but by large numbers of masculine floaters and promiscuous semi-floaters, beggars, bandits, outlaws, kidnappers, militia, and resentful slaves and serfs" [158, p. 427]. It is furthermore only in settled, stratified societies that some rich men can shoulder the costs of women's quarters, guards, and the lost economic production of the confined women themselves. Finally, and perhaps most interestingly, the phenomenon of **hypergynous dowry competition** [158] reinforces status-graded claustration. In stratified societies, it is desirable for a woman to marry up the social ladder, where her children stand to gain and her sons may even grow up to become effectively polygynous. Families therefore compete to marry their daughters to the highest-ranking men. Thus, in India and elsewhere, dowry is practiced in the higher strata and brideprice in the lower [158], and besides competing financially, families compete to offer daughters of the greatest virtue, which means the most severely claustrated.

Besides confining women and threatening them with dire consequences for lapses of virtue, there is another sort of preventative control of female sexuality that is shockingly widespread: genital mutilation designed to destroy the sexual interest and even the penetrability of young women. These mutilations, often euphemistically labeled "female circumcision," range from partial through complete clitoridectomy to surgical removal of most of the external genitalia and the suturing shut of the labia majora (infibulation). In a 1979 exposé, Fran Hosken [280] documented the continued existence of these practices in 23 countries extending across northern and central Africa as well as in Arabia, Indonesia, and Malaysia. She estimated that more than 65 million women and girls presently alive in Africa have been "circumcized." Although various hygienic and religious justifications are proposed by apologists for these practices, it is clear that control of female sexuality is the real goal:

> Infibulation makes sexual intercourse impossible . . . Infibulation is performed to guarantee that a bride is intact—the smaller her opening, the higher the brideprice. She is often inspected before marriage by the husband-to-be or his female relatives. . . .
>
> Women who are infibulated have to be cut open to allow penetration, and more cuts are needed for delivery of a child. Wives, traditionally, are re-infibulated, for instance in the Sudan, after a baby is born; and when the child is weaned, they are opened again for intercourse. During her reproductive life, a woman used to go through this process with each child; and in some areas it still continues today. The decision rests with the husband who has several wives.
>
> It is reported that women often demand re-infibulation after delivery to make intercourse more pleasurable for the husband. A woman must please sexually, or she may be divorced, which means loss of her children, loss of economic support, and disgrace to her family. It is reported that some men have their wives re-infibulated when they leave home for extended periods. [280, p. 2]

Societies differ greatly, and yet the coercive constraint of women by men is a monotonously recurrent theme. The prevalence of this theme is surely a reflection of our selective history of paternal investment and attendant cuckoldry risks. Human marriage tends to polygyny (Figure 11-1), and there is no reason to doubt that the actual mating system is effectively somewhat polygynous too, hence one in which males compete for the limited resource of female reproductive capacity. We shall next consider some rather different sorts of vestigial evidence that polygynous competition has long been characteristic of our ancestors and has thus shaped human nature.

SEX DIFFERENCES IN LIFE SPAN

In Chapter 7, we remarked that the species-characteristic degree of sexual size dimorphism in certain mammalian orders tends to reflect the intensity of polygyny and sexual selection. In the order Primates, monogamous species are notably monomorphic (Figure 7-3, p. 153). *Homo sapiens* is a mildly dimorphic primate: The ratio of mean male body length to mean female body length is 1.08 [12]. We might therefore expect to find a slightly polygynous mating system. That, of course, is just what is indicated by the distribution of modern marriage practices (Figure 11–1), and the fact of sexual size dimorphism suggests that a degree of polygyny has prevailed for many generations. An independent index of a polygynous evolutionary history should be the degree of sexual bimaturism (p. 192); that human males mature somewhat later than females (Figure 10–5, p. 255) again speaks for a polygynous legacy.

A more subtle vestige of our sexual selective history is the sex difference in vulnerability. We have seen (Chapter 7, pp. 92–97, 151–154) that mortality of males is expected to exceed that of females to a degree that again reflects the intensity of polygynous competition, and of course men do suffer higher mortality rates than women. This should follow both from elevated mortality risks attendant upon an escalated expenditure of reproductive effort and from accelerated male senescence consequent upon that expenditure (cf. pp. 187–188). Demographers draw a distinction between "external" and "internal" sources of mortality. The former

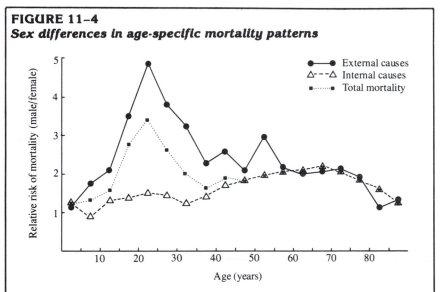

FIGURE 11–4
Sex differences in age-specific mortality patterns

At all ages, males are likelier than females to die as a result of both "external" causes (mainly accidents, as well as suicides, homicides, and poisonings) and "internal" causes (disease and senescence). The sex difference in external mortality risk is maximal in young adulthood, while the sex difference in internal mortality risk is maximal in late middle age. (After Province of Ontario, 1979, Table 24 [512] and Statistics Canada, 1978, *1976 Census of Canada Population: Demographic Characteristics, Table 11.* Ottawa: Statistics Canada.)

category consists primarily of deaths by accident, as well as deaths by suicide, homicide, poisoning, and medical misadventure. To these "external" deaths we shall return shortly. Internal sources of mortality include disease and degenerative processes. Males exceed females, in the Western world, in both sorts of mortality. However, the patterns of sex differences by age are distinct (Figure 11-4). The data in Figure 11-4 are Canadian, but statistics from other industrial countries generate virtually identical curves (see, e.g., Figure 5 in [682]). The sex difference in internal mortality is apparent at all ages but is maximal in later years. Males, in other words, senesce more rapidly than females, and this fact suggests that female life span has exceeded male life span for a significant portion of our evolutionary history. We may also surmise that greater mortality of boys than of girls has a long history from the fact of male-biased sex ratios at birth: Recall Fisher's theory of the sex ratio of investment (Chapter 9) and his explanation of biases in the primary sex ratio as compensatory for sex-differential mortality before the end of the period of parental investment.

SEX DIFFERENCES IN RISK-TAKING

More striking than the "internal" sex difference in Figure 11-4 is of course the "external" sex difference, which is maximal in young adulthood. One way to characterize this difference is to say that men, especially young men, are relatively "risk-prone" and women relatively "risk-averse." This general sex difference is manifest in a variety of spheres [682]. For example, men are more aggressive, risky drivers than women [e.g., 184], and they pay the price (Figure 11-5).

Why should a selective history of polygynous competition have produced a masculine psychology prone to dangerous risk-taking? Where the risks are incurred in direct competition, the answer is apparent. As one theorist has put it, "Sexual competition is demonstrably more intense among males than among females . . . as a general consequence the entire life history strategy of males is a higher-risk, higher-stakes adventure than that of females" [9, p. 241]. However, some categories of risky behavior, such as dangerous driving, are not conspicuously competitive or utilitarian and are nevertheless male-dominated. Perhaps the risk-prone inclination has to be understood as a general masculine attribute that is often dysfunctional in evolutionarily novel settings such as automobiles. However, there may be more utility in apparently senseless risks than meets the eye. Some dangerous driving can be considered a sort of social display that is facilitated by an audience. For example, male drivers are much quicker to hazard a turn into traffic when they have male passengers than when they have

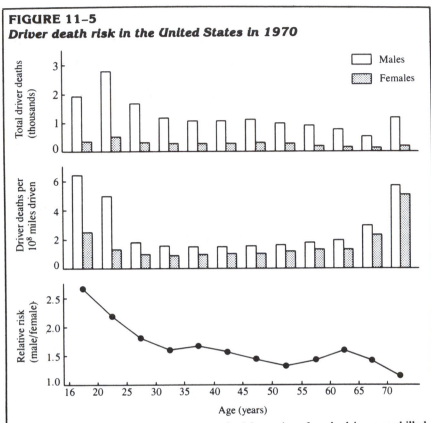

FIGURE 11–5
Driver death risk in the United States in 1970

The top figure shows that many more male drivers than female drivers are killed in accidents in the United States at all ages. (Motorcycles are excluded.) Men drive more than women, so the middle figure presents driver death rates per mile driven; the sex difference is still substantial. The lower figure illustrates that the sex difference is maximal in youth. (After Wilson & Daly, 1984, Figure 6 [682]. Reprinted by permission of the publisher. Copyright 1984 by Elsevier Publishing Co., Inc.)

female passengers or are alone, whereas female drivers are not evidently influenced by passengers [310]. Accident proneness on the part of young men certainly antedates the invention of the automobile. Young men seem powerfully inclined to compete for prestige and status, and these are social "resources" that have surely been indirectly redeemable as fitness throughout human history.

The same sort of social resources are what is at stake in homicidal conflicts in America. Most cases are not robbery-related, nor are the victim and the offender usually strangers. Instead, the leading motive category in American homicides is so-called "trivial altercations" between young men before an audience of their mutual acquaintances. These disputes often begin with overt but minor competition—an argument over who is next in line to play pool or who is tougher—and become ugly when neither party can back down without loss of face [682]. This pattern is by no means unique to America. In fact, homicidal conflict is overwhelmingly a male affair, with victim and offender most often acquainted but not related, in *all* societies.

Much risky behavior is more clearly utilitarian. Ninety-three percent of robberies and 94 percent of burglaries in the United States in 1980, for example, were perpetrated by males [634]. That such crime is a virtual male monopoly is again cross-culturally universal. Men are certainly not poorer than women, yet they help themselves to other people's property more often and are readier to use violence to do so. It may be suggested that the chronic competitive situation among men is ultimately responsible for a greater felt need for disposable—as opposed to subsistence—resources.

ATTRACTION AND MATE SELECTION

One of the corollaries of Bateman's principle is that females, as the sex with more to lose from a poor mating, should be the choosier sex. This prediction can be dealt with at several levels. There is first the matter of utterly misdirected sexual behavior. Women are much less likely than men to engage in sexual contacts with nonhuman animals [290, 328]. They are also less inclined to homosexuality, whether exclusive or not (though it must be remarked that exclusive homosexuality is a rather startling phenotype in either sex, from an evolutionary point of view). Men are also more inclined than women to sexual interactions with altogether inanimate objects; indeed, fetishism is an almost exclusively male phenomenon [328].

The greater selectivity of women is also apparent in comparing their evaluations of members of the opposite sex. A computer dating study of several hundred young Americans is instructive here [115]. People were asked what qualities were really important in determining the acceptability of a date. Women were more insistent than men that the partner be intelligent, a good dancer, of high status, of the right religion, of the right race, and so on. Indeed, they expressed more pronounced preferences on *every* quality but one—physical attractiveness. And it was not just in their prior aspirations that the women were choosier. After the date, men approved the matchings considerably more than did women. Men were two and one half times as likely as women to profess a "strong romantic attraction" to the date. Fifty percent of the men and 39 percent of the women considered the date potential marriage material.

Now, indiscriminate males may be expected when the issue is one of a sexual partner, but a marital partner should be another matter. Both sexes make a heavy commitment to marriage and to children, and so we might expect rather little sex difference in choosiness in this sphere. So who marries whom? In the majority of human societies this decision resides largely or entirely with the parents. We have already mentioned Lévi-Strauss's generalization that male kinship groups *exchange* women, and indeed this practice is widespread. Men who acquire wives from a particular lineage are then under considerable obligation to offer their daughters in return. The exchange of women is part of a larger political reciprocity between the male groups, so that other considerations can be substituted for women. Among the Yanomamö, for example, Napoleon Chagnon has described how one group may offer military aid, refuge from enemies, or agricultural assistance to a hard-pressed ally but will extract payment in women, taking more wives than they give [87]. The arithmetic may be imprecise, and it would be gauche for a Yanomamö man to make a direct reference to the "price," but the principle of exchange is perfectly well understood. Male groups sit down together to negotiate, and with considerable circumlocution each drives the best bargain possible.

Exchange is not often so bald-faced as among the Yanomamö. Parents around the world endeavor to make the best match they can for their offspring. Familial ties and reciprocal obligations, including past interfamilial marriages, are just part of the calculus of costs and benefits that conscientious parents must employ. A major concern in marrying off a daughter is that the man be a good provider. His economic and social status is important. Where inheritance is primarily or entirely a male affair, as it is in most societies, the economic stratum of a prospective husband is of more concern than that of a prospective wife. She should be a good breeder and a hard worker. Among the Yomut Turkmen of Iran, fertility and a propensity to produce sons are highly valued and are thought to be hereditary [305]. Accordingly, the ideal bride has many siblings, most of them brothers.

A near universal of human marriage is that husbands are older than wives. This generalization might seem surprising if cultural diversity were believed to be arbitrary, and it might even be suggested that considerations of sexual compatibility should dictate pairings in which the woman is somewhat older than the man. But when the procreative function of marriage is considered, the age relationship between the sexes becomes comprehensible. Youth in a bride is, after all, a fine predictor of her reproductive potential. Take a woman at puberty and all her reproductive years are yours. If parents seeking a bride for their son were expressly concerned with grandchild maximization (and they often are!), they could hardly do better than to seek a healthy, pubescent virgin.

Parents seeking a husband for their daughter should not be so powerfully drawn to youth. If their daughter is to have many healthy children, the male should be chosen with respect to the resources he commands. Fertility is important, too, but men are fertile into old age. A girl's parents would do well to marry her to the richest, most powerful man they can manage, and that is likely to be someone older than the girl. There is a limit, of course. A really old husband may leave their daughter widowed and unmarriageable. Best to pick a young man of a wealthy, powerful lineage, with a guaranteed inheritance.

Besides direct assessments of wealth, potential grooms may be assessed on the basis of *predictors* of competitive success. That, at least, is one interpretation of this summary statement from a survey of 190 ethnographies, representing every part of the world: "One very interesting generalization is that in most societies the physical beauty of the female receives more explicit consideration than does the handsomeness of the male. The attractiveness of the man usually depends predominantly upon his skills and prowess rather than upon his physical appearance" [195, p. 86].

In Western society we all know that mate selection is supposed to be a matter of falling in love. Some call love a myth, and some insist that it is an objective reality. Most observers agree that it occurs in those societies in which it is expected and is absent or at best an adolescent affair in other societies. In any case, to explain a particular act of mate selection as the result of two people falling in love begs the question. In love or not, why do people choose the mates they choose?

Mainly because they're what's available. This dreary answer is about the only one that finds much empirical support [66, 110, 627]. Personality characteristics are surprisingly difficult to relate to the choice of marriage partners. There are two pieces of contrary folk wisdom on this subject: Opposites attract, and lovers fall for people like themselves. The latter is probably nearer the truth, but neither receives much support from research. In America, to be sure, people tend to marry within their religion, their ethnic group, and their social class. But above all else, they tend to marry within their *neighborhood* [627]. Couples are unlikely to fall in love unless they happen to meet and can continue to meet without a lot of trouble.

Even today, mate selection in the Western world is not entirely a matter of free, individual choice. Parents and other interested parties exert influences that may be profound. Nowadays it may be rare for parents to step in and squash a love match, but influence can be a good deal more subtle than raw veto power. Parents have long understood the value of simple physical separation in extinguishing love affairs of which they disapprove.

Nevertheless, young people clearly select their own mates to a greater extent now than in the past. Has this affected the criteria of mate selection? Apparently not much. A woman's desirability still depends upon her physical attractiveness to a greater extent than does a man's. His worth as a potential mate depends somewhat more upon his status and prospects for success. Several studies of attitudes have confirmed the existence of these values, and there is at least some evidence that particularly attractive women are able to parlay their good looks into high-status marriages [568, 598, 627].

In America, men continue to be older than their brides at first marriage, but this difference has shrunk slowly and steadily during this century, even as the actual age at marriage has declined, bottomed out, and risen again slightly. This trend has been primarily one of a decline in the husband's age at first marriage, which presumably reflects increasing material affluence. Again, a male's marriageability is more dependent upon economic status than is a female's.

There have been a number of social psychological studies of heterosexual attraction [48]. Most have involved paper-and-pencil evaluations of the attractiveness of a potential partner described in terms of a set of hypothetical characteristics. Such studies, in which the rated individual is usually fictitious, are too artificial to be evaluated with much confidence. People are likely to express approval for socially approved characteristics rather than for what actually attracts them.

The advent of computer dating services has provided an opportunity for more realistic research. In one ambitious study men and women were prerated by means of a questionnaire on their "adherence to traditional religious values" and on their "permissive attitudes toward violations of traditional sexual standards." Such questionnaire-based ratings are an acceptable part of the computer dating business, since "personality inventories" are allegedly used to generate "compatible" matches. Matches were then generated in which pairs were similar or dissimilar on each of these scales. Of 300 pairs 239 actually dated and then provided the computer dating service with a retrospective evaluation of their date [618]. The results were intriguing. Men were attracted to women who shared their sexual attitudes. Similarity of religious belief was not important to the men. Women were attracted to men who shared their religious beliefs. One interpretation of these results is that men date largely for sexual reasons, while women are more concerned to evaluate a man's prospects as a stable long-term mate [see also 115].

INCEST AND EXOGAMOUS MARRIAGE RULES

Kinship is relevant to who mates with whom. It defines an ineligible group of relatives with whom sexual contact would be incestuous and also often defines an especially eligible group of somewhat more distant relatives. The universality of incest taboos has long fascinated anthropologists. There are few subjects on which more ink has been spilled, but there is little agreement on any proposition save that incest avoidance is a fundamental organizing principle of every human society. Marriage rules should be distinguished from the incest taboos governing sexual contact [637], but the proscribed categories are usually co-extensive.

What constitutes incest and marital ineligibility varies widely. Mother-son, father-daughter, and brother-sister marriages are forbidden virtually everywhere. (The exceptions involve the inbreeding of special royal or divine lineages and do not apply to any society at large [638].) Beyond this, there is great variability. There may be no taboo group beyond the nuclear family, or there may be literally thousands of (theoretically) forbidden mates tenuously linked by a single common ancestor generations ago. First-cousin marriages are forbidden in some societies but are obligatory in many others. The *type* of first cousin, whether on the mother's or father's side, parallel cousin (through mother's sister or father's brother), or cross cousin (through mother's brother or father's sister), often determines marriageability.

There is a simple biological reason why incest avoidance is a good thing—**inbreeding depression**. Abundant data from many animal species including people demonstrate that mating close relatives with one another leads to reduced fertility and reduced viability of offspring [e.g., 3]. The main reason for this fitness penalty of close inbreeding appears to be that everyone carries a few rare deleterious recessive genes that are not normally expressed, and that some of these rare recessives, duplicated by immediate descent in close kin, become homozygous in the progeny of inbreeding. There is thus a substantial selection pressure in natural populations to avoid close inbreeding, and many mechanisms function to achieve that avoidance: Juvenile dispersal, sex differences in the age of maturation, and a lack of sexual interest in individually recognized relatives are some of the most widespread and effective of these mechanisms.

Social scientists have been curiously reluctant to concede that natural selection against inbreedng might be relevant to the human phenomenon of incest avoidance. Many have claimed that such avoidance is unknown in other animals, a claim that is clearly false. When a male rhesus monkey matures, he will copulate with females of the same age and status as his mother, but not with her [539]. If deer mice are individually caged at weaning and subsequently placed in opposite-sexed pairs at maturity, littermates are considerably more reluctant to mate than are strangers [270]. In nature, a yearling female prairie dog will come into estrus in the absence of her father but will delay first estrus if he is still about [278]. Mechanisms that assure **exogamy** (outbreeding) almost certainly characterize most mammals [54, 56].

It is true that the biological demand for exogamy does not explain the *details* of the incest taboo in particular human societies. Cultural kinship is not perfectly mappable upon biological kinship, and indeed the incest taboo may sometimes forbid sexual contact between classificatory relatives who are no biological kin at all. Moreover, while the details of a particular marriage system are not fully explicable in terms of a function of inbreeding avoidance, they sometimes can be explained as serving other functions. Exogamous marriage rules may distribute women in such a way as to ensure small familial groups against nonavailability of women due to biased sex ratios [307]. Such rules also maintain and strengthen the nexus of interpersonal relations and obligations on which the society rests. About this there can be no quarrel. However, to argue from these facts that the human incest taboo is cultural *rather than* natural, as Freud, Lévi-Strauss, and many others have done, is to fail to grasp the legitimacy, indeed the necessity, of multiple levels of explanation (cf. Chapter 1). Theorists often debate "alternative" explanations that might easily be simultaneously valid and complementary.

Norbert Bischof has effectively argued, with specific reference to the incest taboo, that culturally imposed rules often *reinforce* behavioral tendencies that already exist for other reasons [54]. His proposal is that the explicit taboo formalizes and buttresses an avoidance of incest that would exist to some degree in any event. This suggestion is directly opposed to the belief that taboos are necessary to civilize the socially disruptive fires of incestuous desire burning in every human psyche. We cannot here evaluate the Oedipal scenario of Freudian psychodynamics: that the first sexual desires of the male are directed to his mother with resultant homicidal feelings toward the father, and so forth [but see 263]. But it is worth noting that while many anthropologists influenced by Freud have alluded to the "universal horror of incest," Needham points out that evidence of that horror is hardly to be found in the ethnographies. Many peoples, including some who severely punish the incestuous and others who do not, discuss the subject with indifference or amusement [457].

At least as regards brother-sister incest, however, there is good reason to suppose that Bischof is right: Inbreeding would be unlikely even in the absence of a culturally imposed taboo. The evidence on this point concerns people raised like brothers and sisters but between whom no barriers of biological kinship or cultural incest taboo exist. Studies of young people raised on Israeli kibbutzim have revealed the remarkable fact that they neither marry nor have affairs within their rearing groups [557, 578]. The critical factor is very clearly that the children are raised together from an early age, since the only exceptions involved pairs who were in fact separated from one another for a large part of their childhood. The records of thousands of married adults reveal not a single marriage between a man and a woman reared together throughout childhood! This is all the more remarkable in view of the fact that marriages, in the industrialized West as elsewhere, tend to involve very close neighbors. Nor is there a documented case of any sexual relationship between members of the same rearing group. Many comments by the Israeli youngsters reveal their awareness, sometimes rueful, that rearing together has killed sexual attraction.

If rearing girls and boys together really prevents mutual sexual interest, this should normally function as an effective outbreeding mechanism. The Israeli children were "tricked" into rejecting acceptable mates by the operation of that mechanism among children who were not in fact biological kin. But in most circumstances, including those in which most hominid evolution presumably took place, children reared in very close proximity are likely to be kin; and the mechanism will function to assure exogamy. If familiarity breeds a lack of sexual interest, we may expect that incestuous desires will be especially likely when brothers and sisters are reared with barriers between them, in which case sexual prudery in child rearing may have an effect precisely opposite to that intended.

The notion that incest avoidance occurs naturally as a result of childhood familiarity was proposed by the anthropologist Edward Westermarck as long ago as 1891 [663]. It has been debated ever since. Many have ridiculed the hypothesis, imagining that it required some mystical capacity to recognize biological kin. Of course it does not, as Westermarck made clear. The Israeli example shows that a mechanism that would normally function to achieve incest avoidance can be sidetracked. Further evidence comes from the practice of *Shim-pua* marriage in Taiwan [689, 690]. In this special form of marriage the couple is wed as children and reared together. Such couples have more difficulty consummating their marriages, more extramarital affairs and divorces, and fewer children than do couples whose marriages are arranged postpubertally.

THE SEXUAL MARKETPLACE

"The males are almost always the wooers," said Charles Darwin [137, p. 613] of animals generally. People are no exception. This fact has presented a conundrum to social scientists seeking explanations for the economics of sexual relations.

Consider the practices of the Trobriand Islanders, the foremost subjects for the prolific pen of that great pioneer anthropologist Bronislaw Malinowski. In his much-quoted book of 1929, *The Sexual Life of Savages in North-western Melanesia,* Malinowski made no attempt to conceal his admiration for their relatively relaxed sexuality [412]. They seemed to him much healthier in their attitudes than sexually repressed Europeans. Trobriand women were understood to enjoy sex as much as men, and Malinowski insisted that members of both sexes were free to choose their lovers. But one disturbing fact was hard to reconcile with this idyllic picture of sexual egalitarianism. A Trobriand man gives small presents to his mistress, and if he has nothing to offer, she refuses him. As Malinowski noted, "This custom implies that sexual intercourse . . . is a service rendered by the female to the male" [412, p. 319], an implication that vexed him, since he explicitly expected that sexual relations ought to be treated "as an exchange of services in itself reciprocal" [412, pp. 319–320]. His solution was to dismiss the fact: "But custom, arbitrary and inconsequent here as elsewhere, decrees that it is a service from women to men, and men have to pay" [412, p. 320]. Arbitrary and inconsequent? Our skepticism is necessarily aroused, for it appears that "custom" in *all* societies makes sexual intercourse a female service. In the words of one anthropologist, "Everywhere sex is understood to be something females have that males want" [593, p. 253]. The signs are all around us: in the ubiquity of female prostitution [see 78] and the rarity of its converse, in the male market for pornography, in desires expressed to pollsters, in graffiti, and above all in men's and women's complementary roles in courtship.

That women should generally enjoy this leverage in the sexual marketplace is to be expected from the fact of the male copulatory imperative. In people, as in other mammals, males are more easily aroused, more doggedly interested, less tolerant of sexual abstinence, and less discriminative than females. To the male psyche, women are ever in short supply, both as sexual partners [593] and as wives [380].

SEX DIFFERENCES AMONG HOMOSEXUALS

In order to see clearly the nature of transactions between the sexes, it is, paradoxically, illuminating to consider homosexuals [593]. The rationale is this: By freeing themselves from the necessity of compromises with the alien sex, homosexuals may most clearly reveal the sociosexual proclivities of men and women. We have argued, for example, that male, but not female, psychology should have evolved to value sexual variety and the maximization of numbers of sexual partners. There is a good deal of evidence that male homosexuals have sexual contact with many more partners than do lesbians, and are more inclined to one-night stands, anonymous sexual encounters, and group sex (all of which are rampant but rarely realized fantasies of heterosexual men). As Donald Symons has summed up the argument, "Heterosexual men would be as likely as homosexual men to have sex most often with strangers, to participate in anonymous orgies in public baths, and to stop off in public restrooms for five minutes of fellatio on the way home from work if women were interested in these activities. But women are not interested" [593, p. 300].

It has sometimes been suggested that the dramatic sex differences in homosexual behavior are products of a greater societal discrimination against homosexual men, but the idea is unconvincing. It is precisely their quest for sexual variety that makes homosexual men conspicuous to potential persecutors; they cruise in spite of harassment, not because of it. Other sex differences in the behavior of homosexuals also seem to match or exaggerate those in heterosexuals. The men are ardent consumers of pornography, whereas there is no such market among the women. Men judge the attractiveness of potential partners by their physical beauty, and especially by their youth, to a far greater extent than do women.

The homosexual evidence is important because it provides a clear refutation of the naive but popular view that "society" and "the media" are the sources of the differences between male and female sexuality. Homosexual men embrace the larger society's idealization of monogamy but usually fail in their efforts to achieve it. They do not see anything manly or admirable in fleeting sexual encounters in public places, but they pursue them. To quote Symons again, "That homosexual men are at least as likely as heterosexual men to be interested in pornography, cosmetic qualities, and youth seems to me to imply that these interests are no more the result of advertising than adultery and alcohol consumption are the results of country and western music" [593, p. 304].

PSYCHOLOGY AND THE PURSUIT OF FITNESS

The fact of a substantial homosexual subculture presents an obvious challenge to the proposition that people have been "designed" by natural selection to be fitness-maximizers. And of course, exclusive homosexuals are by no means the only people who are conspicuously abstaining from efforts to replicate their genes. Some sociobiological enthusiasts have attempted to explain away these embarrassing cases by arguing that homosexuals are like sterile worker bees laboring for their kin; that two-child families are a strategy for long-term fitness maximization; and so forth. We find these hypotheses farfetched and unnecessary. Rather than seeking a fitness-promoting rationale for entering the nunnery or for undergoing a vasectomy, we believe that the adaptive logic of the human psyche can be revealed at a more abstract level. If people value honor enough to die for it, for example, then it is not the dying but the valuing that is likely to make adaptive sense. Describing the human psyche at a level of abstraction that is cross-culturally valid and appropriate to the elucidation of its adaptive logic—*that* is the task facing those who would construct an evolutionary theoretical account of human behavior [593]. In this chapter, we have suggested several psychological traits that might qualify as panhuman and adaptively intelligible: male sexual jealousy, female selectivity, male risk proneness, sexual disinterest after childhood intimacy, and so forth.

Our suggestion that adaptation must be sought at the level of psyche is not special pleading. We are not arguing that the application of evolutionary biological theory is somehow more problematic and less direct for *Homo sapiens* than for other creatures. The individual acts of individual organisms are often maladaptive whatever the species; we do not expect positive fitness consequences as a result of every unique behavioral event, but rather as the *average* result of the behavioral output of evolved ("psychological") mechanisms of behavioral control operating within the species' natural environment. This is every bit as true of planaria as it is of people. And when organisms confront evolutionarily novel environmental circumstances, their adaptive behavioral control mechanisms may fail them dramatically. Endogenous opiates in the brain, for example, mediate normal, evolved mechanisms of pleasure and reward; confront the organism with an environment containing opium-derived drugs—an environment for which the organism has not been evolutionarily adapted—and destructive consequences may follow. Often it is just such failures that best reveal how the mechanisms normally function. If rapid technocultural evolution has landed modern people in circumstances for which they are ill-adapted, still we may hope to elucidate the functional design of the human psyche. We might even derive some benefit from so doing.

SUMMARY

Most human cultures permit polygynous marriage, but relatively few men are able to attain it. Polyandrous marriage is much rarer, and nowhere entails a reversal of male and female roles. Marriage often presents the appearance of a trafficking in women: Men commonly pay for wives and acquire rights to their reproductive capacity in the process.

Like other male animals, men compete for reproductive opportunity. This competition takes the form of social contracts including adultery law, of violent conflict between individual men and between kinship groups, and of coercive constraint of the women themselves. People exhibit several characteristics indicative of a history of polygynous competition, including sex differences in body size, life expectancy, age at maturity, and risk proneness.

Incest avoidance—which characterizes many animals—is a universal organizing principle of human societies, and appears to develop naturally among children raised in close proximity. Mates tend to be neighbors of like background, and women are more selective than men. Males are the wooers and are more promiscuously inclined. All these attributes are consistent with the evolutionary theoretical view of organisms as reproductive strategists.

SUGGESTED READINGS

Alexander, R. D. 1979. *Darwinism and human affairs.* Seattle: University of Washington Press.

Chagnon, N. A., & Irons, W., eds. 1979. *Evolutionary biology and human social behavior.* North Scituate, Mass.: Duxbury Press.

Crook, J. H. 1980. *The evolution of human consciousness.* Oxford: Clarendon Press.

Dickemann, M. 1979. The ecology of mating systems in hypergynous dowry societies. *Social Science Information,* 18: 163–195.

Symons, D. 1979. *The evolution of human sexuality.* New York: Oxford University Press.

12

Human reproductive investment

Of all the million and more species of animals on this planet, *Homo sapiens* boasts one of the lowest reproductive rates. Our nine-month gestation period is the longest of any primate, though it is surpassed by some large mammals, including elephants and whales. Lactation may persist for two years or more, and young remain dependent for several years beyond weaning. It takes more than a decade to attain sexual maturity, but any individual who does so has a fairly good chance of living several decades more. Ovulation occurs at a leisurely pace: Usually only one egg is shed at a time, and if it goes unfertilized, there is a four-week wait for the next one.

The problem of explaining our very low reproductive rate is not in principle different from that of explaining an albatross's single egg or the low fecundity of mammals in general. The evolution of low reproductive rates has already been discussed in Chapter 8. In that discussion, we concluded that low reproductive rates evolve where intensive parental nurture enhances the success of offspring sufficiently to compensate for the costs of reduced parental fecundity. This might result from intense competition in saturated environments (*K*-selection), from high vulnerability of unprotected juveniles, or from a need for prolonged learning and parental tutelage. All these and more are probably relevant to the hominid case, and the causal priorities among these considerations remain a topic of debate and active theorizing. In any event, people are very nearly unsurpassed as *K*-strategists. We usually bear but a single offspring at remarkably long intervals—not just annually, as do most of the species that we are inclined to call *K*-strategists, nor even every second year. In a state of nature we are one of a handful of species that bear young at intervals of three years or more.

With such a low reproductive rate we invest a huge proportion of our resources in every child. Parents maintain a benevolent interest in the welfare of their offspring for life—a rare phenomenon among mammals. Intergenerational familial bonds that continue beyond the sexual maturity of the young are almost exclusively limited to primates. This trait is shared to some extent with elephants, some carnivores, and perhaps whales and dolphins. We particularly surpass our primate kin and these other mammalian *K*-strategists in the degree to which males are integrated into those long-term familial bonds.

MAN AMONG THE PRIMATES

It is more than two hundred years since Linnaeus placed our species firmly and correctly among the primates, describing himself as the type specimen of *Homo sapiens,* wise man. That the great apes are our closest relatives has been clear since T. H. Huxley's anatomical treatise in 1863 [294]. Just how close was not appreciated, however, until 1975 when King & Wilson [327] demonstrated that chimps and people were as genetically alike as most congeneric pairs of "sibling species." Traditional Primate taxonomy accords us our own family, Hominidae, while placing orangutans (*Pongo*), gorillas (*Gorilla*), and chimpanzees (*Pan*) together in the family Pongidae. However, the most recent studies of homologous proteins and chromosome banding patterns suggest that *Pan* and *Homo* are more closely related to one another than either is to *Gorilla*, and that these three form a larger taxonomic unit that excludes *Pongo* [705].

Taxonomic classification is a relatively neat affair for extant groups of species classified at one point in time. It is necessarily messier when we deal with temporal lineages. Our ancestors of a few hundred thousand years ago differed sufficiently from us that they are given another specific name, *Homo erectus*. Our ancestors of more than a million years ago were so different as to be given another generic name, *Australopithecus*. Still further back was the last common ancestor of man and chimpanzee, and it is not yet certain just what sort of creature that was. Long assumed to have been relatively chimplike, this last common ancestor may instead have been intermediate between *Homo* and *Pan*, with both modern genera having diverged therefrom.

The various structural specializations of hominid evolution are well known, and we will simply mention a few—the skeletal adaptation to a terrestrial bipedal life, the improved capacity for a precision grip with the fingers, the reduction of the canine teeth, the enlargement of the brain case. Just as well known is the interpretation, based on archaeological remains, that hominid evolution accompanied the invasion of a new ecological niche, namely a semicarnivorous life-style on the open savannas, and that the selection pressures that drove hominid evolution were inherent in that lifestyle. The pelvic modifications necessary for bipedality evidently constricted the birth canal at the same time as the hominid brain case was increasing in size. This conflict demanded that birth take place well before the human baby's head was fully grown; that, at least, is a plausible account of the neural immaturity of the human neonate relative to other primates.

While these structural modifications were in progress, hominids evolved behaviorally too, but the reconstruction of our behavioral history is necessarily much more speculative. Perhaps the extreme immaturity of human babies tipped the scales toward male-female bonding and paternal investment. Certainly the evolution of language, elaborate kinship reckoning, and marital exchange politics set people apart from other primates. Complex culture must have a long pre-history. G. P. Murdock [450] offered an intriguing list of cross-culturally universal human practices including cooking, dancing, family feasting, funeral rites, gift giving, law, mealtimes, religious ritual, and weaving. These practices are socially transmitted and culturally variable in their details, to be sure, but they are no less species-typical for that, and no less the products of evolution. These cultural practices are manifestations of the intense sociality of the human organism, which pursues its fitness interests by long-term strategies within a complexly structured social milieu.

OUR NEAREST RELATIVES

Some clues about the social organization of early hominids can be derived from studies of chimpanzees, who share several interesting characteristics with modern man, characteristics that we may therefore consider ancient in our lineage.

We are still a long way from a full understanding of chimpanzee society. Early observations suggested that chimps lived in an amiable web of friendships and occasional get-togethers [365, 366, 524]. More recently it has been discovered that their easygoing affability is reserved for members of a discrete community of mutual acquaintances. Between adjacent communities the males, at least, are mutually hostile or even warlike, conducting raids and skirmishes that can lead to fatalities [216, 462].

We owe much of our knowledge of the behavior of wild chimpanzees to the bold and fascinating long-term studies conducted by Jane Goodall. She spent many years establishing contact with and observing the animals in Gombe Stream National Park in Tanzania, and her observations are widely known through a delightful popular book [366]. While Goodall has been responsible for a number of significant discoveries, it has gradually become apparent that a certain bias was introduced into her early observations by her methods, particularly her initial practice of setting out bananas to attract the animals to her camp. It now appears that chimpanzees are not as sociable as we once believed; adult females, for instance, spend a majority of their time accompanied only by dependent offspring [243]. There is, however, considerable interaction among adults: Groups of males, particularly, move about together, and animals of both sexes may maintain close lifelong relationships with their mothers.

The community of male chimpanzees is in fact a kinship group. Males apparently never transfer out of their natal communities, whereas many females transfer, sometimes at puberty and sometimes later. This situation contrasts with that in baboons and most other monkeys, where *males* typically move from their natal troop at sexual maturity while *females* stay put [476]. (Differential dispersal by the two sexes is generally believed to have evolved to reduce deleterious inbreeding. In mammals generally, it is usually the males who disperse the greater distance, in birds the females [239].)

Until the community structure of chimpanzees was elucidated, the widespread human tendency to solidary male kinship groups who exchange women seemed a complete departure from the other primates. But the chimpanzee results raised the interesting possibility that female transfer was not a human novelty. Then, hard on the heels of the chimpanzee studies, came a report on gorilla behavior indicating that here too it is generally the females who transfer between troops [257]. Recent evidence suggests that such "patrilineal" troop structure also characterizes a couple of species of monkeys [350, 587]. In all the primates, then, there are just a handful of species in which exogamy is achieved primarily by female movement between groups rather than by male movement, and these include ourselves and our two nearest taxonomic kin. The discovery that we share with *Pan* and *Gorilla* at least a rudimentary form of "female exchange" has reoriented our view of human social evolution, suggesting that male kin-group solidarity is an ancient anthropoid adaptation antedating our invasion of the savannas.

Early observations suggested that chimpanzee mating is a promiscuous free-for-all. Near the site where she provisioned the chimps, Goodall saw males queue up to copulate, whenever a female was in estrus, with surprisingly little male competitiveness. By doggedly tracking individual chimps, however, Richard Wrangham has discovered that a female often consorts with a single male for several days when she is in estrus, and they take considerable pains to avoid other chimps [695]. It has been estimated that about half the conceptions at Gombe take place during such consortships [624], but the consorting male pays a price, becoming the object of attacks by his compatriots when he returns to the male community. Even in the absence of artificial provisioning, many estrous females do not enter a consortship but instead mate with all the community males. In these cases, mating is at first noncompetitive; but in the last few days of estrus, when ovulation is imminent, the top-ranking male (or a pair of brothers!) is likely to establish exclusive mating rights by aggressively guarding the female [624].

There has been considerable interest in the likely mating structure of early hominids, although theories are necessarily speculative. Comparative data provide some clues. It seems unlikely that we are adapted to quite the same sort of within-group sexual competition as chimpanzees. The evidence on this point comes from comparisons of testis size and sperm counts [566]! A chimp's sperm count is much higher than a man's, and his testes are relatively huge (Table 12-1). These facts may be interpreted as reflecting the relative intensity of sperm competition. Field observations of the mating behavior of the great apes indeed suggest that sperm competition is most extreme in chimpanzees where several males sometimes copulate with an estrous female in rapid succession, and probably least in gorillas where there is a stable troop association with exclusive mating bonds between adult females and individual dominant silverback males [256]. Interestingly, the testis size comparisons place man well below the chimp but above the gorilla. (The large size of the male gorilla makes this comparison rather less than conclusive. Note, too, that the intensity of sperm competition is not exactly the same thing as the intensity of sexual selection or of polygyny. The great degree of sexual dimorphism in *Gorilla* is assuredly the result of intense sexual selection, but male competition is not fought out as sperm competition.)

TABLE 12-1
Testis size and sperm count in Hominoidea

The data suggest that sperm competition is most intense in chimpanzees and least intense in gorillas, a conjecture in accord with accounts of mating behavior in the field. Semen samples for the great apes were collected by electroejaculation.

	Approximate weight of testes of mature male		Approximate number of sperm per ejaculate
	Grams	(% of body weight)	
Chimpanzee	120	0.3	60×10^7
Man	25–50	0.04–0.08	25×10^7
Orangutan	35	0.05	7×10^7
Gorilla	35	0.02	5×10^7

Source: After Short, 1981 [566]; Warner *et al.* 1974 [651].

LOSS OF ESTRUS

Unlike chimpanzees (and, to a lesser extent, gorillas), women do not advertise imminent ovulation with a conspicuous vulvar swelling. In fact, the human midcycle seems positively cryptic, although it is by no means uneventful (Figure 12-1). Near the time of ovulation, bodily odors change detectably, the sensitivity of several sensory modalities is enhanced, and there seem to be a number of mood and attitudinal fluctuations [e.g., 154, 162, 436, 531, 671]; whether these phenomena have some functional significance is unknown. Many women experience a temporally discrete pang ("Mittelschmerz") shortly before ovulation [2, 468]. Whether there might be a midcycle peak in sexual behavior in people, as in other primates, has been somewhat controversial. Such a peak has now been observed in several studies and is perhaps especially likely when female-initiated sexual activity is under consideration (Figure 12-1). According to one study, American women entering a discothèque wear more jewelry and makeup and are more often touched by men when near midcycle [269].

A number of authors have recently addressed the question of why there has been a "loss of estrus" or "concealment of ovulation" during human evolution. These phrases refer to both the absence of conspicuous external signs at midcycle and the human tendency to engage in sexual behavior more or less throughout the menstrual cycle. It was once popular to attribute the loss of estrus to the utility of continuous sexual receptivity in maintaining a monogamous pair bond [443]. This view has been criticized by Devra Kleiman, who points out that monogamous mammals are not

FIGURE 12–1
Some phenomena that vary over the human menstrual cycle

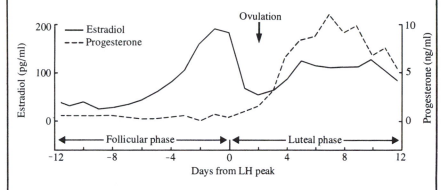

A. Estradiol rises during the follicular phase leading to the surge of luteinizing hormone that triggers ovulation and the subsequent luteal phase. (After Mishell et al., 1971, Figure 1 [438].)

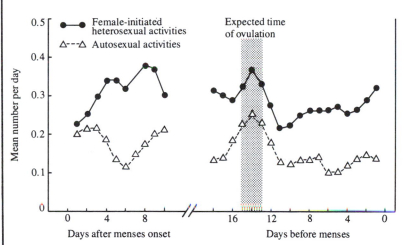

B. Sexual behavior, as recorded daily by 23 American women over 109 woman-cycles. All were using contraceptive practices other than the pill. (After Adams et al., 1978, Figure 2 [2]. Reprinted by permission of the New England Journal of Medicine, 229 (1979): 1145–1150.)

generally more sexually active than related species with other mating systems [332]. Pairs of foxes or gibbons or any other monogamous species maintain their close association through months of asexuality. However, a related argument that seems more tenable has selection favoring any woman who could keep a consorting male in doubt about the exact time of ovulation, thereby obliging him to guard her more prolongedly and in so doing to forsake mating competition for a career of monogamous investment in wife and children [13, 584]. Of course, males need not have been monogamously bonded to females for the latter to have profited by prolonging their receptive period; perhaps early women earned more male attention and gifts of meat that way [593]. A rather different hypothesis is that a great deal of male attention is the *last* thing a female might want; a female with a large estrous swelling can be the center of violent activity that interferes with her maintenance activities and even introduces a risk of injury. If she already has a relationship with a consorting male, a female might prefer to conceal ovulation and avoid the attendant brouhaha (though it is not clear why this argument should apply to people more than to chimps). Another popular hypothesis has women concealing ovulation in order to cuckold their husbands and have their children sired by the best available males [42, 593], although it is hard to understand how continued parental investment by the husbands could be evolutionarily stable if the rate of cuckoldry were very high [584]. Yet another hypothesis has the female primarily adapted to conceal ovulation from *herself*, on the argument that advances in human cognitive capacities led women to an awareness of the pain and perils of motherhood, an awareness that might have inspired opting out if the fertile period were not obscured [77]; this may sound farfetched, but women in many cultures do indeed express a weary wish for relief from successive pregnancies, and when the technology necessary to effect that wish is made available, many use it.

Our suspicion is that the concealment of ovulation is not such a specific human adaptation as is usually assumed. It is not clear to what extent behavior or even odors might advertise midcycle in less antiseptic cultures than our own. And it is not proven that people are so very much less cyclic in their sexual behavior than other primates [287]. As for conspicuous sexual swellings, they are characteristic of only some primates, perhaps especially terrestrial, polygynous, group-living species [587], where they may be designed to incite male-male competition [cf. 118]; females may even be advertising their estrous state to quite distant males outside the troop, for matings with such males during between-troop encounters have been observed in several species of terrestrial monkeys. It seems more to the point to enquire what function conspicuous morphological correlates of ovulation serve in those who have them rather than why we lack them. But there is no question that a reconstruction of ancestral mating systems will be an essential part of any satisfactory account of human origins.

It is true throughout the world, and has probably been true for a very long stretch of hominid evolution, that people form predominantly monogamous mateships that persist through repeated reproductive episodes. There is of course a persistent incidence of extramarital matings, but mateships are sufficiently stable that husbands can invest in their wives' children with reasonable confidence of paternity. The tendency to keep most productive human matings within nuclear families that are most often monogamous means that our breeding system is effectively only mildly polygynous. What is really unusual about *H. sapiens* is the fact that our relatively stable mateships persist within a rich context of highly personalized social interactions with a large number of other people. A pair of gibbons lives in monogamous seclusion on their territory in a state of chronic mild hostility toward the neighbors. That description may be reminiscent of American suburbia, but by and large we differ from all other pair-forming mammals in this regard: People everywhere maintain active, individualized social relationships within the marital household, within larger kin groups, and within like-sexed friendships all at the same time.

HUMAN KINSHIP

Kinship is a basic organizing principle of every human society, although the rules for classifying relatives are fascinating in their variability [197]. Our society's practice of retaining male surnames, for example, has the effect that we can often trace an ancestry through the male line for several generations, whereas we quickly lose track of collateral relatives, with different surnames, who are just as close genetic relatives. Such a patrilineal bias in kinship reckoning is common but by no means universal. Other classificatory rules vary as well. We lump as "cousins" maternal and paternal relatives (not to mention in-laws), who are terminologically distinguished and treated quite distinctly in other cultures. On the other hand, some of the relatives we distinguish are elsewhere conflated into larger terminological categories. Our kinship system is comparatively rather simple, combining an ego-centered ("kindred") terminology of relatively limited extension with patrilineal surnames. Other systems are far more complex. Whole academic careers have been spent pursuing the intricacies of Australian aboriginal kinship.

But as variable as kinship systems may be, they all share certain features. People everywhere understand the concept of a **dimension from close to distant kin**. Furthermore, there is invariably an expectation that there will be a greater commonality of interest the more closely two people are related. This dimension of kinship distance always bears some relationship, more or less exact, to genealogical descent (and hence to genetic relatedness as a biologist would compute it). Furthermore, kinship seems always to be understood by the people themselves in terms of genealogical descent and of metaphors of biological relationship such as shared blood or common ancestors. In every society there are individuals, often older people, who take it upon themselves to learn the detailed ramifications of their own families' genealogies and to pass their expert knowledge along to relatives.

As we saw in Chapter 3, evolution by natural selection is a process that produces organisms with effectively nepotistic behavioral control systems. We also saw that the implementation of nepotistic strategies often depends upon recognizing one's relatives and that this recognition is often the product of experience with individuals who, on circumstantial evidence, can be presumed to be one's offspring, siblings, and so forth. The human preoccupation with kinship is of course just what one might expect of an evolved nepotist with elaborate cognitive skills.

Inevitably, anthropologists and other scientists have begun to apply Hamilton's kinship theory and related evolutionary biological ideas to the study of human sociality, and there have already been a number of exciting results. One example concerns the "avuncular" relationship, according to which the men in certain societies take a benevolent interest in their sisters' children rather than their wives' children. This initially perplexing social practice can be related theoretically and empirically to circumstances that lower a man's confidence of his paternity of his wife's children and thus make the sister's children surer fitness vehicles [8, 192, 352]. Another example concerns marital exchange practices [158]: Brideprices and dowries often involve major resource transfers that are unintelligible under strictly economic models. The apparent inequities can be understood when inclusive fitness prospects are treated as a valued commodity for which people are willing to pay a price. The idea that economic resources are invested where they will yield fitness returns has also been offered as an explanation for male-biased inheritance [263]: Polygyny and the sex difference in fitness variance mean that males are likely to be better able than females to convert extra resources into extra descendants.

The view of organisms as evolved nepotists with a natural solidarity with their genealogical relatives sheds light on a number of aspects of human kinship. For example, an unreciprocated one-way flow of benefits is more readily tolerated when the parties are close kin [49, 182]; between nonrelatives, a failure of strict reciprocity is resented as exploitative. Kinship mitigates conflict; the risk of homicide, for example, is much greater between nonrelatives living together (and between in-laws) than between blood kin [133]. The other side of the coin is that blood relationship is highly predictive of the frequency of cooperative interaction [e.g., 193, 244, 245]. Even the inconsistencies in the genealogical information given to anthropologists by different informants have been interpreted as the results of nepotistically self-interested manipulations of the facts (who is eligible to marry whom; who has what obligations to whom; and so forth) [89, 199]. These are just a few of the recent applications of evolutionary theory to the analysis of human kinship.

The evolutionary biological model of human kinship has been criticized on the grounds that kin terminology rarely if ever corresponds precisely with genetic relatedness; indeed, nominal kinship categories often incorporate "fictive kin" who are not blood relatives at all. But there is really no reason to expect that people must calculate kinship like a population geneticist; after all, ground squirrels and sweat bees (pp. 51–55) manage to be effective nepotists on the basis of simple rules of thumb. It is sometimes argued that human kinship is cultural *rather than* biological, but this is a naive nature-nurture dichotomy of the sort we have criticized before. No one has suggested any real alternative to the evolutionist's explanation for the central role of kinship (indisputably based on the biological relationship between parent and child) in human affairs.

In any event, the fact that individuals of differing relatedness are addressed by the same kin term does not mean that people are insensitive to finer genealogical distinctions. Napoleon Chagnon has shown, for example, that blood relationship is a better predictor of solidarity in conflicts among the Yanomamö Indians than is terminological kinship [88, 90]: Many men are nominal "brothers," but when the chips are down, it is *real* brothers who come to one another's assistance. The fictive use of kin terms is furthermore readily interpreted from a sociobiological view [8]. We address nonrelatives in kin terms when we are trying to promote solidarity or to solicit aid: "Brother, can you spare a dime?" This very usage, which has numerous parallels in other languages and other cultures, testifies to a general expectation of benevolence toward close relatives. The artificial use of a kin term is here designed to exploit the natural solidarity of true relatives. A most interesting topic for detailed study would be the contexts in which different modes of address, including kinship terms, are employed.

THE HUMAN LIFE HISTORY

If human social responses and life historical parameters are evolved adaptations, then they should be primarily adapted to the mixed foraging niche that has been ours for at least a couple of million years and possibly several times that long. A minimum of one hundred thousand generations of differential survival and reproduction have contributed to our adaptation to the foraging way of life. No hominid subsisted otherwise until agriculture was invented about ten thousand years ago. So if human nature is a complex of adaptations functioning to maximize fitness in the ecological niche in which and for which natural selection created it, then we must study that niche to understand the organism.

That, at least, is the rationale for the enormous scientific interest in surviving foraging societies. Well into this century, there remained many peoples in Africa, Australia, Asia, the Americas, and the Arctic who had never planted a crop for harvest, never domesticated an animal for food. Their life-styles, already almost extinguished as a result of contact with technological cultures, have been studied by a host of anthropologists convinced of their special significance. The best-studied such people, and one whose world is perhaps as close as any to ancestral mankind's, is the !Kung San of southwestern Africa's Kalahari Desert.

Most San have been resettled as agriculturalists in recent years, but a few hundred carry on their ancient activities, pursued by anthropologists. The women gather and the men hunt; and as in most foraging societies, it is the women who contribute most of the calories, the men most of the protein [373]. !Kung Sun move their residences frequently but not far, forming camps that commonly contain some twenty to forty people [375]. Richard Lee, a leading expert on their affairs, insists that there is really no such thing as a band, for each camp "is an open aggregate of cooperating persons which changes in size and composition from day to day" [373, p. 31]. The more permanent building blocks from which these temporary aggregations are formed are nuclear families, about 5 percent of which are polygynous [284, 374]. Movements are largely constrained by the location of waterholes. Although these waterholes are in a sense the property of residents, there is no such thing as territorial exclusion.

In the !Kung San population studied by Nancy Howell [283, 284], menarche occurs at a mean age of 16.6 years, and it is at about that age that women marry. The husband is likely to be at least five years older than his wife. A period of adolescent sterility follows menarche, with the first child being born at an average maternal age of 19.5 years. Fertility is remarkably low, only 4.65 live births by menopause in one San population [258], although neither contraception nor abortion is evidently practiced. The effective breeding system is, as in most human societies, somewhat polygynous (Figure 12-2). The males' greater variance in offspring production is due partly to greater mortality in youth and partly to simultaneous polygyny, but mainly to successive marriages.

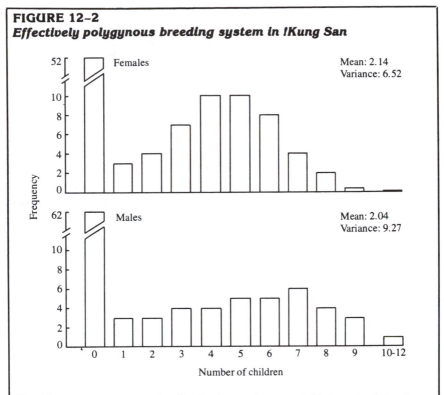

FIGURE 12-2
Effectively polygynous breeding system in !Kung San

The histogram represents the distributions of expected lifetime fertilities for a cohort of 100 females and 100 males, as reconstructed from demographic data. Males are likelier (62 percent) to die childless than females (52 percent), partly because males are likelier (49 percent) to die before marriageable age than females (44 percent). Males are also likelier than females to produce 7 or more children, primarily because of successive marriages. (After Howell, 1979, Tables 16-1 and 16-2 [284].)

Well nourished but thin, San women are unlikely to conceive while nursing infants. The baby is carried in a sling, and suckles essentially at will throughout the day and night. According to one careful study, babies nursed in two-minute bouts at an average interval of 13 minutes, hence four bouts per hour throughout the daylight hours, and this nursing schedule did not change greatly through 139 postpartum weeks [340]. Nursing commonly continues until age four and exceptionally until age six [283]. The child is typically weaned (often with articulate protest!) when the mother discovers that she is again pregnant. The frequent nursing evidently suppresses maternal ovarian function, delaying the next conception [340]. A three-month-old baby spends perhaps 80 percent of the day and all night in direct physical contact with mother; at a year, such contact still occupies half the day [339].

Howell [284] has cautioned against considering the !Kung to be typical of all foraging societies and hence human prototypes. However, there are several reasons to suppose that the patterns of mother-infant interaction exhibited by the !Kung are indeed natural to the human species. Prolonged mother-infant contact, including sleeping together, and demand nursing beyond the age of two are general features of foraging societies. Moreover, the composition of human milk reveals us to be adapted to demand nursing (Table 12-2): Low fat and protein content is typical of the milk of continuous or near-continuous sucklers, whereas species that leave their offspring in a nest and visit them periodically produce richer milk.

Observations by Lee [373] indicate that there is surprisingly little problem in food getting. The women need to go gathering only about every third day, and the men commonly hunt for a week and then relax for two. Both women and men thus spend an average of just a couple of hours a day in

TABLE 12-2
Nursing intervals and milk analyses for various species

Species	Nursing interval	Fat %	Protein %
Red kangaroo	continuous	4.0	3.9
Brown bear	continuous	3.0	3.8
Rhesus macaque	demand	3.9	2.1
Chimpanzee	demand	3.7	1.2
Human	demand	4.0	1.3
Domestic pig	1 hour	4.0	3.7
Rat	2–3 hour	12.6	9.2
Fox	2–3 hour	6.3	6.2
Domestic cat	2–3 hour	4.9	7.1
Hedgehog	3–4 hour	10.1	7.2
Hamster	2–4 hour	12.6*	9.0
Coyote	3–4 hour	10.7	9.9
Rabbit	4 hours	10.4	15.5
Lion	6–8 hours	9.5	9.4
Impala	8–12 hours	20.4	10.8
Virginia deer	12 hours	19.6	10.3

Source: From Ben Shaul, 1962 [41].

food-getting activities, and that is enough to support not just themselves but unproductive oldsters and youngsters, who together comprise some 40 percent of the population. However, since the women carry and closely mind dependent offspring, two children in close succession are burdensome [58]. Indeed, too short a birth interval may occasionally lead a woman to kill the second child at birth. Infanticide is also practiced in the case of birth abnormalities or upon one of twins. Howell [283] reports 6 cases of infanticide out of 500 live births. About one in five infants fails to survive the first year, with malaria and gastrointestinal and respiratory diseases the leading causes of mortality.

Superficially, the leisurely pace of !Kung life is not easy to reconcile with the idea of evolved adaptations for reproduction maximization. As we have seen, however, fitness maximization is not simply a matter of fertility maximization. Furthermore, the assumption that the ecological circumstances of the !Kung are essentially those of prehistoric foragers would surely be wrong. Agriculture has swept the field in more productive areas, forcing foragers into marginal habitats. This makes the ease of their food gathering the more surprising, but water may be more limiting than food, and the costs of carrying children in a hot, arid climate probably preclude more frequent reproduction [58]. In drawing inferences to our selective history, we must also recall the faunal differences between the modern Kalahari and the human evolutionary theater. *Australopithecus* had to cope with a large number of carnivorous mammals that were both competitors and potential predators. For the !Kung, the impact of carnivorous mammals has been modified, though by no means eliminated. This is not entirely a result of recent European influence. The ecosystem of hominid evolution has been progressively disrupted over the several centuries in which pastoralists and sedentary agriculturalists have flourished in Africa. Everywhere in the world farmers tend to exterminate dangerous and competitive animals. No present-day foraging society can now reveal to us exactly what constraints hyenas and lions once imposed upon human dispersion, hunting strategies, or parental vigilance. But special attention to modern foragers remains warranted, though we must be cautious in our inferences therefrom. The !Kung are not just another tribe, more grist for the ethnographic mill. They are as near as we can get to a pristine society living as all humankind lived for most of our evolutionary history.

One thing that studies of the !Kung reveal is just what extreme *K*-strategists people have evolved to be. Interestingly, our nearest relatives, the chimpanzees, are approximately our equals in this regard [601, 624]. They mature slightly earlier, menarche taking place at twelve years of age with adolescent sterility until about fourteen. Birth intervals, however, tend to be even longer than in *Homo*, a mean of almost six years if both infants survive. A female chimp who enjoys a full life span will have had five live births by her death at forty or forty-five years of age. Clearly, the slow reproductive pace of *Homo* and our remarkably intensive nurture cannot be attributed to a uniquely human intellectual development, to the hunting adaptation, nor indeed to any novel characteristics of the hominid line: The basic reproductive life history antedated the evolution of people's peculiarities.

HUMAN FERTILITY

The advent of agriculture changed the game. We have already noted that new kinds of material wealth provided a new medium of male-male competition, greatly exacerbating male fitness variance, extreme polygyny, and the treatment of women as commodities. Domestic animals also made it possible to wean babies sooner and therefore to squeeze more into a reproductive life span. Foragers like the San manage just 4 or 5 live births. Horticulturalists, lacking domestic animals, are seldom much more fertile; Yanomamö women who survive to forty, for example, report a mean of 3.8 live births [407], and though there is assuredly an underreporting of infanticides, the figure is impressively low.

Pastoralists and settled agriculturalists, by contrast, experienced a fertility boom. Lactational inhibition of ovulation is partially dependent upon the mother's nutritional state: Even with a fixed duration of lactation, a better-fed mother is likely to resume ovulation sooner [315]. More important, when the child is weaned partially or entirely, ovulatory probability rises quickly [23, 285, 431, 502] (Figure 12–3). This mechanism, nicely adapted for both variable birth-spacing according to food availability and rapid replacement of lost infants, allowed new heights of fertility in agricultural woman. According to the *Guinness Book of World Records* the greatest number of live births ever recorded for one woman is 69, but even in high-fertility populations the upper limit for all practical purposes is about 24 [432]. This is an extreme, of course; typical numbers are much lower. In Western cultures prior to the "demographic transition" from high to low fertility, the mean number of children born per female reproductive lifetime was commonly 6 to 8 [395].

The highest fertility rate that has been documented for any human population was that recently attained by the Hutterites, pacifist Christians from Eastern Europe who emigrated to become prosperous communal agriculturalists in western North America. In a 1953 study the average number of live births for a Hutterite woman by the time of menopause was 10.4 [170]. Between 1880 and 1950 the North American Hutterite population grew from 443 to 8,542. (Since 1950, Hutterite fertility has declined markedly [361].)

FIGURE 12–3
Lactational suppression of ovulation in women

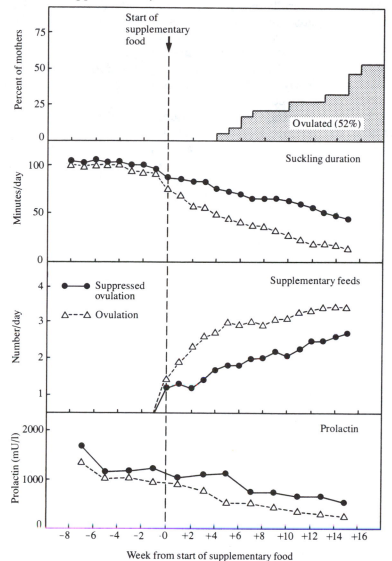

The study involved 27 Scottish mothers, who breastfed for a mean of 40.5 weeks, and introduced supplementary food at a mean of 16.1 weeks postpartum. The top figure shows when ovulation resumed after the introduction of supplementary food. The 14 mothers (52 percent) who had ovulated by week 16 are compared with the 13 who had not in the three lower figures. The two groups did not differ before the introduction of supplementary food. Those who ovulated sooner nursed less, supplemented more, and had lower blood prolactin levels. (After Howie et al., 1981, Figures 2 and 3 [285].)

Even in high-fertility populations, lactational anovulation and maternal nutritional state continue to adjust birth intervals to match maternal capabilities. These physiological controls are frequently reinforced by a taboo against sexual intercourse for some time after birth [452]. The duration of the taboo period varies considerably from one society to another, as does the faithfulness with which the taboo is observed in practice. The most common duration is about six months.

It is important to note that the mechanisms of birth spacing do not entirely eliminate resource competition between the offspring of a single mother. Like the !Kung, many people all over the world routinely sacrifice one of a pair of twins unless unusually favorable circumstances promise that both can be reared, and when successive births occur too close together, mothers often resort to that same drastic measure of infanticide. It seems obvious that very short birth intervals must put infants at risk, although it is not easy to find data bearing directly on this point. There are lots of data indicating that short intervals are associated with high infant mortality, but that relationship does not constitute firm evidence that the short birth interval is a *cause* of infant death. It might instead be a result: Parents often quickly replace a lost infant. A Guatemalan study, for example, showed that a mean birth interval of two and a half years was reduced to less than one and a half when the first child died before one year of age [271]. In order to show effects in the other direction—that is, the effects of the interval upon infant survival—we should like information on the survival prospects of a second infant as a function of the birth interval, *given* the survival of the first infant. The data have been organized in precisely this way in one study of nineteenth-century demographic records from three Bavarian villages [336]. As can be seen in Figure 12-4, the second infant's survival prospects improved markedly, the longer the interval between its birth and that of the previous child.

The Bavarian data can also be used to make another point: Children died. And there is no reason to doubt that they have died throughout the history of the human species. The mechanisms of birth spacing, both physiological and cultural, do not completely eliminate detrimental competition for maternal resources between successive infants. How could they? The "decision" to conceive depends largely upon a prediction of future conditions on the basis of present conditions, and predictions always contain an element of uncertainty. The mother may expect to have adequate resources to cope with two infants nine months and more in the future, but that expectation may fall flat. If a birth-spacing mechanism did not permit conception until *all* risk of detrimental competition between infants had dissipated, it would enforce a suboptimal strategy of excessive birth spacing (Figure 12-4).

FIGURE 12–4
Probability of infant survival as a function of the
preceding birth interval in nineteenth-century Bavaria

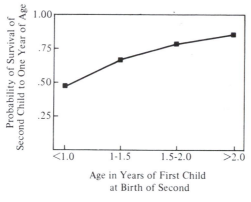

(Data from Knodel, 1968 [336].)

Adaptive birth spacing? An optimal birth-spacing mechanism produces births at such intervals as to maximize the total output of surviving offspring. Very wide spacing will increase the individual probability of survival, yet reduce the total output of survivors. In the Bavarian case a child born after a birth interval of fifteen months has 0.66 probability of survival, while a child born after a twenty-one-month interval has 0.79 probability of survival. Mothers practicing the shorter interval will lose more children, but they will produce more survivors too. Suppose, for example, that A uses a fifteen-month interval after each surviving child and B uses a twenty-one-month interval, while both use a fifteen-month interval after a nonsurvivor. On these assumptions we compute that A can expect to produce on average 3.3 more survivors (while losing 1.2) in the five years after her first surviving child, whereas B can expect to produce an average of 2.4 more survivors (while losing 0.6) in the same time. We have left out of the equations the risk that the mother herself incurs in every confinement, but it should be clear that it is at least possible for the strategy with the higher death rate to enjoy a selective advantage over the slower, safer strategy. There is no reason to expect that an evolved mechanism of birth spacing will necessarily minimize mortality.

The actual intervals that occur between human births (and hence determine a woman's lifetime reproductive output) are influenced by more factors than just lactational anovulation and the woman's nutritional state. Such factors determine whether the woman *can* become pregnant again, but ovulation carries no guarantee of fertilization. If the woman is to conceive, she must copulate at some time quite near to ovulation.

In each twenty-eight-day menstrual cycle the time during which it is physiologically possible for copulation to result in conception is quite brief —almost certainly less than two days and possibly less than one. Therefore, the probability of conception is related to the frequency of sexual intercourse: The more rarely copulation occurs, the more likely it is that the fertile period will be missed, and then another month must elapse before impregnation is again possible [455]. In one American study [664], married women practicing no birth control took an average of 11.0 months to conceive when they reported a coital frequency of less than two per week. Just 7.1 months were required for conception in couples who had coitus twice a week, and 6.6 for those who reported their weekly frequency to be three or more. A large volume of British and American survey data have furthermore demonstrated that couples who have had more children in a fixed duration of marriage reported higher coital frequencies; this is so despite the fact that the couples reported a decline in sexual activity with the addition of each baby [313]. Extremely high coital frequencies, as are sometimes experienced by prostitutes, can nevertheless have a negative effect upon fertility, a phenomenon resulting from a spermicidal immune reaction [455].

Another factor in human fertility is polygyny. The wives in polygynous marriages are less fertile than wives in monogamous unions. We have already presented some data to this effect (see Figure 6-6, p. 127). There are probably several reasons for this trend. In some cases, a second marriage is contracted specifically because of the infertility of the first; moreover, if it is in fact the male who is sterile, then a whole household of polygynous wives will be barren. Actual sterility is rather uncommon, however. Possibly a more important source of the relative infertility of polygynous alliances is a lesser coital frequency. A polygynous wife, particularly one with a young child, is less likely to be exposed to the risk of pregnancy. Where the husband has an alternative sexual outlet, he is more likely to observe a postpartum taboo on sexual intercourse, and indeed that taboo is likely to be longer. One anthropologist used the sample of human societies in Murdock's *Ethnographic Atlas* to search for correlates of prolonged postpartum taboos [544]. About the only strong correlate was the marriage system: Where polygyny was practiced, the taboo tended to be relatively long (Table 12-3).

It is a common prejudice to suppose that the poor reproduce at an especially high rate, but this is far from the truth. Comparing across countries, we find that per capita income is positively related to fertility [266], and the same is usually true within societies. It is of course the affluent men who are polygynous, but that is not the only consideration. Both fertility and mortality are relevant. In one study in central India, for example, the

TABLE 12-3
Relationship between marriage system and duration of postpartum taboo

Marriage system	Duration of postpartum taboo (number of societies)		Taboo greater than 6 months (percent of societies)
	less than 6 months	more than 6 months	
Monogamy	23	3	11.5
Limited Polygyny	50	23	31.5
Polygyny	40	32	44.4

Source: Reprinted from *Current Anthropology*, 13: 238-249 by J. F. Saucier by permission of The University of Chicago Press. © 1972 by The Wenner-Gren Foundation for Anthropological Research.

fertility of (monogamous) couples showed no relationship to the husband's income, but the number of *surviving* children increased with increasing affluence [166]. So despite the lack of a fertility difference, there was differential reproductive success as a function of income. Fertility was furthermore positively related to a different economic variable, namely the size of the couple's landholdings. This was so despite the fact that those with more land had a somewhat greater knowledge of birth control methods and expressed greater interest in their use.

In a study of Yomut Turkmen, a pastoral people in Iran who lack the great degree of economic inequality that is commonly found in agricultural and urban societies, anthropologist William Irons looked at the reproductive performances of the wealthier half and the poorer half of the population [306]. Men from the wealthier half sired surviving offspring at a rate 1.75 times that of men from the poorer half. This depended in part upon the possibility of polygyny, but it also resulted both from higher fertility at all ages in the wealthier half and from their lower mortality. Because of these latter factors women from the wealthier half were outproducing women from the poorer half by the less dramatic ratio of 1.12 to 1.00. Similarly, among Iranian village women, socioeconomic status has been shown to be positively correlated with reproductive rate [5].

The fertility advantage of the affluent is surprising only because of our familiarity with a few industrialized countries that have undergone the demographic transition (the modern phenomenon of a drastically reduced death rate followed by a declining birthrate). Even in post-demographic-transition societies such as the United States, the common belief that low occupational status and poverty are associated with high fertility is false [e.g., 24, 647] (Figure 12-5).

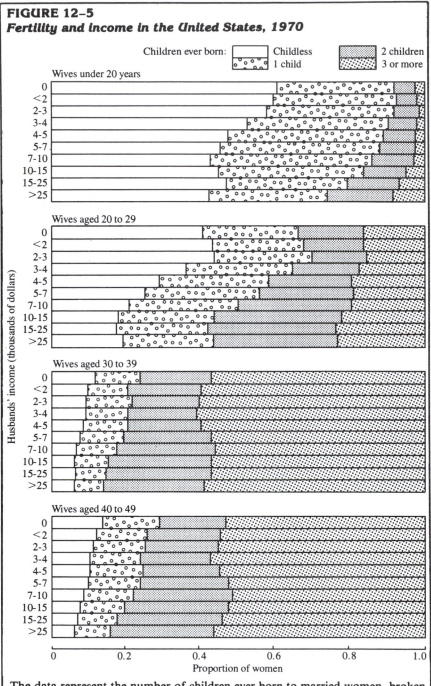

FIGURE 12–5
Fertility and income in the United States, 1970

Children ever born: ☐ Childless ▨ 2 children
○ 1 child ⣿ 3 or more

Wives under 20 years

Wives aged 20 to 29

Wives aged 30 to 39

Wives aged 40 to 49

Husbands' income (thousands of dollars): 0, <2, 2-3, 3-4, 4-5, 5-7, 7-10, 10-15, 15-25, >25

Proportion of women: 0, 0.2, 0.4, 0.6, 0.8, 1.0

The data represent the number of children ever born to married women, broken down according to the woman's age and the husband's income. In each age category, the proportion childless declines with increasing affluence. (After U.S. Bureau of the Census, 1973, Tables 50 and 51 [630].)

Societies that have undergone the demographic transition do, however, pose a puzzle. All about us are people willfully refraining from reproduction! Such behavior seems to contradict the suggestion that human nature is in any sense a product of evolution by natural selection, that process of the competitive ascendancy of whatever traits help some individuals outreproduce others. The solution to this dilemma lies in an appreciation of the evolutionary novelty of modern circumstances. The pill has derailed physiological mechanisms that once assured regular fertilizations. Agriculture, urbanization, and a host of other relatively recent developments may have similarly derailed psychological mechanisms that once were reproductively valuable, but the identification of such basic psychological features is necessarily more speculative. What did it matter before the advent of effective contraception if men were often more interested in copulation than in fatherhood, so long as the former assured the latter? And what did it matter if women were ambivalent about an endless succession of pregnancies, so long as circumstances conspired to assure them that fate?

PARENTAL SOLICITUDE AND DISCRIMINATION

In people, as in all mammals, direct care of the young is predominantly a female concern, and as we saw in the case of the !Kung, women in foraging societies devote a very considerable proportion of their time and energy to infant care. But paternal investment is also clearly characteristic of our species. Men in virtually all societies confer resources upon their mates and children, and they are commonly able to confer status upon sons [452]. In one sample of twenty-three societies, men furthermore shared in direct infant-care activities in thirteen [582]. Fatherless children in Western society and elsewhere suffer social and intellectual disadvantages and elevated mortality risk [e.g., 119]. Biparental care in *H. sapiens* appears to be a fundamental adaptive attribute.

As in other parentally investing animals, human parental solicitude is not indiscriminate. Solicitude seems to be most readily directed toward those children who are the likeliest contributors to parental fitness [130]. Children who are defective or otherwise unlikely to survive and reproduce are killed at birth in many, perhaps most, societies [155, 132]. Where parents are obliged to rear defective children, the risks of physical abuse and destructive neglect are greatly elevated [131].

The risk of cuckoldry presents a special problem. We have already suggested that it is the considerable paternal investment in putative offspring that is responsible for the intensity of male concern about female sexual fidelity. Men must also be sensitive to phenotypic cues indicative of paternity; a staple bit of folk humor in many parts of the world is the ridiculing of a simple-minded man who invests in a child that everyone can see is not his. In North America, new parents and their relatives and friends all pay considerably more attention to paternal resemblances in a newborn than to maternal resemblances, and as we might anticipate, it is the new mother who is most inclined to perceive and remark paternal resemblance [134].

Maternity confidence is not a problem: Like other mammals, human mothers can establish individualized bonds with their children at birth. Or at least they used to do so, for recent hospital practices have seldom permitted the mother-infant attachment process to proceed normally [588]. A series of studies in several different countries has involved comparing control mothers who are given minimal contact with their newborns in the first couple of days (the routine in most hospital births) with experimental mothers given a few hours' more contact [reviewed by 403, 525]. Follow-up studies have shown effects upon both mothers and infants months and even years later. Long after they have left the hospital, the experimental mothers talk to their children more, handle them more affectionately, and express stronger feelings of maternal attachment [331]. The experimental mothers' children are also significantly less likely to become victims of child abuse [467]. Deep maternal feeling develops out of a history of interaction, and early interactions appear to be particularly important. Where adults of either sex are called upon to invest "parentally" in children manifestly not their own and with whom early attachment experiences are lacking, there is a high risk of failure of solicitude; step-parent households, for example, are far riskier environments for child abuse than are natural-parent households [683].

We cannot leave the topic of discriminative parenting without some consideration of the differential treatment of daughters and sons. A preference for sons is widespread and often extreme [677]. Daughters may be so devalued as to be killed at birth, and many women have been divorced or mistreated for producing only daughters (a bitter irony, since the determination of offspring sex resides in the paternal gamete). The effects of female infanticide and destructive neglect may include extremely male-biased sex ratios in the population.

This sort of parental bias seems perplexing from the fitness-maximization perspective. Where women are in short supply, daughters should be *especially* valued (cf. Fisher's sex ratio theory, pp. 224–226). Several partial answers to this paradox have been suggested. Mildred Dickemann [157] has pointed out that female infanticide is status-graded in highly stratified societies, with the highest social classes removing the most daughters. It is in these classes that preferential investment in sons is most likely to yield great returns in political and economic power and in polygynous monopolization of wives and concubines; moreover, there is dowry competition among those a little lower in the social order to marry their daughters to higher-ranking men, which means that marital prospects are scarce for aristocratic women. The extreme patrilineal bias in inheritance of resources and political office leads to an ideology that values only the male line of descent. A Hindu bridegroom prays, "Oh faithful wife, give birth to a son who will live long and perpetuate our line," and mantras are chanted to transform a female fetus into a male; in modern India, amniocentesis is used solely to detect and eliminate female fetuses [517]. Female-selective infanticide occurs in many relatively unstratified societies too, where it is evidently associated with chronic small-scale warfare [159]. Here the preference for sons is commonly expressed as a need for warriors, and the fragile balances of power suggest that this rationale is a very real one. The resultant shortage of women exacerbates the conflict, since women are a large part of what the warfare is all about [91].

CHANGING FERTILITY PATTERNS

The industrialized societies of the West are undergoing a curious demographic change, the results of which can hardly be foreseen. Differential reproduction of the sort we have just described is on the wane. No variable is related to any large amount of variance in reproduction—neither religion nor residence, neither income nor education. The reason is that the variance itself is shrinking. The average number of children borne by an American woman who was herself born in 1870 was 3.9 by the time of menopause. The variance in this reproductive output was 10. For the currently menopausal cohort of women born in 1930, the average lifetime production has fallen to 3.1 and the variance has plummeted to 4 [312]. (These parameters, incidentally, are somewhat closer to those prevailing in "primitive" societies; see Figure 5–6, p. 89, and Figure 12–2). Everyone is aware of the declining occurrence of really large families, but it is less well known that abstention from reproduction has declined too. In 1909, 23 percent of the thirty-five-year-old women in the United States were childless; by 1960 this figure had fallen to 12 percent [329]. More recent data indicate that this trend has continued (Figure 12–6). What this means is that the intensity of selection within the contemporary American population of *Homo sapiens* has recently declined. Selection is a matter of differential reproduction, and there simply isn't much differential.

Another factor that influences human fertility has undergone a striking change in the last century—the age at sexual maturity. This is most

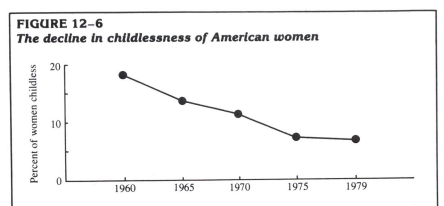

FIGURE 12–6
The decline in childlessness of American women

The percentage of ever-married American women aged 45 to 49 who have had no children has declined substantially since 1960. (Ever-marrieds represented 93 percent of 45- to 49-year-old women in 1960, 95 percent in 1970 and 1980.) (After U.S. Bureau of the Census, 1976, 1981 [631, 632].)

readily determined in women, since there is a conspicuous pubertal event that can be dated—the first menstrual flow. The age at which menarche takes place varies considerably among human populations and conditions. It is much retarded by malnutrition; indeed, menstruation is suspended even after puberty whenever the nutritional state is extremely poor. Harvard demographer Rose Frisch has argued that menstrual function is related to the availability of body fat. A European girl must have deposited fat amounting to about 17 percent of her body weight if menarche is to occur. Fat deposition continues, reaching a mean of about 27 percent of body weight by age sixteen. Before that age girls tend to be relatively infertile— and Frisch considers reproductive maturity to be attained at sixteen to eighteen years. After maturity any weight loss that drops body fat below about 22 percent of body weight is likely to occasion a cessation of menstrual function (secondary amenorrhea). The implication is that a certain minimum level of stored energy is necessary for a woman's physiological systems to make the decision to risk a pregnancy. Lacking that energy store, her best strategy is to hold off and build herself up in anticipation of a more propitious reproductive opportunity [203].

In the Western world the age of menarche has declined steeply and steadily for a hundred years (Figure 12–7). Presumably, this reflects increased caloric intake and a more rapid attainment of the necessary fat deposits. Why has this occurred? An improvement in the material conditions of life may be a major factor, but it seems unlikely that this is the whole story. The simple availability of calories has surely not increased with

FIGURE 12-7
The decline in the average age of menarche in five
Western industrial nations

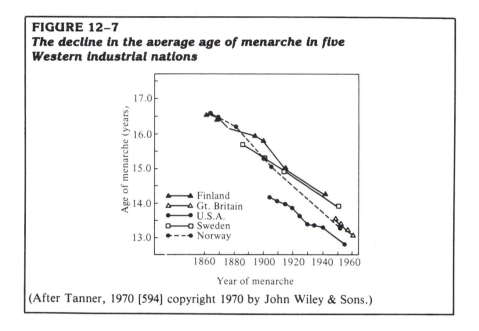

(After Tanner, 1970 [594] copyright 1970 by John Wiley & Sons.)

the same monotonic doggedness with which age of menarche has declined. The data show little effect of war or depression. The trend is remarkably linear.

One intriguing consideration is that there has been a steady increase over the decades in our exposure to artificial lighting. The possible relevance of this phenomenon is suggested by the fact that light exposure (reflecting daylength) is important in the seasonal regulation of endocrine function in many vertebrates. Humans are not so seasonal as some animals but we are not completely liberated from effects of seasonal changes either. An ingenious analysis of Finnish birth records bears this out. Timonen and his colleagues demonstrated an annual cycle with a maximal probability of conception in summer and a minimal probability in the darkest part of winter [612]. This effect is not in itself very convincing evidence of anything, but there is more. The incidence of multiple births due to multiple ovulations, which can be considered an indirect index of ovarian function, exhibits an exaggerated version of the same trend: There is a higher peak of conceptions in summer and a lower trough in winter (Figure 12-8). The effect is also exaggerated in higher latitudes with greater daylength variability. The obvious conclusion is that light exposure has at least some influence on reproductive function in our own species [see also 262]. That light may be involved in the historical change in the age of menarche, however, remains speculative.

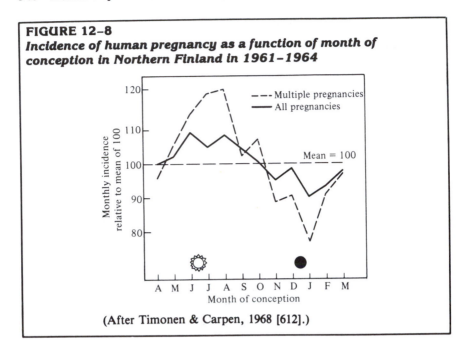

FIGURE 12-8
Incidence of human pregnancy as a function of month of conception in Northern Finland in 1961-1964

(After Timonen & Carpen, 1968 [612].)

There is, moreover, the possibility that the trend reflects a process of biological evolution. This recent hypothesis is supported by the data in Table 12-4. As can be seen, there is a substantial relationship between fertility and early puberty. The slight difference in the age of marriage does not account for this effect. Critescu argues that the selective significance of the relationship between fertility and early puberty was offset in the past by a positive relationship between fertility and mortality that has recently disappeared [122]. She also points to evidence of a substantial degree of

TABLE 12-4
Fertility as a function of age of puberty in Rumanian women

Age at first menses	Mean age at marriage	Mean number of children
12-13	20.2	5.50
14-15	20.0	4.66
16-18	21.7	3.50

Source: After Critescu, 1975 [122] with permission from *Journal of Human Evolution.* Copyright: Academic Press Inc. (London) Ltd.

heritability in the menarchical age, which would have to be true for genetic selection to be able to effect the evolutionary change hypothesized. But if gene-pool changes are involved in the shift, they are assuredly not the whole story. Menarchical age has fallen abruptly in certain cases, as for example in Japanese emigrants to America and in !Kung hunter-gatherers resettled as agriculturalists [283]. In these cases it seems probable that dietary change has effected maturational change irrespective of gene-pool change.

IDEOLOGY AND ATTITUDE

Many of the factors in human reproduction that we have discussed to this point could as readily be discussed for other animals. Mating systems, mechanisms achieving exogamy, physiological and behavioral factors in birth spacing, mate selection, nutritional effects upon the speed of maturation—all these are matters that warrant consideration in a discussion of any species' reproduction. In the special case of people, we have access to something more—the expressions of intent and attitude that "explain" human behavior at one proximate level. We have no intention of getting into the murky subject of psychodynamic theories here. We wish simply to consider the attitudes and intentions of human actors, manifest in speech and action and commonsensically interpreted at their face value.

The most universal and pervasive human attitudes are likely to be so taken for granted as to be rarely perceived. One such is pronatalism. People are supposed to reproduce, and they are supposed to want to. However bizarre an alien culture may seem, it is hard to imagine how it could lack pronatal attitudes. It is a rare social scientist who would think to investigate this directly, but it is apparent incidentally in all sorts of related matters—in a payment of brideprice only after a child is born [215], in the "universal concern" over infertile women [194], in divorce on the grounds of sterility. Why is the announcement of a pregnancy cause for celebration? If culture were truly arbitrary, it might as readily be an occasion for sorrow or indifference!

In industrialized society fertility has declined dramatically, but pronatalism has endured. Rainwater's survey of family design and attitudes in three American cities in the early 1960s found a "universal rejection of the woman who wants no children as either totally self-involved, childish, neurotic or in poor health" [515]. He summarized his informants' statements as a norm to the effect that one should have as many children as one can support. Americans who favor large families feel that they are desirable from every viewpoint: They appeal to personal, social, economic, moral, and patriotic considerations, sometimes all at once [209].

But perhaps pronatalists protest too much. As someone once re-marked, you don't need a taboo against autophagy. Why did Pope Pius XI feel obliged to announce that "Every attempt on the part of the married couple during the conjugal act or during the development of its natural con-sequences, to deprive it of its inherent power and to hinder the procreation of a new life is immoral"? [501]. Is there an antinatalist satan abroad in the world?

It is probably more accurate to say that there is a lot of ambivalence. The human mind is devious enough to want children and not want them at the same time. This fact emerges with particular clarity from a sociological study of poor tenant farm families in North Carolina during the Depression [241]. Asked how many children they had had, mothers were proud of their productivity: "Ten and all a-living," said one, and "Eleven, I done my share, didn't I?" In the researcher's words, "The bearing, 'raising' and 'marrying off' of children are everywhere recognized as being a positive achievement." And yet most women with young children were likely to add, "I hope this is the last one." The author's comment is again significant: "In most cases, the effecting of the wish was limited to 'hoping'."

"Motherhood" is so universally a virtue that the very word has become political slang for any issue with surefire unanimous support. Women's career aspirations and the concern with overpopulation have combined to tarnish the virtue of maternity in recent decades, and yet it is remarkable how general the aspiration to parenthood remains [79]. Cer-tainly there has been no dramatic modern increase in the proportion of adults who remain childless in industrial societies (Figure 12-6).

The most recent surveys still indicate that educated young Americans want to have more than two children on average [147, 209, 571]. They are likely to be well aware that zero population growth demands a lesser output, and they may feel guilty about their excessive desires, but those are their ex-pressed desires nevertheless. If a society that socialized its young to an am-bition of childlessness were ever to arise, what would be its fate?

Many social scientists are still skeptical of an evolutionary biological ap-proach to human behavior. (For an unusually well-informed critique, see [59].) In our view, that skepticism derives largely from a failure to ap-preciate the complementarity of multiple explanatory modes (cf. Chapter 1). Too often, scientists advocate one perspective by denigrating another. In psychology, situational determinants of behavior are still set in opposition to personality traits. In anthropology, economic explanations are still con-sidered alternatives to biopsychological explanations. In sociology, Durkheim's insistence that social facts are explicable only at the level of social facts is still quoted with approval. It is time that we abandoned such one-dimensional views.

Why is that man over there bullying his wife? Because of his testosterone? His upbringing? His Y chromosome? His culture? His social status? His religion? In order that he not be cuckolded? Because he no longer loves her? Because of a frustrating day at the office? Sadism? Poverty? Sunspots?

There may be something to all those answers and there may not. If we arrive at the clinical decision that the man is or is not a sadist, none of the other explanations is obviated: They are still open to affirmation or denial. And if we establish that he bullies her only on days when his boss bullies him, none of the other questions has been answered.

In this example, the simultaneous validity of several explanations in several explanatory modes is easy to grasp. In other examples, such as the explanation of sex differences, people often find it harder to entertain two or more explanatory modes at once. In particular, that of functional explanation encounters resistance. The notion that sex differences might be evolved adaptations can provoke violent opposition. An "alternative" explanation is offered in terms of sexist upbringing and societal stereotypes. These factors are undeniably involved in the development and expression of sex-typed behavior, but the explanation of sex differences in terms of function still needs to be addressed. Moreover, we may seek to explain the stereotypes and sexist rearing practices at the functional level too.

According to our current understanding of the evolutionary process, all aspects of all organisms, including ourselves, are the products of a continuing history of differential reproductive success. Current theory is not inviolate. Our detailed conceptions of "natural selection" and "fitness" have changed before and may change again. Particular evolutionary hypotheses that seem compelling today will seem naive and incomplete tomorrow. But it is of course by rejecting and refining hypotheses that any science progresses. What is certain is that evolutionary biology has profound implications for the social sciences, implications that we have just begun to explore. In the next few years we can look forward to many insights into the selective circumstances that have shaped our species. We can hope to understand how the evolving reproductive strategies appropriate to our unique ecological adaptations have taken us, one species of clever, sociable primates, to the point of critical self-examination.

SUMMARY

People are extreme K-strategists, investing prolongedly and intensively in a few offspring and in collateral relatives.

Attention to patrilineal as well as to matrilineal kinship links is a feature shared to some extent with our closest primate kin, but culture and language have enormously elaborated human kinship systems. Human kinship is profitably viewed from an evolutionary perspective: The effects of kinship upon resource flow and upon cooperation and conflict appear to reflect nepotistic strategies for inclusive-fitness maximization.

In the foraging niche characteristic of human prehistory, women raised perhaps as few as four or five children, born at intervals of several years; both physiological and cultural mechanisms matched reproductive output to maternal capacity. Investing so intensively in a few offspring, human parents are highly discriminative in their solicitude, favoring those children who are most likely to contribute to parental fitness.

Agriculture and related historical developments raised maximal fertility; affluence was converted to reproductive success, and in predemographic transition societies, it still is. In the modern West, fertility has declined but so has childlessness. The aspiration to reproduce remains robust.

SUGGESTED READINGS

Chagnon, N. A. & Irons, W., eds. 1984. *Human sociobiology: new research and theory*. Special issue of *Ethology & Sociobiology*.

Lee, R. B. & DeVore, I., eds. 1976. *Kalahari hunter-gatherers: studies of the !Kung San and their neighbors*. Cambridge, Mass.: Harvard University Press.

References

1. Abele, L. & Gilchrist, S. 1977. Homosexual rape and sexual selection in acanthocephalan worms. *Science,* 197: 81–83.

2. Adams, D. B., Gold, A. R. & Burt, A. D. 1978. Rise in female-initiated sexual activity at ovulation and its suppression by oral contraceptives. *New England Journal of Medicine,* 229: 1145–1150.

3. Adams, M. S. & Neel, J. V. 1967. Children of incest. *Pediatrics,* 40: 55–62.

4. Adkins-Regan, E. K. 1981. Early organizational effects of hormones: an evolutionary perspective. In N. T. Adler, ed. *Neuroendocrinology of reproduction: physiology and behavior.* New York: Plenum Press.

5. Ajami, I. 1976. Differential fertility in peasant communities: a study of six Iranian villages. *Population Studies,* 30: 453–463.

6. Alatalo, R. V., Lundberg, A. & Stahlbrandt, K. 1982. Why do pied flycatcher females mate with already-mated males? *Animal Behaviour,* 30: 585–593.

7. Alcock, J., Jones, C. E. & Buchmann, S. L. 1977. Male mating strategies in the bee *Centris pallida* Fox (Anthophoridae: Hymenoptera). *American Naturalist,* 111: 145–155.

8. Alexander, R. D. 1974. The evolution of social behavior. *Annual Review of Ecology and Systematics,* 5: 325–383.

9. Alexander, R. D. 1979. *Darwinism and human affairs.* Seattle: University of Washington Press.

10. Alexander, R. D. & Borgia, G. 1978. Group selection, altruism, and the levels of organization of life. *Annual Review of Ecology and Systematics,* 9: 449–474.

11. Alexander, R. D. & Borgia, G. 1979. On the origin and the basis of the male-female phenomenon. In M. S. Blum & N. A. Blum, eds. *Sexual selection and reproductive competition in insects.* New York: Academic Press.

12. Alexander, R. D., Hoogland, J. L., Howard, R. D., Noonan, K. M., & Sherman, P. W. 1979. Sexual dimorphisms and breeding systems in pinnipeds, ungulates, primates and humans. In N. A. Chagnon & W. Irons, eds. *Evolutionary biology and human social behavior.* North Scituate, Mass.: Duxbury Press.

13. Alexander, R. D. & Noonan, K. M. 1979. Concealment of ovulation, parental care, and human social evolution. In N. A. Chagnon & W. Irons, eds. *Evolutionary biology and human social behavior.* North Scituate, Mass.: Duxbury Press.

14. Alexander, R. D. & Sherman, P. W. 1977. Local mate competition and parental investment in social insects. *Science,* 196: 494–500.

15. Allen, G. E. 1975. *Life science in the twentieth century.* New York: John Wiley.

16. Andersson, M., Wiklund, C. G. & Rundgren, H. 1980. Parental defence of offspring: a model and an example. *Animal Behaviour,* 28: 536–542.

17. Andrew, R. J. 1969. The effects of testosterone on avian vocalizations. In R. A. Hinde, ed. *Bird vocalizations.* London: Cambridge University Press.

18. Antonovsky, H. F. 1980. *Adolescent sexuality: a study of attitudes and behavior.* Lexington, Mass.: Lexington Books.

19. Arnold, S. J. 1976. Sexual behavior, sexual interference and sexual defence in the salamanders *Ambystoma maculatum, Ambystoma tigrinum* and *Plethodon jordani. Zeitschrift für Tierpsychologie,* 42: 247–300.

20. Ashmole, N. P. 1971. Sea bird ecology and the marine environment. In D. S. Farner & J. R. King, eds. *Avian biology,* vol. 1. London: Academic Press.

21. Askenmo, C. 1979. Reproductive effort and return rate of male pied flycatchers. *American Naturalist,* 114: 748–752.

22. Austin, C. R. 1972. Fertilization. In C. R. Austin & R. V. Short, eds. *Reproduction in mammals, Book 1: germ cells and fertilization.* Cambridge: Cambridge University Press.

23. Baird, D. T., McNeilly, A. S., Sawers, R. S. & Sharpe, R. M. 1979. Failure of estrogen-induced discharge of luteinising hormone in lactating women. *Journal of Clinical Endocrinology and Metabolism,* 49: 500–506.

24. Bajema, C. J. 1968. Relationship of fertility to occupational status, IQ, educational attainment, and size of family of origin: a follow-up study of a male Kalamazoo public school population. *Eugenics Quarterly,* 15: 198–203.

25. Barash, D. P. 1975. Ecology of paternal behavior in the hoary marmot (*Marmota caligata*): an evolutionary interpretation. *Journal of Mammalogy,* 56: 613–618.

26. Barash, D. P. 1975. Evolutionary aspects of parental behavior: the distraction display of the Alpine accentor, *Prunella collaris. Wilson Bulletin,* 87: 367–373.

27. Barash, D. P. 1976. Male response to apparent female adultery in the mountain bluebird (*Sialia currucoides*): an evolutionary interpretation. *American Naturalist,* 110: 1097–1101.

28. Bardin, C. W. & Catterall, J. F. 1981. Testosterone: a major determinant of extragenital sexual dimorphism. *Science,* 211: 1285–1294.

29. Barnes, H. 1962. So-called anecdysis in *Balanus balanoides* and the effect of breeding upon the growth of calcareous shell of some common barnacles. *Limnology and Oceanography,* 7: 462–473.

30. Bateman, A. J. 1948. Intra-sexual selection in *Drosophila. Heredity,* 2: 349–368.

31. Batra, S. W. T. 1966. Life cycle and behavior of the primitively social bee, *Lasioglossum zephyrum. University of Kansas Science Bulletin,* 46: 359–423.

32. Beach, F. A. 1968. Factors involved in the control of mounting behavior by female mammals. In M. Diamond, ed. *Reproduction and sexual behavior.* Lafayette: Indiana University Press.

33. Beach, F. A. 1976. Sexual attractivity, proceptivity, and receptivity in female mammals. *Hormones and Behavior,* 7: 105–138.

34. Beach, F. A. 1981. Historical origins of modern research on hormones and behavior. *Hormones and Behavior,* 15: 325–376.

35. Beach, F. A., Noble, R. G. & Orndoff, R. K. 1969. Effects of perinatal androgen treatment on responses of male rats to gonadal hormones in

adulthood. *Journal of Comparative and Physiological Psychology,* 68: 490–497.

36. Beall, C. M. & Goldstein, M. C. 1981. Tibetan fraternal polyandry: a test of sociobiological theory. *American Anthropologist,* 83: 5–12.

37. Beecher, M. D. & Beecher, I. M. 1979. Sociobiology of bank swallows: reproductive strategy of the male. *Science,* 205: 1282–1285.

38. Bell, G. 1978. The evolution of anisogamy. *Journal of Theoretical Biology,* 73: 247–270.

39. Bell, G. 1980. The costs of reproduction and their consequences. *American Naturalist,* 116: 45–76.

40. Bell, G. 1982. *The masterpiece of nature.* Berkeley: University of California Press.

41. Ben Shaul, D. M. 1962. The composition of the milk of wild animals. *International Zoo Yearbook,* 4: 333–342.

42. Benshoof, L. & Thornhill, R. 1979. The evolution of monogamy and concealed ovulation in humans. *Journal of Social and Biological Structures,* 2: 95–106.

43. Bentley, P. J. 1976. *Comparative vertebrate endocrinology.* Cambridge: Cambridge University Press.

44. Berger, P. J., Negus, N. C., Sanders, E. H. & Gardner, P. D. 1981. Chemical triggering of reproduction in *Microtus montanus. Science,* 214: 69–70.

45. Bernstein, I. S. 1976. Dominance, aggression and reproduction in primate societies. *Journal of Theoretical Biology,* 60: 459–472.

46. Berreman, G. D. 1962. Pahari polyandry: a comparison. *American Anthropologist,* 64: 60–75.

47. Berry, J. F. & Shine, R. 1980. Sexual size dimorphism and sexual selection in turtles (Order Testudines). *Oecologia,* 44: 185–191.

48. Berscheid, E. & Walster, E. H. 1978. *Interpersonal attraction.* 2nd edition. Reading, Mass.: Addison-Wesley.

49. Berté, N. 1984. Some evolutionary implications of K'ekchi' labor transactions. *Ethology and Sociobiology.* To appear.

50. Bertness, M. D. 1981. Pattern and plasticity in tropical hermit crab growth and reproduction. *American Naturalist,* 117: 754–773.

51. Bertram, B. C. R. 1975. Social factors influencing reproduction in wild lions. *Journal of Zoology,* 177: 463–482.

52. Billewicz, W. Z., Fellowes, M. & Thomson, A. M. 1981. Pubertal changes in

boys and girls in Newcastle upon Tyne. *Annals of Human Biology,* 8: 211–219.

53. Birkhead, T. R. 1978. Behavioural adaptations to high density nesting in the commmon guillemot, *Uria aalge. Animal Behaviour,* 26: 321–331.

54. Bischof, N. 1975. Comparative ethology of incest avoidance. In R. Fox, ed. *Biosocial anthropology.* London: Malaby.

55. Bishop, J. A. & Cook, L. M. 1980. Industrial melanism and the urban environment. *Advances in Ecological Research,* 11: 373–404.

56. Bixler, R. 1981. The incest controversy. *Psychological Reports,* 49: 267–283.

57. Blackstone, W. 1803. *Commentaries on the laws of England, in Four Books.* (Edition edited by St. G. Tucker.) Philadelphia: William Young Birch & Abraham Small.

58. Blurton Jones, N. G. & Sibley, R. M. 1978. Testing adaptiveness of culturally determined behaviour: do Bushman women maximise their reproductive success by spacing births widely and foraging seldom? In N. G. Blurton Jones & V. Reynolds, eds. *Human behaviour and adaptation.* Symposium No. 18 of the Society for the Study of Human Biology. London: Taylor & Frances.

59. Bock, K. 1980. *Human nature and history: a response to sociobiology.* New York: Columbia University Press.

60. Bohannan, P., ed. 1960. *African homicide and suicide.* Princeton: Princeton University Press.

61. Borgia, G. 1979. Sexual selection and the evolution of mating systems. In M. S. Blum & N. A. Blum, eds. *Sexual selection and reproductive competition in insects.* New York: Academic Press.

62. Borgia, G. 1980. Sexual competition in *Scatophaga stercoraria:* size- and density-related changes in male ability to capture females. *Behaviour,* 75: 185–206.

63. Bradbury, J. W. 1980. Foraging, social dispersion and mating systems. In G. W. Barlow & J. Silverberg, eds. *Sociobiology: beyond nature/nurture?* Boulder, Colo.: Westview Press.

64. Bradbury, J. W. 1981. The evolution of leks. In R. D. Alexander & D. W. Tinkle, eds. *Natural selection and social behavior.* New York: Chiron Press.

65. Bray, O. E., Kennelly, J. J. & Guarino, J. L. 1975. Fertility of eggs produced on territories of vasectomized red-winged blackbirds. *Wilson Bulletin,* 87: 187–195.

66. Brennan, E. R. & Dyke, B. 1980. Assortative mate choice and mating opportunity on Sanday, Orkney Islands. *Social Biology,* 27: 199–210.

67. Brian, M. V. 1979. Habitat differences in sexual production by two co-existent ants. *Journal of Animal Ecology,* 48: 943–953.

68. Bronson, F. H. 1971. Rodent pheromones. *Biology of Reproduction,* 4: 344–357.

69. Brooks, R. J. & Falls, J. B. 1975. Individual recognition by song in white-throated sparrows. I, II & III. *Canadian Journal of Zoology,* 53: 879–888, 1412–1420, 1749–1761.

70. Brown, J. K. 1982. Cross-cultural perspectives on middle-aged women. *Current Anthropology,* 23: 143–156.

71. Brown, J. L. 1978. Avian communal breeding systems. *Annual Review of Ecology and Systematics,* 9: 123–156.

72. Bruce, H. M. 1959. An exteroceptive block to pregnancy in the mouse. *Nature,* 184: 105–106.

73. Bryant, D. M. 1979. Reproductive costs in the house martin (*Delichon urbica*). *Journal of Animal Ecology,* 48: 655–675.

74. Buckle, G. R. & Greenberg, L. 1981. Nestmate recognition in sweat bees (*Lasioglosssum zephyrum*): does an individual recognize its own odour or only odours of its nestmates? *Animal Behaviour,* 29: 802–809.

75. Bull, J. J. 1980. Sex determination in reptiles. *Quarterly Review of Biology,* 55: 3–21.

76. Buntin, J. D., Cheng, M.-F., & Hansen, E.W. 1977. Effect of parental feeding on squab-induced crop sac growth in ring doves (*Streptopelia risoria*). *Hormones and Behavior,* 8: 297–309.

77. Burley, N. 1979. The evolution of concealed ovulation. *American Naturalist,* 114: 835–858.

78. Burley, N. & Symanski, R. 1981. Women without: an evolutionary and cross-cultural perspective on prostitution. In R. Symanski. *The immoral landscape.* .Toronto: Butterworths.

79. Busfield, J. 1974. Ideologies and reproduction. In M. P. M. Richards, ed. *The integration of a child into a social world.* London: Cambridge University Press.

80. Busse, C. & Hamilton, W. J. 1981. Infant carrying by male chacma baboons. *Science,* 212: 1281–1283.

81. Butler, S. 1878. *Life and habit.* London: Trubner.

82. Cade, W. 1979. The evolution of alternative male reproductive strategies in field crickets. In M. S. Blum & N. A. Blum, eds. *Sexual selection and reproductive competition in insects.* New York: Academic Press.

83. Cade, W. H. 1981. Alternative male strategies: genetic differences in crickets. *Science,* 212: 563–564.

84. Calow, P. 1979. The cost of reproduction—a physiological approach. *Biological Reviews,* 54: 23–40.

85. Carey, M. & Nolan, V. 1979. Population dynamics of indigo buntings and the evolution of avian polygyny. *Evolution,* 33: 1180–1192.

86. Carter, C. S. & Porges, S. W. 1974. Ovarian hormones and the duration of sexual receptivity in the female golden hamster. *Hormones and Behavior,* 5: 303–315.

87. Chagnon, N. A. 1968. *Yanomamö: the fierce people.* New York: Holt, Rinehart and Winston.

88. Chagnon, N. A. 1981. Terminological kinship, genealogical relatedness and village fissioning among the Yanomamö Indians. In R. D. Alexander & D. W. Tinkle, eds. *Natural selection and social behavior.* New York: Chiron Press.

89. Chagnon, N. A. 1982. Sociodemographic attributes of nepotism in tribal populations: man the rule-maker. In King's College Sociobiology Group, Cambridge, eds. *Current problems in sociobiology.* Cambridge: Cambridge University Press.

90. Chagnon, N. A. & Bugos, P. E. 1979. Kin selection and conflict: an analysis of a Yanomamö ax fight. In N. A. Chagnon & W. Irons, eds. *Evolutionary biology and human social behavior.* North Scituate, Mass.: Duxbury Press.

91. Chagnon, N. A., Flinn, M. V. & Melancon, T. F. 1979. Sex-ratio variation among the Yanomamö Indians. In N. A. Chagnon & W. Irons, eds. *Evolutionary biology and human social behavior.* North Scituate, Mass.: Duxbury Press.

92. Chagnon, N. A. & Irons, W., eds. 1979. *Evolutionary biology and human social behavior.* North Scituate, Mass.: Duxbury Press.

93. Chamie, J. & Nsuly, S. 1981. Sex differences in remarriage and spouse selection. *Demography,* 18: 335–348.

94. Charlesworth, B. 1978. The population genetics of anisogamy. *Journal of Theoretical Biology,* 73: 347–357.

95. Charnov, E. L. 1979. Simultaneous hermaphroditism and sexual selection. *Proceedings of the National Academy of Sciences (USA),* 76: 2480–2484.

96. Charnov, E. L. 1982. *The theory of sex allocation.* Princeton: Princeton University Press.

97. Charnov, E. L. & Bull, J. 1977. When is sex environmentally determined? *Nature,* 266: 828–830.

98. Charnov, E. L., Gotshall, D. W. & Robinson, J. G. 1978. Sex ratio: adaptive response to population fluctuations in pandalid shrimp. *Science,* 200: 204–206.

99.   Charnov, E. L., Los-den Hartogh, R. L., Jones, W. T. & van den Assem, J. 1981. Sex ratio evolution in a variable environment. *Nature,* 289: 27–33.

100. Charnov, E. L., Maynard Smith, J. & Bull, J. J. 1976. Why be an hermaphrodite? *Nature,* 263: 125–126.

101. Cheng, M.-F. 1979. Progress and prospects in ring dove research: a personal view. *Advances in the Study of Behavior,* 9: 97–129.

102. Clark, A. B. 1978. Sex ratio and local resource competition in a prosimian primate. *Science,* 201: 163–165.

103. Cline, D. R., Siniff, D. B. & Erickson, A. W. 1971. Underwater copulation of the Weddell seal. *Journal of Mammalogy,* 52: 216–218.

104. Clutton-Brock, T. H. 1974. Primate social organization and ecology. *Nature,* 250: 539–542.

105. Clutton-Brock, T. H. & Albon, S. D. 1979. The roaring of red deer and the evolution of honest advertisement. *Behaviour,* 69: 145–170.

106. Clutton-Brock, T. H. & Harvey, P. H. 1976. Evolutionary rules and primate societies. In P. P. G. Bateson & R. A. Hinde, eds. *Growing points in ethology.* Cambridge: Cambridge University Press.

107. Clutton-Brock, T. H. & Harvey, P. H. 1977. Primate ecology and social organization. *Journal of Zoology,* 183: 1–39.

108. Cody, M. L. 1966. A general theory of clutch size. *Evolution,* 20: 174–184.

109. Cole, L. C. 1954. The population consequences of life history phenomena. *Quarterly Review of Biology,* 29: 103–137.

110. Coleman, D. A. 1979. A study of the spatial aspects of partner choice from a human biological viewpoint. *Man,* 14: 414–435.

111. Conaway, C. H. 1971. Ecological adaptation and mammalian reproduction. *Biology of Reproduction,* 4: 239–247.

112. Concannon, P., Hodgson, B. & Lein, D. 1980. Reflex LH release in estrous cats following single and multiple copulations. *Biology of Reproduction,* 23: 111–117.

113. Connor, J. R. & Davis, H. N. 1980. Postpartum estrus in Norway rats. II. Physiology. *Biology of Reproduction,* 23: 1000–1006.

114. Conover, D. O. & Kynard, B. E. 1981. Environmental sex determination: interaction of temperature and genotype in a fish. *Science,* 213: 577–579.

115. Coombs, R. H. & Kenkel, W. F. 1966. Sex differences in dating aspirations and satisfaction with computer-selected partners. *Journal of Marriage and the Family,* 28: 62–66.

116. Cosmides, L. M. & Tooby, J. 1981. Cytoplasmic inheritance and intragenomic conflict. *Journal of Theoretical Biology,* 89: 83-129.

117. Cox, C. R. 1981. Agonistic encounters among male elephant seals:frequency, context and the role of female preference. *American Zoologist,* 21: 197-209.

118. Cox, C. R. & LeBoeuf, B. J. 1977. Female incitation of male competition: a mechanism in sexual selection. *American Naturalist,* 111: 317-335.

119. Crellin, E., Pringle, M. L. K. & West, P. 1971. *Born illegitimate: social and educational implications.* London: National Foundation for Education Research.

120. Crews, D. & Greenberg, N. 1981. Function and causation of social signals in lizards. *American Zoologist,* 21: 273-294.

121. Crisp, D. J. & Patel, B. 1961. The interaction between breeding and growth rate in the barnacle *Elminus modestus* Darwin. *Limnology and Oceanography,* 6: 105-115.

122. Critescu, M. 1975. Differential fertility depending on the age of puberty. *Journal of Human Evolution,* 4: 521-524.

123. Crook, J. H. 1970. Social organization and the environment: aspects of contemporary social ethology. *Animal Behaviour,* 18: 197-209.

124. Crook, J. H. 1980. *The evolution of human consciousness.* Oxford: Clarendon Press.

125. Crook, J. H. & Gartlan, J. S. 1966. Evolution of primate societies. *Nature,* 210: 1200-1203.

126. Dahlgren, B. T. 1981. Impact of different dietary protein contents on fecundity and fertility in the female guppy, *Poecilia reticulata* (Peters). *Biology of Reproduction,* 24: 734-746.

127. Daly, M. 1978. The cost of mating. *American Naturalist,* 112: 771-774.

128. Daly, M. 1979. Why don't male mammals lactate? *Journal of Theoretical Biology,* 78: 325-345.

129. Daly, M. & Daly, S. 1974. Spatial distribution of a leaf-eating Saharan gerbil (*Psammomys obesus*) in relation to its food. *Mammalia,* 38: 591-603.

130. Daly, M. & Wilson, M. 1980. Discriminative parental solicitude: a biologial perspective. *Journal of Marriage and the Family,* 42: 277-288.

131. Daly, M. & Wilson, M. I. 1981. Abuse and neglect of children in evolutionary perspective. In R. D. Alexander & D. W. Tinkle, eds. *Natural selection and social behavior.* New York: Chiron Press.

132. Daly, M. & Wilson, M. I. 1981. Child maltreatment from a sociobiological perspective. *New Directions for Child Development,* 11: 93-112.

133. Daly, M. & Wilson, M. I. 1982. Homicide and kinship. *American Anthropologist,* 84: 372–378.

134. Daly, M. & Wilson, M. 1982. Whom are newborn babies said to resemble? *Ethology and Sociobiology,* 3: 69–78.

135. Daly, M., Wilson, M. & Weghorst, S. J. 1982. Male sexual jealousy. *Ethology and Sociobiology,* 3: 11–27.

136. Darling, F. F. 1938. *Bird flocks and the breeding cycle.* London: Cambridge University Press.

137. Darwin, C. 1871. *The descent of man, and selection in relation to sex.* (1887 edition.) New York: D. Appleton and Co.

138. Darwin, C. & Wallace, A. R. 1858. *Evolution by natural selection.* (1958 edition.) London: Cambridge University Press.

139. Davies, N. B. & Halliday, T. R. 1978. Deep croaks and fighting assessment in toads *Bufo bufo. Nature,* 274: 683–685.

140. Dawkins, R. 1976. *The selfish gene.* Oxford: Oxford University Press.

141. Dawkins, R. 1978. Replicator selection and the extended phenotype. *Zeitschrift für Tierpsychologie,* 47: 61–76.

142. Dawkins, R. 1982. *The extended phenotype.* Oxford: W. H. Freeman & Co.

143. de Camp, L. S. 1969. The end of the monkey war. *Scientific American,* 220 (2): 15–21.

144. de Steven, D. 1980. Clutch size, breeding success, and parental survival in the tree swallow (*Iridoprocne bicolor*). *Evolution,* 34: 278–291.

145. DeVoogd, T. & Nottebohm, F. 1981. Gonadal hormones induce dendritic growth in the adult avian brain. *Science,* 214: 202–204.

146. De Vos, G. J. 1983. Social behaviour of black grouse. An observational and experimental field study. *Ardea,* 71: 1–103.

147. De Vos, S. 1980. Women's role orientations and expected fertility: evidence from the Detroit area, 1978. *Social Biology,* 27: 130–137.

148. de Winter, F. R. & Steendijk, R. 1975. The effect of a low-calcium diet in lactating rats; observations on the rapid development and repair of osteoporosis. *Calcified Tissue Research,* 174: 303–316.

149. Dewsbury, D. A. 1972. Patterns of copulatory behavior in male mammals. *Quarterly Review of Biology,* 47: 1–33.

150. Dewsbury, D. A. 1981. Effects of novelty on copulatory behavior: the Coolidge effect and related phenomena. *Psychological Bulletin,* 89: 464–482.

151. Dewsbury, D. A. 1982. Dominance, rank, copulatory behavior, and differential reproduction. *Quarterly Review of Biology,* 57: 135–159.

152. Dewsbury, D. A. 1982. Ejaculate cost and male choice. *American Naturalist,* 119: 601–610.

153. Diamond, M. 1982. Sexual identity, monozygotic twins reared in discordant sex roles and the BBC follow-up. *Archives of Sexual Behavior,* 11: 181–186.

154. Diamond, M., Diamond, A. L. & Mast, M. 1972. Visual sensitivity and sexual arousal levels during the menstrual cycle. *Journal of Nervous and Mental Disease,* 155: 170–176.

155. Dickeman, M. 1975. Demographic consequences of infanticide in man. *Annual Review of Ecology and Systematics,* 6: 107–137.

156. Dickemann, M. 1979. The ecology of mating systems in hypergynous dowry societies. *Social Science Information,* 18: 163–195.

157. Dickemann, M. 1979. Female infanticide, reproductive strategies, and social stratification: a preliminary model. In N. A. Chagnon & W. Irons, eds. *Evolutionary biology and human social behavior.* North Scituate, Mass.: Duxbury Press.

158. Dickemann, M. 1981. Paternal confidence and dowry competition: a biocultural analysis of purdah. In R. D. Alexander & D. W. Tinkle, eds. *Natural selection and social behavior.* New York: Chiron Press.

159. Divale, W. T. & Harris, M. 1976. Population, warfare, and the male supremacist complex. *American Anthropologist,* 78: 521–538.

160. Dominey, W. J. 1980. Female mimicry in male bluegill sunfish—a genetic polymorphism? *Nature,* 284: 546–548.

161. Dorjahn, V. R. 1958. Fertility, polygyny and their interrelations in Temne society. *American Anthropologist,* 60: 838–860.

162. Doty, R. L. 1981. Olfactory communication in humans. *Chemical Senses,* 6: 351–376.

163. Dougherty, E. G. 1955. Comparative evolution and the origin of sexuality. *Systematic Zoology,* 4: 145–169.

164. Downhower, J. F. & Armitage, K. B. 1971. The yellow-bellied marmot and the evolution of polygamy. *American Naturalist,* 105: 355–370.

165. Drent, R. 1975. Incubation. In D. S. Farner & J. R. King, eds. *Avian biology,* vol. 5. New York: Academic Press.

166. Driver, E. D. 1963. *Differential fertility in central India.* Princeton: Princeton University Press.

167. Dunn, L. C. 1965. *A short history of genetics.* New York: McGraw-Hill.

168. Duvall, S. W., Bernstein, I. S. & Gordon, T. P. 1976. Paternity and status in a rhesus monkey group. *Journal of Reproduction and Fertility,* 47: 25-31.

169. Dyer, R. G., MacLeod, N. K. & Ellendorf, F. 1976. Electrophysiological evidence for sexual dimorphism and synaptic convergence in the preoptic anterior hypothalamic areas of the rat. *Proceedings of the Royal Society (B),* 193: 421-440.

170. Eaton, J. W. & Mayer, A. J. 1953. The social biology of very high fertility among the Hutterites. *Human Biology,* 25: 206-264.

171. Eberhard, W. G. 1980. Evolutionary consequences of intracellular organelle competition. *Quarterly Review of Biology,* 55: 231-249.

172. Edwards, C. P. & Whiting, B. B. 1980. Differential socialization of girls and boys in the light of cross-cultural research. *New Directions for Child Development,* 8: 45-57.

173. Egboh, E. O. 1972. Polygamy in Iboland. *Civilisations,* 22: 431-444.

174. Ehrhardt, A. A. & Baker, S. W. 1974. Fetal androgens, human central nervous system differentiation, and behavior sex differences. In R. C. Friedman, R. M. Richart, & R. L. Van de Wiele, eds. *Sex differences in behavior.* New York: Wiley.

175. Ehrman, L. 1972. Genetics and sexual selection. In B. Campbell, ed. *Sexual selection and the descent of man 1871-1971.* Chicago: Aldine.

176. Eisenberg, J. F. 1981. *The mammalian radiations.* Chicago: University of Chicago Press.

177. Elliott, P. F. 1975. Longevity and the evolution of polygamy. *American Naturalist,* 109: 281-287.

178. Emlen, S. T. 1978. The evolution of cooperative breeding in birds. In J. R. Krebs & N. B. Davies, eds. *Behavioural ecology.* Oxford: Blackwell.

179. Emlen, S. T. & Oring, L. W. 1977. Ecology, sexual selection and the evolution of mating systems. *Science,* 197: 215-223.

180. Erickson, C. J. 1970. Induction of ovarian activity in female ring doves by androgen treatment of castrated males. *Journal of Comparative and Physiological Psychology,* 71: 210-215.

181. Eshel, I. 1978. On the handicap principle—a critical defence. *Journal of Theoretical Biology,* 70: 245-250.

182. Essock-Vitale, S. M. & McGuire, M. T. 1980. Predictions derived from the theories of kin selection and reciprocation assessed by anthropological data. *Ethology and Sociobiology,* 1: 233-243.

183. Estes, R. D. 1976. The significance of breeding synchrony in the wildebeest. *East African Wildlife Journal,* 14: 135-152.

184. Evans, L. & Wasielewski, P. 1981. Risky driving related to driver and vehicle characteristics. *General Motors Research Publication,* GMR-3897.

185. Faaborg, J. & Patterson, C. B. 1981. The characteristics and occurrence of cooperative polyandry. *Ibis,* 123: 477-484.

186. Faux, S. F. 1981. A sociobiological perspective of the doctrinal development of Mormon polygyny. Sunstone Theological Symposium.

187. Feder, H. H. 1981. Estrus cyclicity in mammals. In N. T. Adler, ed. *Neuroendocrinology of reproduction.* New York: Plenum Press.

188. Ferguson, M. W. J. & Joanen, T. 1982. Temperature of egg incubation determines sex in *Alligator mississippiensis. Nature,* 296: 850-853.

189. Fiedler, K. 1954. Vergleichende Verhaltensstudien an Seenadeln, Schlangennadeln und Seepferdchen. *Zeitschrift für Tierpsychologie,* 11: 358-416.

190. Fischer, E. A. 1981. Sexual allocation in a simultaneously hermaphroditic coral reef fish. *American Naturalist,* 117: 64-82.

191. Fisher, R. A. 1958. *The genetical theory of natural selection,* 2nd revised edition. New York: Dover Press (originally published in 1930).

192. Flinn, M. 1981. Uterine vs. agnatic kinship variability and associated cousin marriage preferences: an evolutionary biological analysis. In R. D. Alexander & D. W. Tinkle, eds. *Natural selection and social behavior.* New York: Chiron Press.

193. Flinn, M. V. 1984. Behavioral interactions in a Trinidadian village: testing predictions from kin selection theory. *Ethology and Sociobiology.* To appear.

194. Ford, C. S. 1952. Control of conception in cross-cultural perspective. *Annals of the New York Academy of Science,* 54: 763-776.

195. Ford, C. S. & Beach, F. A. 1951. *Patterns of sexual behavior.* New York: Harper and Row.

196. Foster, D. L. 1981. Mechanism for delay of first ovulation in lambs born in the wrong season (fall). *Biology of Reproduction,* 25: 85-92.

197. Fox, R. 1967. *Kinship and marriage.* London: Penguin.

198. Frame, L. H., Malcolm, J. R., Frame, G. W. & van Lawick, H. 1979. The social organization of African wild dogs (*Lycaon pictus*) on the Serengeti Plains, Tanzania, 1967-1978. *Zeitschrift für Tierpsychologie,* 50: 225-249.

199. Fredlund, E. V. 1984. Incest and biological relatedness in the Shitari Yanomamö population. *Ethology and Sociobiology*. To appear.

200. Freedman, D. G. 1974. *Human infancy: an evolutionary perspective.* Hillsdale, N. J.: Lawrence Erlbaum Associates.

201. Freeman, D. S., Harper, K. T. & Charnov, E. L. 1980. Sex change in plants: old and new observations and new hypotheses. *Oecologia,* 47: 222-232.

202. Fricke, H. & Fricke, S. 1977. Monogamy and sex change by aggressive dominance in coral reef fish. *Nature,* 266: 830-832.

203. Frisch, R. E. & McArthur, J. W. 1974. Menstrual cycles: fatness as a determinant of minimum weight for height necessary for their maintenance or onset. *Science,* 185: 949-951.

204. Frith, H. J. 1962. *The mallee-fowl: the bird that builds an incubator.* Sydney: Angus and Robertson.

205. Gadgil, M. 1972. Male dimorphism as a consequence of sexual selection. *American Naturalist,* 106: 574-580.

206. Gadgil, M. & Bossert, W. H. 1970. Life historical consequences of natural selection. *American Naturalist,* 104: 1-24.

207. Gadgil, M. & Solbrig, O. T. 1972. The concept of *r*- and *K*-selection: evidence from wild flowers and some theoretical considerations. *American Naturalist,* 106: 14-31.

208. Garson, P. J., Pleszczynska, W. K. & Holm, C. H. 1981. The 'polygyny threshold' model: a reassessment. *Canadian Journal of Zoology,* 59: 902-910.

209. Gaughran, E., Struening, E. L., Raabe, G. & Muhlin, G. 1976. A factor analytic study of attitudes toward human fertility. *Adult Education,* 26: 86-100.

210. Ghiselin, M. T. 1969. The evolution of hermaphroditism among animals. *Quarterly Review of Biology,* 44: 189-208.

211. Ghiselin, M. T. 1969. *The triumph of the Darwinian method.* Berkeley: University of California Press.

212. Ghiselin, M. T. 1974. *The economy of nature and the evolution of sex.* Berkeley: University of California Press.

213. Gladstone, D. E. 1979. Promiscuity in monogamous colonial birds. *American Naturalist,* 114: 545-557.

214. Glesener, R. R. & Tilman, D. 1978. Sexuality and the components of environmental uncertainty: clues from geographic parthenogenesis in terrestrial animals. *American Naturalist,* 112: 659-673.

215. Gluckman, M. 1950. Kinship and marriage among the Lozi of Northern Rhodesia and the Zulu of Natal. In A. R. Radcliffe-Brown & D. Ford, eds. *African systems of kinship and marriage.* London: Oxford University Press.

216. Goodall, J., Bandora, A., Bergmann, E., Busse, C., Matama, H., Mpongo, E., Pierce, A. & Riss, D. 1979. Intercommunity interactions in the chimpanzee population in the Gombe National Park. In D. A. Hamburg & E. R. Mc-Cown, eds. *The great apes.* Menlo Park, Calif.: Benjamin/Cummings.

217. Goodman, D. 1974. Natural selection and a cost ceiling on reproductive effort. *American Naturalist,* 108: 247–268.

218. Gordon, J. W. & Ruddle, F. H. 1981. Mammalian gonadal determination and gametogenesis. *Science,* 211: 1265–1271.

219. Gorski, R. A., Gordon, J. H., Shryne, J. E. & Southam, A. M. 1978. Evidence for a morphological sex difference within the medial preoptic area of the rat brain. *Brain Research,* 148: 333–346.

220. Goss-Custard, J. D. 1977. The energetics of prey selection by redshank, *Tringa totanus* (L.), in relation to prey density. *Journal of Animal Ecology,* 46: 1–19.

221. Goss-Custard, J. D., Dunbar, R. I. M. & Aldrich-Blake, F. P. G. 1972. Survival, mating and rearing strategies in the evolution of primate social structure. *Folia Primatologica,* 17: 1–19.

222. Gowaty, P. A. 1981. Aggression of breeding eastern bluebirds (*Sialia sialis*) toward their mates and models of intra- and interspecific intruders. *Animal Behaviour,* 29: 1013–1027.

223. Goy, R. W. & McEwen, B. S. 1980. *Sexual differentiation of the brain.* Cambridge, Mass.: MIT Press.

224. Graham, C. A. & McGrew, W. C. 1980. Menstrual synchrony in female undergraduates living on a coeducational campus. *Psychoneuroendocrinology,* 5: 245–252.

225. Graham, J. M. & Desjardins, C. 1980. Classical conditioning: induction of luteinizing hormone and testosterone secretion in anticipation of sexual activity. *Science,* 210: 1039–1041.

226. Green, R. 1978. Sexual identity of 37 children raised by homosexual or transsexual parents. *American Journal of Psychiatry,* 135: 692–697.

227. Green, R. F., Gordh, G. & Hawkins, B. A. 1982. Precise sex ratios in highly inbred parasitic wasps. *American Naturalist,* 120: 653–665.

228. Greenberg, L. 1979. Genetic component of bee odor in kin recognition. *Science,* 206: 1095–1097.

229. Greenblatt, R. B. 1972. Inappropriate lactation in men and women. *Medical Aspects of Human Sexuality* 6(5): 25-33.

230. Greenwood, P. J., Harvey, P. H. & Perrins, C. M. 1979. The role of dispersal in the great tit (*Parus major*): the causes, consequences and heritability of natal dispersal. *Journal of Animal Ecology,* 48: 123-142.

231. Greig-Smith, P. W. 1980. Parental investment in nest defence by stonechats (*Saxicola torquata*). *Animal Behaviour,* 28: 604-619.

232. Gross, M. R. 1979. Cuckoldry in sunfishes (*Lepomis:* Centrarchidae). *Canadian Journal of Zoology,* 57: 1507-1509.

233. Gross, M. R. & Charnov, E. L. 1980. Alternative male life histories in bluegill sunfish. *Proceedings of the National Academy of Sciences (USA),* 77: 6937-6940.

234. Gross, M. R. & Shine, R. 1981. Parental care and mode of fertilization in ectothermic vertebrates. *Evolution,* 35: 775-793.

235. Grubb, P. 1974. Social organization of Soay sheep and the behaviour of ewes and lambs. In P. A. Jewell, C. Milner, & J. M. Boyd, eds. *Island survivors.* London: Athlone Press.

236. Gurdon, J. B. 1974. *The control of gene expression in animal development.* Cambridge, Mass.: Harvard University Press.

237. Gwynne, D. T. 1981. Sexual difference theory: Mormon crickets show role reversal in mate choice. *Science,* 213: 779-780.

238. Haartman, L. von. 1969. Nest-site and evolution of polygamy in European passerine birds. *Ornis Fennica,* 46: 1-12.

239. Haartman, L. von. 1971. Population dynamics. In D. S. Farner & J. R. King, eds., *Avian biology,* vol. 1. London: Academic Press.

240. Hadjiyannakis, C. 1969. *Les tendences contemporaines concernant la répression du délit d'adultére.* Thessalonika: Association Internationale de Droit Pénal.

241. Hagood, M. 1939. *Mothers of the south.* Chapel Hill: University of North Carolina Press.

242. Hails, C. J. & Bryant, D. M. 1979. Reproductive energetics of a free-living bird. *Journal of Animal Ecology,* 48: 471-482.

243. Halperin, S. D. 1979. Temporary association patterns in free ranging chimpanzees. In D. A. Hamburg & E. R. McCown, eds. *The great apes.* Menlo Park, Calif.: Benjamin/Cummings.

244. Hames, R. B. 1979. Relatedness and interaction among the Ye'kwana: A preliminary analysis. In N. A. Chagnon & W. Irons, eds. *Evolutionary biology and human social behavior.* North Scituate, Mass.: Duxbury Press.

245. Hames, R. B. 1984. *Ethology and Sociobiology.* To appear.

246. Hamilton, J. B. 1948. The role of testicular secretions as indicated by the effects of castration in man and by studies of pathological conditions and the short lifespan associated with maleness. *Recent Progress in Hormone Research,* 3: 257–322.

247. Hamilton, J. B. & Mestler, G. E. 1969. Mortality and survival: comparison of eunuchs with intact men and women in a mentally retarded population. *Journal of Gerontology,* 24: 395–411.

248. Hamilton, J. B., Hamilton, R. S. & Mestler, G. E. 1969. Duration of life and causes of death in domestic cats: influence of sex, gonadectomy and inbreeding. *Journal of Gerontology,* 24: 427–437.

249. Hamilton, W. D. 1964. The genetical evolution of social behaviour. I and II. *Journal of Theoretical Biology,* 7: 1–52.

250. Hamilton, W. D. 1966. The moulding of senescence by natural selection. *Journal of Theoretical Biology,* 12: 12–45.

251. Hamilton, W. D. 1967. Extraordinary sex ratios. *Science,* 156: 477–488.

252. Hamilton, W. D. 1979. Wingless and fighting males in fig wasps and other insects. In M. S. Blum & N. A. Blum, eds. *Sexual selection and reproductive competition in insects.* New York: Academic Press.

253. Hamilton, W. D. 1980. Sex versus non-sex versus parasite. *Oikos,* 35: 282–290.

254. Hamilton, W. D., Henderson, P. A. & Moran, N. A. 1981. Fluctuation of environment and coevolved antagonist polymorphism as factors in the maintenance of sex. In R. D. Alexander & D. W. Tinkle, eds. *Natural selection and social behavior.* New York: Chiron Press.

255. Hanken, J. & Sherman, P. W. 1981. Multiple paternity in Belding's ground squirrel litters. *Science,* 212: 351–353.

256. Harcourt, A. H. 1981. Intermale competition and the reproductive behavior of the great apes. In C. E. Graham, ed. *Reproductive biology of the great apes.* New York: Academic Press.

257. Harcourt, A. H., Stewart, K. S. & Fossey, D. 1976. Male emigration and female transfer in wild mountain gorilla. *Nature,* 263: 226–227.

258. Harpending, H. 1976. Regional variation in !Kung populations. In R. B. Lee & I. DeVore, eds. *Kalahari hunter-gatherers.* Cambridge, Mass.: Harvard University Press.

259. Harris, G. W. 1964. Sex hormones, brain development and brain function. *Endocrinology,* 74: 627–648.

260. Hart, A. & Begon, M. 1982. The status of general reproductive-strategy theories, illustrated in winkles. *Oecologia,* 52: 37–42.

261. Hart, C. W. M. & Pilling, A. R. 1960. *The Tiwi of North Australia.* New York: Holt, Rinehart and Winston.

262. Hartung, J. 1978. Light, puberty, and aggression: a proximal mechanism hypothesis. *Human Ecology,* 6: 273–297.

263. Hartung, J. 1982. Polygyny and inheritance of wealth. *Current Anthropology,* 23: 1–12.

264. Haseltine, F. P. & Ohno, S. 1981. Mechanism of gonadal differentiation. *Science,* 211: 1272–1278.

265. Hausfater, G. 1975. *Dominance and reproduction in baboons (Papio cynocephalus): a quantitative analysis.* (Contributions to primatology, 7.) Basel: Karger.

266. Hawthorn, G. 1970. *The sociology of fertility.* London: Collier-Macmillan.

267. Herskowitz, I. H. 1977. *Principles of genetics.* 2d edition. New York: Macmillan.

268. Hiatt, L. R. 1980. Polyandry in Sri Lanka: a test case for parental investment theory. *Man,* 15: 583–602.

269. Hill, E. & Wenzl, P. 1981. Variation in ornamentation and behavior in a discotheque for females observed at differing menstrual phases. Paper presented to the Animal Behavior Society, Knoxville, Tenn.

270. Hill, J. L. 1974. *Peromyscus:* effect of early pairing on reproduction. *Science,* 186: 1042–1044.

271. Hinshaw, R., Pyeatt, P. & Habicht, J.-P. 1972. Environmental effects on child-spacing and population increase in highland Guatemala. *Current Anthropology,* 13: 216–230.

272. Hogan-Warburg, L. 1966. Social behavior of the ruff, *Philomachus pugnax* (L.). *Ardea,* 54: 109–229.

273. Högstedt, G. 1980. Evolution of clutch size in birds: adaptive variation in relation to territory quality. *Science,* 210: 1148–1150.

274. Högstedt, G. 1981. Should there be a positive or negative correlation between survival of adults in a bird population and their clutch size? *American Naturalist,* 118: 568–571.

275. Höhn, E. O. 1967. Observations on the breeding biology of Wilson's phalarope (*Steganopus tricolor*) in central Alberta. *Auk,* 84: 220–244.

276. Höhn, E. O. 1969. The phalarope. *Scientific American,* 220(6): 104–111.

277. Holmes, W. G. & Sherman, P. W. 1982. The ontogeny of kin recognition in two species of ground squirrels. *American Zoologist,* 22: 491–517.

278. Hoogland, J. L. 1982. Prairie dogs avoid extreme inbreeding. *Science,* 215: 1639–1641.

279. Horn, H. S. 1978. Optimal tactics of reproduction and life-history. In J. R. Krebs & N. B. Davies, eds. *Behavioural ecology.* Oxford: Blackwell.

280. Hosken, F. P. 1979. *The Hosken Report. Genital and sexual mutilation of females,* 2d revised edition. Lexington, Mass.: Women's International Network News.

281. Howard, R. D. 1981. Male age-size distribution and male mating success in bullfrogs. In R. D. Alexander & D. W. Tinkle, eds. *Natural selection and social behavior.* New York: Chiron.

282. Howe, M. A. 1975. Behavioral aspects of the pair bond in Wilson's phalarope. *Wilson Bulletin,* 87: 248–270.

283. Howell, N. 1976. The population of the Dobe area !Kung. In R. B. Lee & I. DeVore, eds. *Kalahari hunter-gatherers.* Cambridge, Mass.: Harvard University Press.

284. Howell, N. 1979. *Demography of the Dobe !Kung.* New York: Academic Press.

285. Howie, P. W., McNeilly, A. S., Houston, M. J., Cook, A. & Boyle, H. 1981. Effect of supplementary food on suckling patterns and ovarian activity during lactation. *British Medical Journal,* 283: 757–759.

286. Hrdy, S. B. 1979. Infanticide among animals: a review, classification, and examination of the implications for the reproductive strategies of females. *Ethology and Sociobiology,* 1: 13–40.

287. Hrdy, S. B. 1981. *The woman that never evolved.* Cambridge, Mass.: Harvard University Press.

288. Hull, D. L. 1980. Individuality and selection. *Annual Review of Ecology and Systematics,* 11: 311–332.

289. Hunt, G. L. 1980. Mate selection and mating systems in seabirds. In J. Burger, B. L. Olla, & H. E. Winn, eds. *Behavior of marine animals. Vol. 4: Marine birds.* New York: Plenum.

290. Hunt, M. 1974. *Sexual behavior in the 1970's.* Chicago: Playboy Press.

291. Hunter, J. 1861. *Essays and observations on natural history, anatomy, physiology, psychology, and geology,* vol. 1. London.

292. Hutt, C. 1972. *Males and females.* London: Penguin.

293. Huxley, L. 1900. *Life and letters of Thomas Henry Huxley.* New York: Appleton.

294. Huxley, T. H. 1863. *Evidence as to man's place in nature.* London: William & Norgate.

295. Hytten, F. E. & Leitch, I. 1971. *The physiology of human pregnancy,* 2d edition. Oxford: Blackwell Scientific Publications.

296. Iersel, J. J. A., van. 1953. An analysis of the parental behaviour of the male three-spined stickleback (*Gasterosteus aculeatus L.*). *Behaviour,* supplement 3: 1–159.

297. Imber, M. J. 1976. Breeding biology of the grey-faced petrel *Pterodroma macroptera gouldi. Ibis,* 118: 51–64.

298. Immelmann, K. 1963. Drought adaptations in Australian desert birds. In *Proceedings of the 13th International Ornithological Congress:* 649–657.

299. Immelmann, K. 1971. Ecological aspects of periodic reproduction. In D. S. Farner & J. R. King, eds. *Avian biology,* Vol. 1. New York: Academic Press.

300. Imperato-McGinley, J. & Peterson, R. E. 1976. Male pseudohermaphroditism: the complexities of male phenotypic development. *American Journal of Medicine,* 61: 251–272.

301. Imperato-McGinley, J., Peterson, R. E., Gautier, T. & Sturla, E. 1979. Androgens and the evolution of male-gender identity among male pseudohermaphrodites with 5α -reductase deficiency. *New England Journal of Medicine,* 300: 1233–1237.

302. Imperato-McGinley, J., Peterson, R. E., Leshin, M., Griffin, J. E., Cooper, G., Draghi, S., Berenyi, M. & Wilson, J. D. 1980. Steroid 5α -reductase deficiency in a 65-year-old male pseudohermaphrodite: the natural history, ultrastructure of the testes, and evidence for inherited enzyme heterogeneity. *Journal of Clinical Endocrinology and Metabolism,* 50: 15–22.

303. Inglis, J. & Lawson, J. S. 1981. Sex differences in the effects of unilateral brain damage on intelligence. *Science,* 181: 693–695.

304. Ingram, J. C. 1977. Interactions between parents and infants, and the development of independence in the common marmoset (*Callithrix jacchus*). *Animal Behaviour,* 25: 811–827.

305. Irons, W. 1975. The Yomut Turkmen: a study of social organization among a central Asian Turkic-speaking population. *Anthropological Papers of the Michigan University Museum of Anthropology,* 58: 1–193.

306. Irons, W. 1979. Emic and reproductive success. In N. A. Chagnon & W. Irons, eds. *Evolutionary biology and human social behavior: an anthropological perspective.* North Scituate, Mass.: Duxbury Press.

307. Irons, W. 1981. Why lineage exogamy? In R. D. Alexander & D. W. Tinkle, eds. *Natural selection and social behavior.* New York: Chiron Press.

308. Irons, W. 1983. Human female reproductive strategies. In S. K. Wasser, ed. *Social behavior of female vertebrates.* New York: Academic Press.

309. Isaac, B. L. 1980. Female fertility and marital form among the Mende of rural upper Bambara chiefdom, Sierra Leone. *Ethnology,* 19: 297–313.

310. Jackson, T. T. & Gray, M. 1976. Field study of risk-taking behavior of automobile drivers. *Perceptual and Motor Skills,* 43: 471–474.

311. Jacobson, C. D., Shryne, J. E., Shapiro, F. & Gorski, R.A. 1980. Ontogeny of the sexually dimorphic nucleus of the preoptic area. *Journal of Comparative Neurology,* 193: 541–548.

312. Jacquard, A. & Ward, R. H. 1976. The genetic consequences of changing reproductive behavior. *Journal of Human Evolution,* 5: 139–154.

313. James, W. H. 1974. Marital coital rates, spouses' ages, family size and social class. *Journal of Sex Research,* 10: 205–218.

314. Jarman, P. J. 1974. The social organisation of antelope in relation to their ecology. *Behaviour,* 48: 215–267.

315. Jelliffe, D. B. & Jelliffe, E. F. P. 1978. *Human milk in the modern world. Psychosocial, nutritional, and economic significance.* Oxford: Oxford University Press.

316. Jenni, D. A. 1974. Evolution of polyandry in birds. *American Zoologist,* 14: 129–144.

317. Jenni, D. A. & Collier, G. 1972. Polyandry in the American jaçana (*Jacana spinosa*). *Auk,* 89: 743–765.

318. Johnson, R. E. 1970. Some correlates of extramarital coitus. *Journal of Marriage and the Family,* 32: 449–456.

319. Johnson, S. R. & West, G. C. 1973. Fat content, fatty acid composition and estimates of energy metabolism of Adélie penguins (*Pygoscelis adeliae*) during the early breeding season fast. *Comparative Biochemistry and Physiology B,* 45: 709–719.

320. Jolly, A. 1967. Breeding synchrony in wild *Lemur catta*. In S. A. Altmann, ed. *Social communication among primates.* Chicago: University of Chicago Press.

321. Josso, N., Picard, J.-Y. & Tran, D. 1980. A new testicular glycoprotein: anti-Müllerian hormone. In A. Steinberger & E. Steinberger, eds. *Testicular development, structure, and function.* New York: Raven.

322. Jost, A. 1953. Problems of fetal endocrinology: the gonadal and hypophyseal hormones. *Recent Progress in Hormone Research,* 8: 379–413.

323. Keenleyside, M. H. 1972. Intraspecific intrusions into nests of spawning longear sunfish (Pisces: Centrarchidae). *Copeia,* 1972: 272–278.

324. Keenleyside, M. H. A. 1979. *Diversity and adaptation in fish behavior.* New York: Springer-Verlag.

325. Kern, M. D. & King, J. R. 1972. Testosterone-induced singing in female white-crowned sparrows. *Condor,* 74: 204–209.

326. Kettlewell, B. 1973. *The evolution of melanism.* Oxford: Clarendon.

327. King, M.-C. & Wilson, A. C. 1975. Evolution at two levels in humans and chimpanzees. *Science,* 188: 107–116.

328. Kinsey, A. C., Pomeroy, W. B., Martin, C. E. & Gebhard, P. H. 1953. *Sexual behavior in the human female.* Philadelphia: Saunders.

329. Kirk, D. 1966. Demographic factors affecting the opportunity for natural selection in the United States. *Eugenics Quarterly,* 13: 270–273.

330. Kirkwood, T. B. L. & Holliday, R. 1979. The evolution of ageing and longevity. *Proceedings of the Royal Society of London (B),* 205: 531–546.

331. Klaus, M. H. & Kennell, J. H. 1976. *Maternal-infant bonding.* St. Louis: C. V. Mosby.

332. Kleiman, D. G. 1977. Monogamy in mammals. *Quarterly Review of Biology,* 52: 39–69.

333. Kleiman, D. G. & Malcolm, J. R. 1981. The evolution of male parental investment in mammals. In D. J. Gubernick & P. H. Klopfer, eds. *Parental care in mammals.* New York: Plenum Press.

334. Klomp, H. 1970. The determination of clutch size in birds: a review. *Ardea,* 58: 1–124.

335. Klopfer, P. 1971. Mother love: what turns it on? *American Scientist,* 59: 404–407.

336. Knodel, J. 1968. Infant mortality and fertility in three Bavarian villages: an analysis of family histories from the 19th century. *Population Studies,* 22: 297–318.

337. Kolodny, R. C., Jacobs, L. S. & Daughaday, W. H. 1972. Mammary stimulation causes prolactin secretion in non-lactating women. *Nature,* 238: 284–286.

338. Konishi, M. 1965. The role of auditory feedback in the control of vocalization in the white-crowned sparrow. *Zeitschrift für Tierpsychologie,* 22: 770–783.

339. Konner, M. J. 1976. Maternal care, infant behavior, and development among the !Kung. In R. B. Lee & I. DeVore, eds. *Kalahari hunter-gatherers.* Cambridge, Mass.: Harvard University Press.

340. Konner, M. J. & Worthman, C. 1980. Nursing frequency, gonadal function, and birth spacing among !Kung hunter-gatherers. *Science,* 207: 788–791.

341. Korschgen, C. E. 1977. Breeding stress of female eiders in Maine. *Journal of Wildlife Management,* 41: 360–373.

342. Krebs, C. J. & Myers, J. H. 1974. Population cycles in small mammals. *Advances in Ecological Research,* 8: 267–399.

343. Krebs, J. R., Ryan, J. C. & Charnov, E. L. 1974. Hunting by expectation or optimal foraging? A study of patch use by chickadees. *Animal Behaviour,* 22: 953–964.

344. Krekorian, C. O. 1976. Field observations in Guyana on the reproductive biology of the spraying characid, *Copeina arnoldi* Regan. *American Midland Naturalist,* 96: 88–97.

345. Kressel, G. M. 1981. Sororicide/filiacide: homicide for family honour. *Current Anthropology,* 22: 141–158.

346. Krombein, K. V. 1967. *Trap nesting wasps and bees—life histories, nests and associates.* Washington, D.C.: Smithsonian Press.

347. Kruijt, J. P. & Hogan, J. A. 1967. Social behavior on the lek in black grouse, *Lyrurus tetrix tetrix (L.). Ardea,* 55: 203–239.

348. Kruuk, H. 1978. Spatial organization and territorial behaviour of the European badger, *Meles meles. Journal of Zoology,* 184: 1–19.

349. Kummer, H. 1968. *Social organization of hamadryas baboons.* Chicago: University of Chicago Press.

350. Kummer, H. 1977. Generation turnover in a band of hamadryas baboons. Keynote address to the annual meeting of the Animal Behavior Society, June 1977.

351. Kurland, J. A. 1977. *Kin selection in the Japanese monkey.* (Contributions to primatology, 12.) Basel: Karger.

352. Kurland, J. A. 1979. Matrilines: the primate sisterhood and the human avunculate. In N. A. Chagnon & W. Irons, eds. *Evolutionary biology and human social behavior: an anthropological perspective.* North Scituate, Mass.: Duxbury Press.

353. Labov, J. B. 1981. Pregnancy blocking in rodents: adaptive advantages for females. *American Naturalist,* 118: 361–371.

354. Lack, D. 1954. *The natural regulation of animal numbers.* Oxford: Oxford University Press.

355. Lack, D. 1966. *Population studies of birds.* Oxford: Oxford University Press.

356. Lack, D. 1968. *Ecological adaptations for breeding in birds.* London: Methuen.

357. Lack, D. 1970. *The life of the robin,* 4th edition. London: Fontana.

358. Lack, D., Gibb, J. A., & Owen, D. F. 1957. Survival in relation to brood-size in tits. *Proceedings of the Zoological Society of London,* 128: 313–326.

359. Lacoste-Utamsing, C. de & Holloway, R. L. 1982. Sexual dimorphism in the human corpus callosum. *Science,* 216: 1431–1432.

360. Lacy, R. C. & Sherman, P. W. 1983. Kin recognition by phenotype matching. *American Naturalist,* 121: 489–512.

361. Laing, L. M. 1980. Declining fertility in a religious isolate: the Hutterite population of Alberta, Canada, 1951–1971. *Human Biology,* 52: 288–310.

362. Lande, R. 1980. Sexual dimorphism, sexual selection, and adaptation in polygenic characters. *Evolution,* 34: 292–305.

363. Langman, J. 1975. *Medical embryology,* 3d edition. Baltimore: Williams and Wilkins.

364. Lawick, H. van & Lawick-Goodall, J. van. 1971. *Innocent killers.* Boston: Houghton Mifflin.

365. Lawick-Goodall, J. van. 1968. The behaviour of free-living chimpanzees in the Gombe Stream Reserve. *Animal Behaviour Monographs,* 1: 161–311.

366. Lawick-Goodall, J. van. 1971. *In the shadow of man.* London: Book Club Associates.

367. LeBoeuf, B. J. 1972. Sexual behaviour in the northern elephant seal *Mirounga angustirostris. Behaviour,* 41: 1–26.

368. LeBoeuf, B. J. 1974. Male-male competition and reproductive success in elephant seals. *American Zoologist,* 14: 163–176.

369. LeBoeuf, B. J. & Kaza, S., eds. 1981. *The natural history of Año Nuevo.* Pacific Grove, Calif.: Boxwood Press.

370. LeBoeuf, B. J. & Peterson, R. S. 1969. Social status and mating activity in elephant seals. *Science,* 163: 91–93.

371. LeBoeuf, B. J., Whiting, R. J. & Gantt, R. F. 1972. Perinatal behaviour of northern elephant seal females and their young. *Behaviour,* 43: 121–156.

372. Lee, A. K., Bradley, A. J. & Braithwaite, R. W. 1977. Corticosteroid levels and male mortality in *Antechinus stuartii.* In B. Stonehouse & D. Gilmore, eds. *The biology of marsupials.* Baltimore: University Park Press.

373. Lee, R. B. 1968. What hunters do for a living, or, how to make out on scarce resources. In R. B. Lee & I. DeVore, eds. *Man the hunter.* Chicago: Aldine.

374. Lee, R. B. 1972. The !Kung bushmen of Botswana. In M. G. Bicchieri, ed. *Hunters and gatherers today.* New York: Holt, Rinehart and Winston.

375. Lee, R. B. 1979. *The !Kung San: men, women and work in a foraging society.* Cambridge: Cambridge University Press.

376. Lehrman, D. S. 1965. Interaction between internal and external environments

in the regulation of the reproductive cycle of the ring dove. In F. A. Beach, ed. *Sex and behavior*. New York: Wiley.

377.   Leutert, R. 1975. Sex-determination in *Bonellia*. In R. Reinboth, ed. *Intersexuality in the animal kingdom*. New York: Springer-Verlag.

378.   Levin, D. A. 1975. Pest pressure and recombination systems in plants. *American Naturalist,* 109: 437–451.

379.   Levine, N. E. 1980. Nyinba polyandry and the allocation of paternity. *Journal of Comparative Family Studies,* 11: 283–298.

380.   Lévi-Strauss, C. 1969. *The elementary structures of kinship*. Boston: Beacon.

381.   Licht, P. 1971. Regulation of the annual testis cycle by photoperiod and temperature in the lizard, *Anolis carolinensis. Ecology,* 52: 240–252.

382.   Liley, N. R. 1966. Ethological isolating mechanisms in four sympatric species of poeciliid fishes. *Behaviour,* supplement 13: 1–197.

383.   Liley, N. R. 1980. Patterns of hormonal control in the reproductive behavior of fish, and their relevance to fish management and culture programs. In J. E. Bardach, J. J. Magnuson, R. C. May & J. M. Reinhart, eds. *Fish behavior and its use in the capture and culture of fishes. ICLARM Conference Proceedings 5*. Manila, Philippines: Internal Center for Living Aquatic Resources Management.

384.   Liley, N. R. & Wishlow, W. 1974. The interaction of endocrine and experiential factors in the regulation of sexual behaviour in the female guppy *Poecilia reticulata. Behaviour,* 48: 185–214.

385.   Lincoln, G. A. 1971. The seasonal reproductive changes in the red deer stag (*Cervus elaphus*). *Journal of Zoology* (London), 163: 105–123.

386.   Lincoln, G. A. 1981. Seasonal aspects of testicular function. In H. Burger and D. deKretser, eds. *The testis*. New York: Raven Press.

387.   Lincoln, G. A. & Short, R. V. 1980. Seasonal breeding: nature's contraceptive. *Recent Progress in Hormone Research,* 36: 1–52.

388.   Lindburg, D. G. 1971. The rhesus monkey in north India: an ecological and behavioral study. In L. A. Rosenblum, ed. *Primate behavior: developments in field and laboratory research*, vol. 2. New York: Academic Press.

389.   Lisk, R. D. 1973. Hormonal regulation of sexual behavior in polyestrous mammals common to the laboratory. In R. O. Greep & E. B. Astwood, eds. *Handbook of physiology*, section 7, volume 2, part 1. Washington, D.C.: American Physiological Society.

390.   Lloyd, D. G. 1980. Benefits and handicaps of sexual reproduction. *Evolutionary Biology,* 13: 69–111.

391. Lloyd, J. E. 1966. Studies on the flash communication system in *Photinus* fireflies. *Miscellaneous Publications of the Museum of Zoology*, No. 130, Ann Arbor: University of Michigan.

392. Lobban, C. F. 1972. *Law and anthropology in the Sudan (an analysis of homicide cases in Sudan)*. African Studies Seminar Series #13, Sudan Research Unit, Khartoum University.

393. Longair, R. W. 1981. Sex ratio variations in xylophilous aculeate Hymenoptera. *Evolution*, 35: 597–600.

394. Lorenz, K. Z. 1966. *On aggression*. New York: Harcourt Brace Jovanovich.

395. Lorimer, F. 1954. *Culture and human fertility*. Zurich: UNESCO.

396. Lott, D. F. 1979. Dominance relations and breeding rate in mature male American bison. *Zeitschrift für Tierpsychologie*, 49: 418–432.

397. Lott, D. F. 1981. Sexual behavior and intersexual strategies in American bison. *Zeitschrift für Tierpsychologie*, 56: 97–114.

398. Lott, D., Scholz, S. D. & Lehrman, D. S. 1967. Exteroceptive stimulation of the reproductive system of the female ring dove (*Streptopelia risoria*) by the mate and by the colony milieu. *Animal Behaviour*, 15: 433–437.

399. Lovari, S. & Hutchison, J. B. 1975. Behavioural transitions in the reproductive cycle of Barbary doves (*Streptopelia risoria*). *Behaviour*, 53: 126–150.

400. Løvtrup, S. 1974. *Epigenetics*. London: Wiley.

401. Low, B. S. 1978. Environmental uncertainty and the parental strategies of marsupials and placentals. *American Naturalist*, 112: 197–213.

402. Low, B. S. 1979. Sexual selection and human ornamentation. In N. A. Chagnon & W. Irons, eds. *Evolutionary biology and human social behavior: an anthropological perspective*. North Scituate, Mass.: Duxbury Press.

403. Lozoff, B., Brittenham, G. M., Trause, M. A., Kennell, J. H. & Klaus, M. H. 1977. The mother-newborn relationship: limits of adaptability. *Journal of Pediatrics*, 91: 1–12.

404. Lumpkin, S. 1981. Avoidance of cuckoldry in birds: the role of the female. *Animal Behaviour*, 29: 303–304.

405. Lumpkin, S., Kessel, K., Zenone, P. G. & Erickson, C. J. 1982. Proximity between the sexes in ring doves: social bonds or surveillance? *Animal Behaviour*, 30: 506–513.

406. MacArthur, R. H. 1962. Some generalized theorems of natural selection. *Proceedings of the National Academy of Sciences (USA)*, 48: 1893–1897.

407. MacCluer, J. W., Neel, J. V. & Chagnon, N. A. 1971. Demographic structure of a primitive population: a simulation. *American Journal of Physical Anthropology*, 35: 193–207.

408. Maccoby, E. E. & Jacklin, C. N. 1974. *The psychology of sex differences.* Palo Alto: Stanford University Press.

409. Mack, W. S. 1964. Ruminations on the testis. *Proceedings of the Royal Society of Medicine,* 57: 47–51.

410. MacLusky, N. J. & Naftolin, F. 1981. Sexual differentiation of the central nervous system. *Science,* 211: 1294–1303.

411. Madison, D. M. 1978. Behavioral and sociochemical susceptibility 'of meadow voles (*Microtus pennsylvanicus*) to snake predators. *American Midland Naturalist,* 100: 23–28.

412. Malinowski, B. 1929. *The sexual life of savages in north-western Melanesia.* London: Routledge.

413. Manning, J. T. 1980. Sex ratio and optimal male time investment strategies in *Asellus aquaticus* (L.) and *A. meridianus* Racovitza. *Behaviour,* 74: 264–273.

414. Martin, S. G. 1974. Adaptations for polygynous breeding in the bobolink, *Dolichonyx oryzivorus. American Zoologist,* 14: 109–119.

415. Martinez-Vargas, M. C. & Erickson, C. J. 1973. Some social and hormonal determinants of nest-building behaviour in the ring dove (*Streptopelia risoria*). *Behaviour,* 45: 12–37.

416. Massey, A. 1977. Agonistic aids and kinship in a group of pigtail macaques. *Behavioral Ecology and Sociobiology,* 2: 31–40.

417. Massey, A. & Vandenbergh, J. G. 1980. Puberty delay by a urinary cue from female house mice in feral populations. *Science,* 209: 821–822.

418. Matsumoto, A. & Arai, Y. 1981. Effect of androgen on sexual differentiation of synaptic organization in the hypothalamic arcuate nucleus: an ontogenetic study. *Neuroendocrinology,* 33: 166–169.

419. Maxson, S. J. & Oring, L. W. 1980. Breeding season time and energy budgets of the polyandrous spotted sandpiper. *Behaviour,* 74: 200–263.

420. Maynard Smith, J. 1956. Fertility, mating behaviour and sexual selection in *Drosophila subobscura. Journal of Genetics,* 54: 261–279.

421. Maynard Smith, J. 1958. The effects of temperature and of egg-laying on the longevity of *Drosophila subobscura. Journal of Experimental Biology,* 35: 832–842.

422. Maynard Smith, J. 1964. Group selection and kin selection. *Nature,* 201: 1145–1147.

423. Maynard Smith, J. 1971. What use is sex? *Journal of Theoretical Biology,* 30: 319–335.

424. Maynard Smith, J. S. 1978. *The evolution of sex.* Cambridge: Cambridge University Press.

425. McClintock, M. K. 1971. Menstrual synchrony and suppression. *Nature,* 229: 244–245.

426. McClure, P. A. 1981. Sex-biased litter reduction in food-restricted wood rats (*Neotoma floridana*). *Science,* 211: 1058–1060.

427. McEwen, B. S. 1981. Neural gonadal steroid actions. *Science,* 211: 1303–1311.

428. McGlone, J. 1980. Sex difference in human brain asymmetry: a critical survey. *The Behavioral and Brain Sciences,* 3: 215–263.

429. McKinney, F., Derrickson, S. R. & Mineau, P. 1983. Forced copulation in waterfowl. *Behaviour,* 86: 250–294.

430. McLaren, A. 1972. The embryo. In C. R. Austin & R. V. Short, eds. *Reproduction in mammals. Book 2: embryonic and fetal development.* Cambridge: Cambridge University Press.

431. McNeilly, A. S. 1979. Effects of lactation on fertility. *British Medical Bulletin,* 35: 151–154.

432. McWhirter, N. & McWhirter, R. 1975. *Guinness book of world records.* New York: Sterling.

433. Meggitt, M. J. 1962. *Desert people: a study of the Walbiri aborigines of central Australia.* Sydney: Angus and Robertson.

434. Meikle, D. B. & Vessey, S. H. 1981. Nepotism among rhesus monkey brothers. *Nature,* 294: 160–161.

435. Metcalf, R. A. 1980. Sex ratios, parent-offspring conflict, and local competition for mates in the social wasps *Polistes metricus* and *Polistes variatus. American Naturalist,* 116: 642–654.

436. Michael, R. P., Bonsall, R. W. & Warner, P. 1974. Human vaginal secretions: volatile fatty acid content. *Science,* 186: 1217–1219.

437. Miller, J. W. 1975. Much ado about starlings. *Natural History,* 84 (7): 38–45.

438. Mishell, D. R., Nakamura, R. M., Crosignani, P. G., Stone, S., Kharma, K., Nagata, Y. & Thorneycroft, I. H. 1971. Serum gonadotropin and steroid patterns during the normal menstrual cycle. *American Journal of Obstetrics and Gynecology,* 111: 60–65.

439. Moehlman, P. D. 1979. Jackal helpers and pup survival. *Nature,* 277: 382–383.

440. Money, J. 1969. Sex reassignment as related to hermaphroditism and transsexualism. In R. Green & J. Money, eds. *Transsexualism and sex reassignment.* Baltimore: Johns Hopkins Press.

441. Money, J. & Ehrhardt, A. A. 1972. *Man and woman, boy and girl.* Baltimore: Johns Hopkins Press.

442. Money, J. & Mathews, D. 1982. Prenatal exposure to virilizing progestins: an adult follow-up study of twelve women. *Archives of Sexual Behavior,* 11: 73–83.

443. Morris, D. 1967. *The naked ape: a zoologist's study of the human animal.* New York: McGraw-Hill.

444. Mrosovsky, N. 1980. Thermal biology of sea turtles. *American Zoologist,* 20: 531–547.

445. Mrosovsky, N. & Sherry, D. F. 1980. Animal anorexias. *Science,* 207: 837–842.

446. Mrosovsky, N. & Yntema, C. L. 1980. Temperature dependence of sexual differentiation in sea turtles: implications for conservation practices. *Biological Conservation,* 18: 271–280.

447. Muller, H. J. 1932. Some genetic aspects of sex. *American Naturalist,* 66: 118–138.

448. Muller, H. J. 1964. The relation of recombination to mutational advance. *Mutation Research,* 1: 2–9.

449. Muller, J.-C. 1980. On the relevance of having two husbands: contribution to the study of polygynous/polyandrous marital forms of the Jos Plateau. *Journal of Comparative Family Studies,* 11: 359–369.

450. Murdock, G. P. 1945. The common denominator of culture. In R. Linton, ed. *The science of man in the world crisis.* New York: Columbia University Press.

451. Murdock, G. P. 1965. *Culture and society.* Pittsburgh: University of Pittsburgh Press.

452. Murdock, G. P. 1967. *Ethnographic atlas.* Pittsburgh: University of Pittsburgh Press.

453. Murton, R. K. & Westwood, N. J. 1975. Integration of gonadotropin and steroid secretion, spermatogenesis and behaviour in the reproductive cycle of male pigeon species. In P. G. Caryl & D. M. Vowles, eds. *Neural and endocrine aspects of behaviour in birds.* Amsterdam: Elsevier.

454. Myers, J. H. 1978. Sex ratio adjustment under food stress: maximization of quality or numbers of young? *American Naturalist,* 112: 381–388.

455. Nag, M. 1972. Sex, culture, and human fertility: India and the United States. *Current Anthropology,* 13: 231–237.

456. Naroll, R. 1961. Two solutions to Galton's problem. *Philosophy of Science,* 28: 15–39.

457. Needham, R. 1971. Introduction. In R. Needham, ed. *Rethinking kinship and marriage.* London: Tavistock.

458. Netboy, A. 1980. *The Columbia River salmon and steelhead trout.* Seattle: University of Washington Press.

459. Newton, I., Marquiss, M. & Moss, D. 1981. Age and breeding in sparrow-hawks. *Journal of Animal Ecology,* 50: 839–853.

460. Newton, N. 1977. Breeding strategies in birds of prey. *Living Bird,* 16: 51–82.

461. Nisbet, I. C. T. 1973. Courtship-feeding, egg-size and breeding success in common terns. *Nature,* 241: 141–142.

462. Nishida, T. 1979. The social structure of chimpanzees of the Mahale Mountains. In D. A. Hamburg & E. R. McCown, eds. *The great apes.* Menlo Park, Calif.: Benjamin/Cummings.

463. Nishizuka, M. & Arai, Y. 1981. Organizational action of estrogen on synaptic pattern in the amygdala: implications for sexual differentiation of the brain. *Brain Research,* 213: 422–426.

464. Noonan, K. M. 1978. Sex ratio of parental investment in colonies of the social wasp *Polistes fuscipes. Science,* 199: 1354–1356.

465. Nottebohm, F. 1980. Testosterone triggers growth of brain vocal control nuclei in adult female canaries. *Brain Research,* 189: 429–436.

466. Nottebohm, F. 1981. A brain for all seasons: cyclical anatomical changes in song control nuclei of the canary brain. *Science,* 214: 1368–1370.

467. O'Connor, S., Vietze, P. M., Sherrod, K. B., Sandler, H. M. & Altemeier, W. A. 1979. Reduced incidence of parenting disorders following rooming-in. Unpublished manuscript. Nashville General Hospital.

468. O'Herlihy, C. & Robinson, H. P. 1980. Mittelschmerz is a preovulatory phenomenon. *British Medical Journal,* 280: 986.

469. Olive, P. J. W. 1980. Environmental control of reproduction in Polychaeta: experimental studies of littoral species in northeast England. In W. H. Clark & T. S. Adams, eds. *Advances in invertebrate reproduction.* New York: Elsevier-North Holland.

470. Ollason, J. C. & Dunnet, G. M. 1978. Age, experience and other factors affecting the breeding success of the fulmar, *Fulmarus glacialis,* in Orkney. *Journal of Animal Ecology,* 47: 961–976.

471. Olmsted, J. M. D. 1946. *Charles Édouard Brown-Séquard: a nineteenth century neurologist and endocrinologist.* Baltimore: Johns Hopkins Press.

472. Opie, I. & Opie, P. 1959. *The lore and language of schoolchildren.* Oxford: Clarendon.

473. Orians, G. H. 1969. On the evolution of mating systems in birds and mammals. *American Naturalist,* 103: 589–603.

474. Overton, W. R. 1982. Creationism in schools: the decision in McLean versus the Arkansas Board of Education. *Science,* 215: 934–943.

475. Packer, C. 1979. Male dominance and reproductive activity in *Papio anubis. Animal Behaviour,* 27: 37–45.

476. Packer, C. 1979. Inter-troop transfer and inbreeding avoidance in *Papio anubis. Animal Behaviour,* 27: 1–36.

477. Packer, C. & Pusey, A. E. 1982. Cooperation and competition within coalitions of male lions: kin selection or game theory? *Nature,* 296: 740–742.

478. Paige, K. E. & Paige, J. M. 1981. *The politics of reproductive ritual.* Berkeley: University of California Press.

479. Parker, G. A. 1970. Sperm competition and its evolutionary consequences in the insects. *Biological Reviews,* 45: 525–568.

480. Parker, G. A. 1974. Courtship persistence and female guarding as male time investment strategies. *Behaviour,* 48: 157–184.

481. Parker, G. A. 1978. Searching for mates. In J. R. Krebs & N. B. Davies, eds. *Behavioural ecology: an evolutionary approach.* Oxford: Blackwell Scientific Publications.

482. Parker, G. A., Baker, R. R. & Smith, V. G. F. 1972. The origin and evolution of gamete dimorphism and the male-female phenomenon. *Journal of Theoretical Biology,* 36: 529–553.

483. Partridge, L. 1980. Mate choice increases a component of offspring fitness in fruit flies. *Nature,* 283: 290–291.

484. Partridge, L. & Farquhar, M. 1981. Sexual activity reduces lifespan of male fruitflies. *Nature,* 294: 580–582.

485. Patterson, C. B., Erckmann, W. J. & Orians, G. H. 1980. An experimental study of parental investment and polygyny in male blackbirds. *American Naturalist,* 116: 757–769.

486. Patterson, I. J. 1965. Timing and spacing of broods in the black-headed gull *Larus ridibundus. Ibis,* 107: 433–459.

487. Patterson, T. L., Petrinovich, L. & James, D. K. 1980. Reproductive value and appropriateness of response to predators by white-crowned sparrows. *Behavioral Ecology and Sociobiology,* 7: 227–231.

488. Payne, R. B. 1974. The evolution of clutch size and reproductive rates in parasitic cuckoos. *Evolution,* 28: 169–181.

489. Payne, R. B. 1977. The ecology of brood parasitism in birds. *Annual Review of Ecology and Systematics,* 8: 1–28.

490. Perry, J. S. 1971. *The ovarian cycle of mammals.* Edinburgh: Oliver and Boyd.

491. Peter, R. E. 1981. Gonadotropin secretion during reproductive cycles in teleosts: influences of environmental factors. *General and Comparative Endocrinology,* 45: 294–305.

492. Peter, R. E. & Crim, L. W. 1979. Reproductive endocrinology of fishes: gonadal cycles and gonadotropin in teleosts. *Annual Review of Physiology,* 41: 323–335.

493. Peters, H. & McNatty, K. P. 1980. *The ovary: a correlation of structure and function in mammals.* London: Granada.

494. Peterson, R. E., Imperato-McGinley, J., Gautier, T. & Sturla, E. 1977. Male pseudohermaphroditism due to steroid 5α-reductase deficiency. *American Journal of Medicine,* 62: 170–191.

495. Pfaff, D. 1970. Nature of sex hormone effects on rat sex behavior: specificity of effects and individual patterns of response. *Journal of Comparative and Physiological Psychology,* 73: 349–358.

496. Phillips, C. L. 1887. Egg-laying extraordinary in *Colaptes auratus. Auk,* 4: 346.

497. Pianka, E. R. 1970. On r- and K-selection. *American Naturalist,* 104: 592–597.

498. Pienkowski, M. W. & Greenwood, J. J. D. 1979. Why change mates? *Biological Journal of the Linnaean Society,* 12: 85–94.

499. Pitelka, F. A., Holmes, R. T. & MacLean, S. F. 1974. Ecology and evolution of social organization in arctic sandpipers. *American Zoologist,* 14: 185–204.

500. Pittard, E. 1934. La castration chez l'homme: recherches sur les adeptes d'une secte d'eunuques mystiques, les Skoptzy. *Archives Suisses d'Anthropologie Générale,* 6: 213–536.

501. Pius XI. 1930. Papal encyclical *Casti connubii.* Vatican.

502. Plant, T. M., Schallenberger, E., Hess, D. L., McCormack, J. T., Dufy-Barbe, L. & Knobil, E. 1980. Influence of suckling on gonadotropin secretion in the female rhesus monkey (*Macaca mulatta*). *Biology of Reproduction,* 23: 760–766.

503. Pleszczynska, W. K. 1978. Microgeographic prediction of polygyny in the lark bunting. *Science,* 201: 935–937.

504. Pleszczynska, W. & Hansell, R. I. C. 1980. Polygyny and decision theory: testing of a model in lark buntings (*Calamospiza melanocorys*). *American Naturalist,* 116: 821–830.

505. Popp, J. L. & DeVore, I. 1979. Aggressive competition and social dominance

theory: synopsis. In D. A. Hamburg & E. R. McCown, eds. *The great apes.* Menlo Park, Calif.: Benjamin/Cummings.

506. Power, H. W. 1980. A model of parental care. I. Types of parental care and the unit of cost and benefit. Manuscript.

507. Power, H. W. & Doner, C. G. P. 1980. Experiments on cuckoldry in the mountain bluebird. *American Naturalist,* 116: 689–704.

508. Power, H. W., Litovich, E. & Lombardo, M. P. 1981. Male starlings delay incubation to avoid being cuckolded. *Auk,* 98: 386–389.

509. Pressley, P. H. 1981. Parental effort and the evolution of nest-guarding tactics in the threespine stickleback, *Gasterosteus aculeatus* L. *Evolution,* 35: 282–295.

510. Price, M. V. & Waser, N. M. 1982. Population structure, frequency-dependent selection, and the maintenance of sexual reproduction. *Evolution,* 36: 35–43.

511. Procter-Gray, E. & Holmes, R. T. 1981. Adaptive significance of delayed attainment of plumage in male American redstarts: tests of two hypotheses. *Evolution,* 35: 742–751.

512. Province of Ontario, Office of Registrar General. [1979]. *Vital Statistics for 1975 and 1976.* Toronto: Office of Registrar General.

513. Pryor, S. & Bronson, F. H. 1981. Relative and combined effects of low temperature, poor diet, and short daylength on the productivity of wild house mice. *Biology of Reproduction,* 25: 734–743.

514. Pugesek, B. H. 1981. Increased reproductive effort with age in the California gull (*Larus californicus*). *Science,* 212: 822–823.

515. Rainwater, L. 1965. *Family design: marital sexuality, family size and contraception.* Chicago: Aldine.

516. Ralls, K. 1977. Sexual dimorphism in mammals: avian models and unanswered questons. *American Naturalist,* 111: 917–938.

517. Ramanamma, A. & Bambawale, U. 1980. The mania for sons: an analysis of social values in South Asia. *Social Science and Medicine,* 14B: 107–110.

518. Rasmussen, D. R. 1981. Pair-bond strength and stability and reproductive success. *Psychological Review,* 88: 274–290.

519. Rathbun, G. B. 1979. The social structure and ecology of elephant-shrews. *Zeitschrift für Tierpsychologie,* supplement 20: 1–77.

520. Reichman, O. J. & Van de Graaff, K. M. 1975. Association between ingestion of green vegetation and desert rodent reproduction. *Journal of Mammalogy,* 56: 503–506.

521. Reinisch, J. M. 1981. Prenatal exposure to synthetic progestins increases potential for aggression in humans. *Science,* 211: 1171–1173.

522. Reiter, J., Panken, K. J. & LeBoeuf, B. J. 1981. Female competition and reproductive success in northern elephant seals. *Animal Behaviour,* 29: 670–687.

523. Reyes, F. I., Winter, J. S. D. & Faiman, C. 1973. Studies on human sexual development. I. Fetal gonadal and adrenal sex steroids. *Journal of Clinical Endocrinology and Metabolism,* 74: 74–78.

524. Reynolds, V. & Reynolds, F. 1965. Chimpanzees of the Budongo Forest. In I. DeVore, ed. *Primate behavior: field studies of monkeys and apes.* New York: Holt, Rinehart and Winston.

525. Richards, M. P. M. 1978. Possible effects of early separation on later development of children: a review. In F. S. W. Brimblecombe, M. P. M. Richards & N. C. R. Robertson, eds. *Separation and special-care baby units.* Philadelphia: Lippincott.

526. Ridpath, M. G. 1972. The Tasmanian native hen, *Tribonyx mortierii. CSIRO Wildlife Research,* 17: 1–118.

527. Ridley, M. 1978. Paternal care. *Animal Behaviour,* 26: 904–932.

528. Rivière, P. G. 1971. Marriage: a reassessment. In R. Needham, ed. *Rethinking kinship and marriage.* London: Tavistock.

529. Robertson, D. R. 1972. Social control of sex reversal in a coral-reef fish. *Science,* 177: 1007–1009.

530. Robinette, W. L., Gashwiler, J. S., Low, J. B. & Jones, D. A. 1957. Differential mortality by sex and age among mule deer. *Journal of Wildlife Management,* 21: 1–16.

531. Robinson, J. E. & Short, R. V. 1977. Changes in breast sensitivity at puberty, during the menstrual cycle, and at parturition. *British Medical Journal,* 1: 1188–1191.

532. Roger, S. C. 1975. Female forms of power and the myth of male dominance: a model of male/female interaction in peasant society. *American Ethnologist,* 2: 727–756.

533. Rohwer, S. 1977. Status signaling in Harris sparrows: some experiments in deception. *Behaviour,* 61: 107–129.

534. Rohwer, S., Fretwell, S. D. & Niles, D. M. 1980. Delayed maturation in passerine plumages and the deceptive acquisition of resources. *American Naturalist,* 115: 400–437.

535. Rowell, T. E. 1974. The concept of social dominance. *Behavioral Biology,* 11: 131–154.

536. Royama, T. 1966. A re-interpretation of courtship feeding. *Bird Study,* 13: 116–129.

537. Ruse, M. 1979. *The Darwinian revolution: science red in tooth and claw.* Chicago: University of Chicago Press.

538. Ryder, J. P. 1980. The influence of age on the breeding biology of colonial nesting seabirds. In J. Burger, B. L. Olla & H. E. Winn, eds. *Behavior of marine animals, vol. 4. Marine birds.* New York: Plenum Press.

539. Sade, D. S. 1968. Inhibition of son-mother mating among free-ranging rhesus monkeys. *Science and Psychoanalysis,* 12: 18–37.

540. Sadleir, R. M. F. S. 1969. *The ecology of reproduction in wild and domestic mammals.* London: Methuen.

541. Safilios-Rothschild, C. 1969. 'Honor' crimes in contemporary Greece. *British Journal of Sociology,* 20: 205–218.

542. Safilios-Rothschild, C. 1969. Attitudes of Greek spouses toward marital infidelity. In G. Neubeck, ed. *Extramarital relations.* Englewood Cliffs, N.J.: Prentice-Hall.

543. Salzano, F. M., Neel, J. V. & Maybury-Lewis, D. 1967. Further studies on the Xavante Indians. I. Demographic data on two additional villages: genetic structure of the tribe. *American Journal of Human Genetics,* 19: 463–489.

544. Saucier, J.-F. 1972. Correlates of the long postpartum taboo: a cross-cultural study. *Current Anthropology,* 13: 238–249.

545. Savage, M. O., Preece, M. A., Jeffcoate, S. L., Ransley, P. G., Rumsby, G., Mansfield, M. D. & Williams, D. I. 1980. Familial male pseudohermaphroditism due to deficiency of 5α-reductase. *Clinical Endocrinology* (Oxford), 12: 397–406.

546. Schadler, M. H. 1981. Postimplantation abortion in pine voles (*Microtus pinetorum*) induced by strange males and pheromones of strange males. *Biology of Reproduction,* 25: 295–297.

547. Schaffer, W. M. & Elson, P. F. 1975. The adaptive significance of variations in life history among local populations of Atlantic salmon in North America. *Ecology,* 56: 577–590.

548. Schaller, G. B. 1972. *The Serengeti lion: a study of predator-prey relations.* Chicago: University of Chicago Press.

549. Schamel, D. & Tracy, D. 1977. Polyandry, replacement clutches, and site tenacity in the red phalarope (*Phalaropus fulicarius*) at Barrow, Alaska. *Bird Banding,* 48: 314–324.

550. Schultz, R. J. 1971. Special adaptive problems associated with unisexual fishes. *American Zoologist,* 11: 351–360.

551. Schwagmeyer, P. L. 1979. The Bruce effect: an evaluation of male/female advantages. *American Naturalist,* 114: 932–938.

552. Selander, R. K. 1966. Sexual dimorphism and differential niche utilization in birds. *Condor,* 68: 113–151.

553. Selander, R. K. 1972. Sexual selection and dimorphism in birds. In B. Campbell, ed. *Sexual selection and the descent of man 1871–1971.* Chicago: Aldine.

554. Shapiro, D. Y. 1979. Social behaviour, group structure, and the control of sex reversal in hermaphroditic fish. *Advances in the Study of Behaviour,* 10: 43–102.

555. Shapiro, D. Y. 1980. Serial female sex changes after simultaneous removal of males from social groups of a coral reef fish. *Science,* 209: 1136–1137.

556. Sharman, G. B. 1970. Reproductive physiology of marsupials. *Science,* 167: 1221–1228.

557. Shepher, J. 1971. Mate selection among second generation kibbutz adolescents and adults: incest avoidance and negative imprinting. *Archives of Sexual Behavior,* 1: 293–307.

558. Sherman, P. W. 1977. Nepotism and the evolution of alarm calls. *Science,* 197: 1246–1253.

559. Sherman, P. W. 1980. The limits of ground squirrel nepotism. In G. W. Barlow & J. Silverberg, eds. *Sociobiology: beyond nature/nurture?* Boulder, Colo.: Westview Press.

560. Sherman, P. W. 1981. Reproductive competition and infanticide in Belding's ground squirrels and other animals. In R. D. Alexander & D. W. Tinkle, eds. *Natural selection and social behavior.* New York: Chiron Press.

561. Sherry, D. F. 1981. Adaptive changes in body weight. In L. A. Cioffi, ed. *The body weight regulatory system.* New York: Raven Press.

562. Sherry, D. F., Mrosovsky, N. & Hogan, J. A. 1980. Weight loss and anorexia during incubation in birds. *Journal of Comparative and Physiological Psychology,* 94: 89–98.

563. Shields, W. M. 1982. Inbreeding and the paradox of sex: a resolution? *Evolutionary Theory,* 5: 245–279.

564. Shine, R. 1978. Sexual size dimorphism and male combat in snakes. *Oecologia,* 33: 269–277.

565. Shine, R. 1979. Sexual selection and sexual dimorphism in the amphibia. *Copeia,* 1979: 297–306.

566. Short, R. V. 1981. Sexual selection in man and the great apes. In C. E. Graham, ed. *Reproductive biology of the great apes.* New York: Academic Press.

567. Shuster, S. M. 1981. Sexual selection in the socorro isopod, *Thermosphaeroma thermophilum* (Cole) (Crustacea: Peracarida). *Animal Behaviour,* 29: 698–707.

568. Sigusch, V. & Schmidt, G. 1971. Lower-class sexuality: some emotional and social aspects in West German males and females. *Archives of Sexual Behavior,* 1: 29–44.

569. Siiteri, P. K. & Wilson, J. D. 1974. Testosterone formation and metabolism during male sexual differentiation in the human embryo. *Journal of Clinical Endocrinology,* 38: 113–125.

570. Silk, J. B. 1980. Adoption and kinship in Oceania. *American Anthropologist,* 82: 799–820.

571. Simon, W. E. & Primavera, L. H. 1976. Attitudes toward ideal family size: some preliminary data. *Psychological Reports,* 38: 1282.

572. Simpson, G. G. 1951. *Horses.* New York: Oxford University Press.

573. Sizonenko, P. C. 1978. Endocrinology in preadolescents and adolescents. I. Hormonal changes during normal puberty. *American Journal of Diseases of Children,* 132: 704–712.

574. Smith, D. G. 1980. Paternity exclusion in six captive groups of rhesus monkeys (*Macaca mulatta*). *American Journal of Physical Anthropology,* 53: 243–249.

575. Smith, J. E. & Kunz, P. R. 1976. Polygyny and fertility in nineteenth-century America. *Population Studies,* 30: 465–480.

576. Smith, J. N. M. 1981. Does high fecundity reduce survival in song sparrows? *Evolution,* 35: 1142–1148.

577. Smith, R. L. 1979. Paternity assurance and altered roles in the mating behaviour of a giant water bug, *Abedus herberti* (Heteroptera: Belostomatidae). *Animal Behaviour,* 27: 716–725.

578. Spiro, M. E. 1958. *Children of the kibbutz.* Cambridge, Mass.: Harvard University Press.

579. Stacey, N. E. 1981. Hormonal regulation of female reproductive behavior in fish. *American Zoologist,* 21: 305–316.

580. Stearns, S. C. 1976. Life-history tactics: a review of the ideas. *Quarterly Review of Biology,* 51: 3–47.

581. Stearns, S. C. 1977. The evolution of life history traits: a critique of the theory and a review of the data. *Annual Review of Ecology and Systematics,* 8: 145–171.

582. Stephens, W. N. 1963. *The family in cross-cultural perspective.* New York: Holt, Rinehart and Winston.

583. Stern, C. & Sherwood, E. R. eds. 1966. *The origin of genetics: a Mendel source book.* San Francisco: Freeman.

584. Strassmann, B. I. 1981. Sexual selection, paternal care, and concealed ovulation in humans. *Ethology and Sociobiology,* 2: 31–40.

585. Strathmann, R. R. & Strathmann, M. F. 1982. The relationship between adult size and brooding in marine invertebrates. *American Naturalist,* 119: 91–101.

586. Struhsaker, T. T. 1969. Correlates of ecology and social organization among African cercopithecines. *Folia Primatologica,* 11: 80–118.

587. Struhsaker, T. T. & Leland, L. 1979. Socioecology of five sympatric monkey species in the Kibale Forest, Uganda. *Advances in the Study of Behavior,* 9: 159–228.

588. Sugarman, M. 1977. Paranatal influences on maternal-infant attachment. *American Journal of Orthopsychiatry,* 47: 407–421.

589. Sugiyama, Y. 1967. Social organization of Hanuman langurs. In S. A. Altmann, ed. *Social communication among primates.* Chicago: University of Chicago Press.

590. Suzuki, D. T., Griffiths, A. J. F. & Lewontin, R. C. 1981. *An introduction to genetic analysis,* 2d edition. San Francisco: W. H. Freeman.

591. Swerdloff, R. S. & Heber, D. 1981. Endocrine control of testicular function from birth to puberty. In H. Burger & D. de Kretser, eds. *The testis.* New York: Raven Press.

592. Symons, D. 1978. *Play and aggression.* New York: Columbia University Press.

593. Symons, D. 1979. *The evolution of human sexuality.* New York: Oxford University Press.

594. Tanner, J. M. 1970. Physical growth. In P. H. Mussen, ed. *Carmichael's manual of child psychology,* vol. 1, 3d edition. New York: John Wiley & Sons.

595. Tanner, J. M. 1981. Postnatal growth of gonads and genital tracts, and development of secondary sex characteristics. In C. R. Austin & R. G. Edwards, eds. *Mechanisms of sex differentiation in animals and man.* New York: Academic Press.

596. Tata, J. R. 1980. The action of growth and developmental hormones. *Biological Reviews,* 55: 285–319.

597. Tauber, E. S. 1940. Effects of castration upon the sexuality of the adult male. *Psychosomatic Medicine,* 2: 74–87.

598. Taylor, P. A. & Glenn, N. D. 1976. The utility of education and attractiveness for females' status attainment through marriage. *American Sociological Review,* 41: 484–498.

599. Taylor, P. D. 1981. Intra-sex and inter-sex sibling interactions as sex-ratio determinants. *Nature,* 291: 64–66.

600. Teismann, M. W. 1975. Jealous conflict: a study of verbal interaction and labeling of jealousy among dating couples involved in jealousy improvisations. Doctoral dissertation, University of Connecticut.

601. Teleki, G., Hunt, E. E. & Pfifferling, J. H. 1976. Demographic observations (1963–1973) on the chimpanzees of Gombe National Park, Tanzania. *Journal of Human Evolution,* 5: 559–598.

602. Tepperman, J. 1973. *Metabolic and endocrine physiology,* 3d edition. Chicago: Year Book Medical Publishers.

603. Thompson, S. K. 1975. Gender labels and early sex role development. *Child Development,* 46: 339–347.

604. Thornes, B. & Collard, J. 1979. *Who divorces?* London: Routledge & Kegan Paul.

605. Thornhill, R. 1976. Sexual selection and nuptial feeding behavior in *Bittacus apicalis* (Insecta: Mecoptera). *American Naturalist,* 110: 529–548.

606. Thornhill, R. 1976. Sexual selection and paternal investment in insects. *American Naturalist,* 110: 153–163.

607. Thornhill, R. 1980. Rape in *Panorpa* scorpionflies and a general rape hypothesis. *Animal Behaviour,* 28: 52–59.

608. Thornhill, R. 1981. Panorpa (*Mecoptera: Panorpidae*) scorpionflies: systems for understanding resource-defense polygyny and alternative male reproductive efforts. *Annual Review of Ecology and Systematics,* 12: 355–386.

609. Thornhill, R. & Alcock, J. 1983. *The evolution of insect mating systems.* Cambridge, Mass.: Harvard University Press.

610. Thornhill, R. & Thornhill, N. W. 1983. Human rape: an evolutionary analysis. *Ethology and Sociobiology* 4: 137–173.

611. Thurber, J. & White, E. B. 1957. *Is sex necessary? or, why you feel the way you do.* New York: Harper and Row.

612. Timonen, S. & Carpen, E. 1968. Multiple pregnancies and photoperiodicity. *Annales Chirurgiae Gynaecologiae Fenniae,* 57: 135–138.

613. Tinbergen, N. 1963. On aims and methods of ethology. *Zeitschrift für Tierpsychologie,* 20: 410–433.

614. Tinbergen, N. 1968. On war and peace in animals and man. *Science,* 160: 1411–1418.

615. Tinkle, D. W. 1969. The concept of reproductive effort and its relation to the evolution of life histories in lizards. *American Naturalist,* 103: 501–516.

616. Tokarz, R. R. & Crews, D. 1981. Effects of prostaglandins on sexual receptivity in the female lizard, *Anolis carolinensis. Endocrinology,* 109: 451–457.

617. Tooby, J. 1982. Pathogens, polymorphism, and the evolution of sex. *Journal of Theoretical Biology,* 97: 557–576.

618. Touhey, J. C. 1972. Comparison of two dimensions of attitude similarity on heterosexual attraction. *Journal of Personality and Social Psychology,* 23: 8–10.

619. Trivers, R. L. 1972. Parental investment and sexual selection. In B. Campbell, ed. *Sexual selection and the descent of man 1871–1971.* Chicago: Aldine.

620. Trivers, R. L. 1974. Parent-offspring conflict. *American Zoologist,* 14: 249–264.

621. Trivers, R. L. & Hare, H. 1976. Haplodiploidy and the evolution of the social insects. *Science,* 191: 249–263.

622. Trivers, R. L. & Willard, D. E. 1973. Natural selection of parental ability to vary the sex ratio of offspring. *Science,* 179: 90–92.

623. Turkington, R. W. 1972. Serum prolactin levels in patients with gynecomastia. *Journal of Clinical Endocrinology and Metabolism,* 34: 62–66.

624. Tutin, C. E. G. & McGinnis, P. R. 1981. Chimpanzee reproduction in the wild. In C. E. Graham, ed. *Reproductive biology of the great apes.* New York: Academic Press.

625. Tutor, B. M. 1962. Nesting studies of the boat-tailed grackle. *Auk,* 79: 77–84.

626. Tyler, M. J. & Carter, D. B. 1981. Oral birth of the young of the gastric brooding frog *Rheobatrachus silus. Animal Behaviour,* 29: 280–282.

627. Udry, J. R. 1974. *The social context of marriage,* 3d edition. Philadelphia: Lippincott.

628. Ukaegbu, A. O. 1977. Fertility of women in polygynous unions in rural Eastern Nigeria. *Journal of Marriage and the Family,* 39: 397–404.

629. U.S. Bureau of the Census. 1972. *Statistical abstract of the United States: 1972.* Washington, D.C.: U.S. Government Printing Office.

630. U.S. Bureau of the Census. 1973. Census of Population: 1970. Subject Reports. Final Report PC(2)-3A. *Women by number of children everborn.* Washington, D.C.: U.S. Government Printing Office.

631. U.S. Bureau of the Census. 1976. *Current population reports.* Series P-20, no. 288. Fertility history and prospects of American women: June 1975. Washington, D.C.: U.S. Government Printing Office.

632. U.S. Bureau of the Census. 1981. *Current population reports.* Series P-20, no. 363. Population profile of the United States: 1980. Washington, D.C.: U.S. Government Printing Office.

633. U.S. Department of Health, Education and Welfare. 1972. *Cohort mortality and survivorship: United States death-registration, states, 1900–1968.* Vital and Health Statistics, Series 3, No. 16. Washington, D.C.: U.S. Government Printing Office.

634. U.S. Department of Justice, Federal Bureau of Investigation. 1981. *Uniform crime reports for the United States 1980.* Washington, D.C.: U.S. Government Printing Office.

635. Vandenbergh, J. G. 1973. Environmental influences on breeding in rhesus monkeys. Symposium of the IVth International Congress of Primatology, vol. 2: *Primate reproductive behavior.* Basel: Karger.

636. van den Berghe, P. L. 1979. *Human family systems.* New York: Elsevier.

637. van den Berghe, P. L. 1980. Incest and exogamy: a sociobiological reconsideration. *Ethology and Sociobiology,* 1: 151–162.

638. van den Berghe, P. L. & Mesher, G. M. 1980. Royal incest and inclusive fitness. *American Ethnologist,* 7: 300–317.

639. Van Valen, L. 1973. A new evolutionary law. *Evolutionary Theory,* 1: 1–30.

640. Vehrencamp, S. L. 1977. Relative fecundity and parental effort in communally nesting anis, *Crotophaga sulcirostres. Science,* 197: 403–405.

641. Verme, L. J. & Ozoga, J. J. 1981. Sex ratio of white-tailed deer and the estrus cycle. *Journal of Wildlife Management,* 45: 710–715.

642. Verner, J. & Willson, M. F. 1966. The influence of habitats on mating systems of North American passerine birds. *Ecology,* 47: 143–147.

643. Waage, J. K. 1982. Sib-mating and sex ratio strategies in scelionid wasps. *Ecological Entomology,* 7: 103–112.

644. Wachtel, S. S. & Koo, G. C. 1981. H-Y antigen in gonadal differentiation. In C. R. Austin & R. G. Edwards, eds. *Mechanisms of sex differentiation in animals and man.* London: Academic Press.

645. Waddington, C. H. 1966. *Principles of development and differentiation.* New York: Macmillan.

646. Walker, I. 1967. Effect of population density on the viability and fecundity in *Nasonia vitripennis* Walker (Hymenoptera, Pteromalidae). *Ecology,* 48: 294–301.

647. Waller, J. H. 1971. Differential reproduction: its relation to IQ test score, education, and occupation. *Social Biology,* 18: 122–136.

648. Ward, I. L. 1969. Differential effects of pre- and postnatal androgen on the sexual behavior of intact and spayed female rats. *Hormones and Behavior,* 1: 25–36.

649. Ward, P. 1965. The breeding biology of the black-faced dioch *Quelea quelea* in Nigeria. *Ibis,* 107: 326–349.

650. Ware, H. 1979. Polygyny: women's views in a transitional society, Nigeria, 1975. *Journal of Marriage and the Family,* 141: 185–195.

651. Warner, H., Martin, D. E. & Keeling, M. E. 1974. Electroejaculation of the Great Apes. *Annals of Biomedical Engineering,* 2: 419–432.

652. Warner, R. R. 1980. The coevolution of behavioral and life-history characteristics. In G. W. Barlow & J. Silverberg, eds. *Sociobiology: beyond nature/nurture?* Boulder, Colo.: Westview Press.

653. Warner, R. R. & Harlan, R. K. 1982. Sperm competition and sperm storage as determinants of sexual dimorphism in the dwarf surfperch, *Micrometrus minimus. Evolution,* 36: 44–55.

654. Warner, R. R., Robertson, D. R. & Leigh, E. G. 1975. Sex change and sexual selection. *Science,* 190: 633–638.

655. Weatherhead, P. J. & Robertson, R. J. 1979. Offspring quality and the polygyny threshold: "the sexy son hypothesis." *American Naturalist,* 113: 201–208.

656. Weir, B. J. & Rowlands, I. W. 1973. Reproductive strategies of mammals. *Annual Review of Ecology and Systematics, 4: 139*–163.

657. Wells, K. D. 1981. Parental behavior of male and female frogs. In R. D. Alexander & D. W. Tinkle, eds. *Natural selection and social behavior.* New York: Chiron Press.

658. Werren, J. H. 1980. Sex ratio adaptations to local mate competition in a parasitic wasp. *Science,* 208: 1157–1159.

659. Werren, J. H., Gross, M. R. & Shine, R. 1980. Paternity and the evolution of male parental care. *Journal of Theoretical Biology,* 82: 619–631.

660. West, M. J., King, A. P. & Eastzer, D. H. 1981. Validating the female bioassay of cowbird song: relating differences in song potency to mating success. *Animal Behaviour,* 29: 490–501.

661. West-Eberhard, M. J. 1975. The evolution of social behavior by kin selection. *Quarterly Review of Biology,* 50: 1–33.

662. West-Eberhard, M. J. 1981. Intragroup selection and the evolution of insect societies. In R. D. Alexander & D. W. Tinkle, eds. *Natural selection and social behavior.* New York: Chiron Press.

663. Westermarck, E. 1891. *The history of human marriage.* New York: Macmillan.

664. Westoff, C. F., Potter, R. G. & Sagi, P. C. 1963. *The third child.* Princeton: Princeton University Press.

665. Weygoldt, P. 1980. Complex brood care and reproductive behaviour in cap-

tive poison-arrow frogs, *Dendrobates pumilio* O. Schmidt. *Behavioral Ecology and Sociobiology,* 7: 329–332.

666. Whalen, R. E. & Edwards, D. A. 1967. Hormonal determinants of the development of masculine and feminine behavior in male and female rats. *Anatomical Record,* 157: 173–180.

667. Whitten, W. K. & Champlin, A. K. 1973. The role of olfaction in mammalian reproduction. In R. O. Greep & E. B. Astwood, eds. *Handbook of physiology,* section 7, volume 2, part 1. Washington, D.C.: American Physiological Society.

668. Whyte, M. K. 1978. *The status of women in preindustrial societies.* Princeton: Princeton University Press.

669. Widdowson, E. M. 1981. The role of nutrition in mammalian reproduction. In D. Gilmore & B. Cook, eds. *Environmental factors in mammal reproduction.* London: Macmillan.

670. Wiese, L. 1981. On the evolution of anisogamy from isogamous monoecy and on the origin of sex. *Journal of Theoretical Biology,* 89: 573–580.

671. Wilcoxon, L. A., Schrader, S. L. & Sherif, C. W. 1976. Daily self-reports on activities, life events, moods, and somatic changes during the menstrual cycle. *Psychosomatic Medicine,* 38: 399–417.

672. Wiley, R. H. 1973. Territoriality and non-random mating in sage grouse, *Centrocercus urophasianus. Animal Behaviour Monographs,* 6: 85–169.

673. Wiley, R. H. 1974. Evolution of social organization and life history patterns among grouse (Aves: Tetraonidae). *Quarterly Review of Biology,* 49: 201–227.

674. Williams, G. C. 1966. *Adaptation and natural selection.* Princeton: Princeton University Press.

675. Williams, G. C. 1966. Natural selection, the costs of reproduction, and a refinement of Lack's principle. *American Naturalist,* 100: 687–690.

676. Williams, G. C. 1975. *Sex and evolution.* Princeton: Princeton University Press.

677. Williamson, N. E. 1976. *Sons or daughters. A cross-cultural survey of parental preferences,* vol. 31. Sage Library of Social Research. Beverly Hills, Calif.: Sage Publications.

678. Wilson, A. P. & Boelkins, R. C. 1970. Evidence for seasonal variation in aggressive behaviour by *Macaca mulatta. Animal Behaviour,* 18: 719–724.

679. Wilson, E. O. 1971. *The insect societies.* Cambridge, Mass.: Harvard University Press.

680. Wilson, E. O. 1975. *Sociobiology: the new synthesis.* Cambridge, Mass.: Harvard University Press.

681. Wilson, J. D., George, F. W. & Griffin, J. E. 1981. The hormonal control of sexual development. *Science*, 211: 1278-1284.

682. Wilson, M. & Daly, M. 1984. Competitiveness, risk-taking and violence: the young male syndrome. *Ethology and Sociobiology*. To appear.

683. Wilson, M. I., Daly, M. & Weghorst, S. J. 1980. Household composition and the risk of child abuse and neglect. *Journal of Biosocial Science*, 12: 333-340.

684. Wimsatt, W. A. 1975. Some comparative aspects of implantation. *Biology of Reproduction*, 12: 1-40.

685. Witt, R., Schmidt, C. & Schmitt, J. 1981. Social rank and Darwinian fitness in a multimale group of Barbary macaques (*Macaca sylvana* Linnaeus 1758). Dominance reversals and male reproductive success. *Folia Primatologica*, 36: 201-211.

686. Wittenberger, J. F. 1976. The ecological factors selecting for polygyny in altricial birds. *American Naturalist*, 110: 779-799.

687. Wittenberger, J. F. 1978. The evolution of mating systems in grouse. *Condor*, 80: 126-137.

688. Wittenberger, J. F. 1981. *Animal social behavior*. Boston: Duxbury Press.

689. Wolf, A. P. 1970. Childhood association and sexual attraction: a further test of the Westermarck hypothesis. *American Anthropologist*, 72: 503-515.

690. Wolf, A. P. & Huang Chieh-shan. 1980. *Marriage and adoption in China, 1845-1945*. Stanford: Stanford University Press.

691. Wolf, L. L. 1975. Prostitution behavior in a tropical hummingbird. *Condor*, 77: 140-144.

692. Wolf, U. 1981. Genetic aspects of H-Y antigen. *Human Genetics*, 58: 25-28.

693. Woodside, B., Wilson, R., Chee, P. & Leon, M. 1981. Resource partitioning during reproduction in the Norway rat. *Science*, 211: 76-77.

694. Woolfenden, G. E. 1975. Florida scrub jay helpers at the nest. *Auk*, 92: 1-15.

695. Wrangham, R. W. 1975. The behavioural ecology of chimpanzees in Gombe National Park, Tanzania. Doctoral dissertation, Cambridge University.

696. Wrangham, R. W. 1979. On the evolution of ape social systems. *Social Science Information*, 18: 335-368.

697. Wrangham, R. W. 1980. Female choice of least costly males: a possible factor in the evolution of leks. *Zeitschrift für Tierpsychologie*, 54: 357-367.

698. Wright, S. 1977. *Evolution and the genetics of populations. Vol. 3. Experimental results and evolutionary deductions*. Chicago: University of Chicago Press.

699. Wylie, H. G. 1966. Some mechanisms that affect the sex ratio of *Nasonia vitripennis* (Walk.) (Hymenoptera: Pteromalidae) reared from superparasitized housefly pupae. *Canadian Entomologist,* 98: 645–653.

700. Wynne-Edwards, V. C. 1962. *Animal dispersion in relation to social behaviour.* Edinburgh: Oliver and Boyd.

701. Yasukawa, K. 1981. Male quality and female choice of mate in the red-winged blackbird (*Agelaius phoeniceus*). *Ecology,* 62: 922–929.

702. Yom-Tov, Y. 1980. Intraspecific nest parasitism in birds. *Biological Reviews,* 55: 93–108.

703. Young, J. Z. 1975. *The life of mammals: their anatomy and physiology,* 2d edition. Oxford: Clarendon.

704. Young, K. 1970. *Isn't one wife enough?* Newport, Conn.: Greenwood Press.

705. Yunis, J. Y. & Prakash, O. 1982. The origin of man: a chromosomal pictorial legacy. *Science,* 215: 1525–1530.

706. Zahavi, A. 1975. Mate-selection—a selection for a handicap. *Journal of Theoretical Biology,* 53: 205–214.

707. Zahavi, A. 1977. The cost of honesty (further remarks on the handicap principle). *Journal of Theoretical Biology,* 67: 603–605.

708. Zajonc, R. B. 1968. Attitudinal effects of mere exposure. *Journal of Personality and Social Psychology,* Monograph Supplement, Part 2: 1–29.

709. Zenone, P. G., Sims, M. E. & Erickson, C. J. 1979. Male ring dove behavior and the defense of genetic paternity. *American Naturalist,* 114: 615–626.

Additional Credits:

Page 34, Fig. 2.3. Reprinted by permission of the publishers from *Sociobiology: The New Synthesis,* by E. O. Wilson, Cambridge, Mass.: The Belknap Press of Harvard University Press, Copyright © 1975 by the President and Fellows of Harvard College.

Page 88, Fig. 5.6. Data from the *American Journal of Human Genetics,* 19: 463–489 by F. M. Salzano, J. V. Neel and D. Maybury-Lewis by permission of The University of Chicago Press. © 1967 by the American Society of Human Genetics.

Page 119, Fig. 6.2. After Verner & Willson. Copyright © 1966 by the Ecological Society of America [642]; and from the *American Naturalist,* 103: 589–603 by G. H. Orians by permission of The University of Chicago Press. © 1969 by The University of Chicago.

Pages 122–123, Fig. 6.3. After the *American Naturalist*, 116: 821–830 by W. Pleszczynska and R. I. C. Hansell by permission of The University of Chicago Press. © 1980 by The University of Chicago; and Garson, Pleszczynska & Holm, 1981, Table 1 [208].

Page 125, Fig. 6.5. After the *American Naturalist*, 105: 355–370 by J. F. Downhower and K. B. Armitage by permission of The University of Chicago Press. © 1971 by The University of Chicago.

Page 127, Fig. 6.6. Reproduced by permission of the American Anthropological Association from *American Anthropologist* 60 (5): 838–860, 1958.

Page 214, Fig. 8.10. After Patterson, 1965; *Ibis*, Journal the British Ornithologists' Union [486].

Glossary

allele Any of a set of alternative forms of a gene.

androgen Any of several steroid hormones, primarily of testicular origin but also produced by adrenals and ovaries, generally producing mainly masculine effects (cf. estrogen).

asexual reproduction Any mode of reproduction that does not entail fusion of gametes and thus produces offspring genetically identical to the parent.

avunculate A relationship characteristic of certain human societies in which men hold authority over, and take benevolent interest in, their sisters' offspring.

bride price Goods or money the groom or his kin pay the bride's kin at marriage in certain human societies.

carrying capacity The size or density of a population that can be supported in stable equilibrium with other biota in a locale or habitat type.

chiasma (pl. chiasmata) A point of contact between homologous chromosomes, the boundary of genetic material exchanged in crossing-over.

clone A group of genetically identical organisms descended asexually from a single individual.

clutch A set of eggs incubated together in a single nest.

coefficient of relatedness A measure of kinship varying from 0.0 (unrelated) to 1.0 (genetically identical), representing the probability that a focal allele will be identical in the two individuals by descent from a recent common ancestor.

cohort A like-aged subset of a population.

convergence The independent evolution of functionally similar characteristics in two or more unrelated populations.

corpus luteum A mass of endocrine cells formed after ovulation in the ruptured ovarian follicle of a mammal.

cytoplasm The contents of a cell other than its nucleus.

Darwinian fitness Personal reproductive success (cf. inclusive fitness).

demographic transition The historical process, in industrialized nations, in which the death rate declines before there is a decline in the birth rate.

diploid Containing a dual set of chromosomes, twice the number in a normal gamete (cf. haploid).

disruptive selection Natural selection operating against intermediate values of a trait, hence producing a bimodal (or multimodal) trait distribution.

dowry The goods or money the bride's kin provide the groom at marriage in certain human societies.

electrophoresis The motion of charged particles through a viscous liquid medium under the influence of an electric field; used to discriminate differentially charged (and hence differently structured) forms of closely similar molecules such as the enzymes coded by alleles of a single gene.

endocrine Pertaining to any of several internally secreting ductless glands and their products.

epigenesis The developmental process of structural elaboration of an organism through the continuous interaction of organism and environment.

estradiol An estrogen (see below), $C_{18}H_{24}O_2$, produced primarily in the ovarian follicles.

estrogen Any of several steroid hormones, primarily of ovarian origin, generally producing mainly feminine effects (cf. androgen).

estrus The period of sexual receptivity in certain female mammals (adjective: estrous).

eusocial Of a social system in which obligate sterility characterizes certain individuals who enhance their fitness by contributions to collateral kin.

evolution Any gradual change, especially change over generations in a gene pool and its population of phenotypes.

exogamy The tendency, by whatever mechanism, to mate selectively with individuals other than kin.

fecundity The realized or potential numerical production of ova or offspring by a female in a reproductive episode, season, or lifetime.

fertilization The process by which two gametes combine to produce a zygote.

fitness The reproductive success of a gene, organism, behavioral attribute, and so forth; commonly expressed relative to either an equilibrial replacement level fitness of 1.0 or a maximal fitness of 1.0.

gene The basic particulate unit of inheritance that is passed intact from parent to offspring.

gamete A haploid sperm cell (spermatozoan or ovum).

gene pool The set of all genes of the members of a population of (potentially) interbreeding conspecific organisms.

genotype The genetic constitution of an individual organism (cf. phenotype).

gonadotropin Any of several hormones of predominantly pituitary origin acting to stimulate the gonads.

gonochorism The sexual system in which each individual is either a female or male (cf. hermaphroditism).

haplodiploidy A genetic system in which unfertilized ova develop into haploid males while fertilized ova develop into diploid females.

haploid Containing a single set of chromosomes, half the number in a typical somatic cell (cf. diploid).

heritability The degree, ranging from 0.0 to 1.0, to which the phenotypic variance of a trait among the members of a population is related to genotypic variance.

hermaphroditism The sexual system in which both female and male reproductive functions are exercised by an individual organism, whether simultaneously or sequentially (cf. gonochorism).

heterogametic Of the sex that is determined by the presence of two different sex chromosomes (XY or ZW; cf. homogametic).

heterozygous Bearing different alleles at one or more chromosomal loci under consideration; said of individuals or populations (cf. homozygous).

homogametic Of the sex that is determined by the presence of two identical sex chromosomes (XX or ZZ; cf. heterogametic).

homozygous Bearing identical alleles at one or more chromosomal loci under consideration; said of individuals or populations (cf. heterozygous).

hormone A chemical substance secreted by an endocrine gland and influencing other tissues within the same organism.

hybrid The offspring of genetically dissimilar parents, especially parents of different species.

hypergyny Upward mating, especially marriage of a woman to a man of higher social standing.

implantation The process of attachment of a mammalian conceptus to the uterine lining from which it will draw sustenance.

inclusive fitness The sum of an individual's Darwinian fitness (personal reproductive success) and his or her influence upon the Darwinian fitnesses of relatives, weighted according to their coefficients of relatedness to the focal individual.

isogamy The circumstance in which all gametes are of a single size class (as opposed to the size distinction of ovum versus sperm).

iteroparity The strategy of reproducing repeatedly (cf. semelparity).

K-strategy A set of reproductive characteristics that tends to maximize resource utilization largely by emphasizing intensive nurture and a slow reproductive rate (cf. *r*-strategy).

lactational anovulation The tendency in female mammals for lactation to inhibit ovulation.

lek A fixed geographic site where males of a species aggregate for courtship displays to females.

linkage The circumstance in which two or more genetic factors segregate nonindependently.

lordosis A posture of forward spinal curvature assumed for sexual intercourse by estrous females of several mammalian species.

matrilineal Of or pertaining to a kinship system in which descent is reckoned or affiliative bonds exist primarily through the maternal line.

meiosis The process of cell division in diploid sexually reproducing organisms that leads to haploid gametes (cf. mitosis).

mitosis The process of cell division typical of diploid somatic cells by which further diploid cells are produced (cf. meiosis).

monogamy An exclusive mateship of one female with one male, whether for a reproductive cycle, a season, or a lifetime (cf. polygamy).

nepotism Any discriminative behavior tending to favor the actor's relatives and hence to contribute to the actor's inclusive fitness.

ontogeny The course of life-span development of an individual organism (cf. phylogeny).

oviparous Egg-laying (cf. viviparous, ovoviviparous).

oviposition The deposition by a female of one or more of her ova into the environment.

ovoviviparous Producing eggs that hatch within the mother's body (cf. viviparous, oviparous).

ovulation The (production and) release of ova by the ovary.

parental investment Any investment by a parent in an offspring that increases the recipient's fitness at the cost of parental capacity to invest elsewhere.

parthenogenesis The development of an unfertilized ovum into an independent organism.

patrilineal Of or pertaining to a kinship system in which descent is reckoned or affiliative bonds exist primarily through the paternal line.

phenotype The manifest nature of an organism, including morphological, physiological, and behavioral attributes, especially when considering genetic variation relevant to those attributes (cf. genotype).

pheromone A chemical substance secreted into the external environment by one organism to influence another, usually conspecific, organism.

phylogeny Evolutionary development over generations (cf. ontogeny).

polyandry The mating of one female with more than one male (cf. polygyny, monogamy).

polygamy Polygyny and/or polyandry (cf. monogamy).

polygynandry   A marriage system by which the marital unit consists of more than one husband and more than one wife.

polygyny   The mating of one male with more than one female (cf. polyandry, monogamy).

polygyny threshold   The difference in quality of male territories (or of the males themselves) necessary for it to be preferable for a female to mate with an already mated male rather than with an available bachelor.

polymorphism   The existence, within a single population, of two or more alternative forms.

preadaptation   Any characteristic of a population that suits it to a niche it does not yet occupy and hence facilitates the population's evolving to exploit that niche.

primordial germ cell   Embryologically primitive sex cells (versus somatic cells) that give rise to eggs and spermatozoa.

progesterone   A steroid hormone, $C_{21}H_{30}O_2$, produced primarily by the ovarian corpus luteum.

protandry   Sequential hermaphroditism with the male stage preceding the female stage (cf. protogyny).

protogyny   Sequential hermaphroditism with the female stage preceding the male stage (cf. protandry).

r-strategy   A set of reproductive characteristics that tends to maximize the potential rate of population increase at the expense of intensive nurture of young and efficient resource utilization (cf. *K*-strategy).

recombination   The production of a novel genotype by the combination of genetic elements from two parental genotypes; the statistical shuffling of genotypic combinations from one generation to another as a result of this process.

reproductive effort   Any activity that diminishes reproductive value while affording some prospect of contributing to Darwinian fitness.

reproductive value   An organism's expected lifetime reproductive output; the sum of present reproductive output and residual reproductive value (see below).

residual reproductive value   An organism's expected future reproductive output; based on the average reproductive performance of individuals of the same sex, age, parity, and so forth.

semelparity   The "big bang" strategy of reproducing only once in a lifetime; practiced by Pacific salmon, bamboo, and some other organisms (cf. iteroparity).

sexual reproduction   Any mode of reproduction that entails fusion of gametes to produce a genetically novel offspring.

species   A set of organisms (in practice typologically identified) capable of interbreeding without major (nongeographic) barriers.

steroid   Any of several organic compounds based on a particular 17-carbon structure, including the principle gonadal and adrenal cortical hormones.

sympatric   Geographically overlapping; said of (at least partially) reproductively isolated populations.

taxonomy   Classification, especially that categorizing organisms into species, genera, families, orders, and so forth.

testosterone   An androgen (see above), $C_{19}H_{28}O_2$, produced primarily in the testes.

variance   A measure of the dispersion of a set of scores; equal to the average squared deviation from the mean.

vasectomy   The surgical removal or obstruction of the vas deferens, rendering a male sterile by preventing sperm from entering the ejaculate.

viviparous   Bearing living offspring that develop within the mother's body (cf. oviparous, ovoviviparous).

zygote   The single cell produced by the fusion of gametes, before mitotic divisions have begun.

Index

397